Information, Computers, Machines and Man

INFORMATION, COMPUTERS, MACHINES AND MAN

An Introductory Text to Systems Engineering

Edited by
A. E. Karbowiak, Professor of Electrical Engineering and
R. M. Huey, Associate Professor of Electrical Engineering
University of New South Wales

John Wiley & Sons Australasia Pty. Ltd.
Sydney
New York London Toronto

ISBN and National Library of Australia Card
Number: Cloth 0 471 45853-8
Paper 0 471 45854-6
Library of Congress Catalog Card Number: 70-152338

Registered at the General Post Office, Sydney for
transmission through the post as a book
Printed by Offset Printing Coy. Pty. Limited, Chatswood.

Dedicated to the students of today:
the architects of tomorrow . . .

Contributors

M. W. Allen
Professor and Head of Department of Electronic Computation, U.N.S.W.

L. W. Davies
Chief Physicist. A.W.A. Physical Laboratory; Visiting Professor of Electrical Engineering and Head, Department of Solid-State Electronics, U.N.S.W.

G. A. Horridge, F.R.S.
Professor of Neurobiology, The Australian National University.

R. M. Huey
Associate Professor of Electrical Engineering, U.N.S.W.

A. E. Karbowiak
Professor of Electrical Engineering and Head, Department of Communications, U.N.S.W.

R. C. Olsson
Professor and Head of Department of Finance, U.N.S.W.

C. B. Speedy
Technical Director, Gresham Lion Group, U. K. formerly Professor of Electrical Engineering, U.N.S.W.

A. A. Thompson
Manager, Computer Services, U.N.S.W.

R. E. Vowels
Professor of Electrical Engineering and Head Department of Electric Power Engineering, U.N.S.W.

Contents

Contributors vi
Preface xiii

PART I SYSTEMS–BACKGROUND AND CONCEPTS

Foreword 3

Chapter 1 5
On Patterns, Systems and Creation—*A. E. Karbowiak*
A. Introduction 5
B. What is a System 7
C. A Framework 8
D. Examples 9
E. Exercises 11

Chapter 2 12
Energy, Power and Intuitive Notions of Information—*A. E. Karbowiak*
A. The Meaning of Words 12
B. Energy and Power, the Two Fundamental Quantities of Applied Science ... 13
C. Energy and Information 15
D. Exercises 16
E. A Solved Example 16

Chapter 3 18
Patterns, Waveforms and Scientific Measure of Information—*A. E. Karbowiak*
A. Concepts and Definitions 18
B. Attempts at Measuring Information 19
C. Pattern Generating Capacity of a Set: A Possible Basis for a Scientific Measure of Information 22
D. Waveforms: A Class of Patterns 25
E. Exercises 27

Chapter 4 28
The Semiconductor Revolution—*L. W. Davies*
A. Introduction 28
B. Components and Devices in Systems 28
C. Vacuum Tubes and Transistors 29
D. The Revolution in Electronics 33
E. Problems 34

Chapter 5 ... 35
Building Blocks for Systems—*R. M. Huey*
A. Introduction ... 35
B. How can a System be Described? ... 36
C. Some Kinds of Systems ... 36
D. Conservation Laws or Principles ... 38
E. Inequalities and Constraints ... 40
F. System Statics ... 41
G. System Dynamics—Block Diagrams ... 41
H. Input and Output ... 42
I. Signals ... 43
J. Methods Available to Calculate the Response ... 45
K. Amplifiers, Machines and Transducers ... 45
L. Summary of Highlights in Chapter 5 ... 49
M. Exercises ... 50
N. Endnotes ... 50

Chapter 6 ... 51
History and Applications of Computers—*M. W. Allen*
A. Analogue and Digital Computers ... 51
B. History of Digital Computers ... 53
C. Applications of Computers ... 56
D. Exercises ... 57

Chapter 7 ... 59
Man-Made Systems and Their Description—*R. M. Huey*
A. Control Systems ... 59
B. Communications Systems ... 60
C. Measurements on Systems ... 61
D. Simulation ... 62
E. Mathematical Models ... 64
F. Exercises ... 65
G. Endnotes ... 65

Chapter 8 ... 66
Biological Systems—*G. A. Horridge*
A. A Mixed Diet ... 66
B. Exercises ... 68

Chapter 9 ... 69
Human Systems—*R. C. Olsson*
A. Introduction ... 69
B. The Nature of Human Systems ... 71
C Factors leading to Complexity in Organisations ... 73
D. Traditional Theories of Organisation and Management ... 74
E. Development of Modern Views ... 77
F. The Systems Approach ... 78
G. The Organisation Viewed as a Total System ... 80
H. Conclusion and Areas for Further Enquiry ... 82
I. Review Questions ... 84
J. Exercises ... 84

Chapter 10 85
Learning, Adaptive and Goal-Seeking Systems—*A. E. Karbowiak and R. M. Huey*
A. Introduction 85
B. Adaptive Control Systems 86
C. Predictors and Learning Machines 88
D. Variable Systems and Loss of Information 89
E. Patterns and Semantics 90
F. Future Uses of Learning Systems 92
G. Epilogue 93

PART II ELEMENTS OF SYSTEMS

Foreword 97

Chapter 11 98
Signals, Information and Languages—*A. E. Karbowiak*
A. Introduction 98
B. Patterns, Constraints, Redundancy 98
C. Waveform: A Sub-set of a Chequer-Board Pattern 100
D. Pattern and Constraints 102
E. Some Information Theoretic Characteristics of the English Language 104
F. Discrete and Continuous Sources 105
G. Languages, Concept of Entropy 108
H. Experiments 109
I. Problem 109

Chapter 12 111
Networks of Building Blocks: Dynamic Systems—*R. M. Huey*
A. Introduction 111
B. Methods Available to Calculate the Response of a Dynamic System 111
C. The Assembly of Blocks into Block Diagrams 115
D. Linear Circuit Theory 120
E. Highlights of Chapter 12 125
F. Exercises 125
G. Endnotes 126

Chapter 13 127
Networks of Building Blocks—Static Systems—*R. M. Huey*
A. Static Systems 127
B. Block Diagrams 128
C. Task Networks and Time Models 128
D. Evaluation of the Performance of the Mathematical Model 131
E. Exercises 133

Chapter 14 134

Distributed Systems and Fields—*R. M. Huey*

A. Introduction 134
B. The Uniform Transmission Line 135
C. The Telegrapher's Equation 135
D. Methods Available for Solution 136
E. An Historical Viewpoint 136
F. Field Theory 138
G. The Maxwell Equations 139
H. Highlights of Chapter 14 140
I. Exercises 140
J. Endnotes 140

Chapter 15 142

Materials—*L. W. Davies*

A. Introduction 142
B. Classification of Crystals 143
C. Material Properties 147
D. Problems 150

Chapter 16 152

Semiconductors—*L. W. Davies*

A. Intrinsic Semiconductors 152
B. Extrinsic Semiconductors 152
C. *p-n* Junctions 155
D. Junction Transistors 158
E. Problems 160

Chapter 17 161

Integrated Circuits—*L. W. Davies*

A. Introduction 161
B. Solid-State Diffusion in Silicon 161
C. Oxide Masking 163
D. Planar Technology 165
E. Monolithic Bipolar Integrated Circuits 166
F. Conclusion 167

Chapter 18 168

Switching Circuits, Number Systems and Basic Computer Organisation—*M. W. Allen*

A. Switching Circuits and Logic 169
B. Number Representations and Arithmetic Operations 173
C. Basic Computer Organisation 179
D. Exercises 185

Chapter 19 186

Introduction to Computer Programming—*A. A. Thompson*

A. Introduction 186
B. General Structure of A Computer System 189
C. Computer Instructions and Programs 195
D. Machine Language Programming 202
E. Exercises 220

Chapter 20 221
Introduction to Computer Software—*A. A. Thompson*
A. Introduction 221
B. Symbolic Programming Languages 223
C. Formal Definition of Programming Languages 230
D. Translation of High Level Programming Languages 234
E. Computer Operating Systems 236
F. Exercises 242

PART III ACTUAL SYSTEMS

Foreword 247

Chapter 21 248
The Crab Eye—*G. A. Horridge*
A. Introduction 248
B. Stabilistation against Tilt 248
C. Eye Structure 249
D. Acuity 252
E. Motion Perception 253
F. Eye Movement 256
G. Conclusion 259
H. Exercises 259

Chapter 22 266
Control of Large Systems—*C. B. Speedy*
A. Introduction 266
B. Mathematical Framework 268
C. The Control Problem 272
D. Satellite Attitude Positioning System 273
E. Satellite Launch System 276
F. Electric Power System 276
G. Vehicular Traffic System 278
H. Hydraulic System 279
I. The Economic System 280
J. Conclusions 281
K. Exercises 281

Chapter 23 282
Communication Systems—*A. E. Karbowiak*
A. Types of Systems 282
B. Characteristics of Systems 284
C. Communication Systems: A Brief Outline of Developments 285
D. Some Future Possibilities 289
E. Examples of Information Theoretic Problems Encountered with Communication Systems 291

Chapter 24 293
Computer Systems—*A. A. Thompson*
A. Introduction 293
B. Time-Shared, Multi-Access, Interactive Computer Systems 294
C. Computer Networks and Utilities 297
D. Computers in Engineering 300
E. Computers in Education 303
F. Computers in Industry 306
G. Computers in Medicine 309
H. Exercises 312

Chapter 25 313
Electric Power and Computers—*R. E. Vowels*
A. Introduction 313
B. Early History 313
C. Sources of Energy 313
D. Generator Unit Size 317
E. Transmission of Power 317
F. Nuclear Power 323
G. Linear Motion Electrical Machines 324
H. Magneto-hydrodynamic Generators 324
I. Automatic Control 324
J. Exercises 326
K. Problems for Discussion 327

Chapter 26 328
Modelling and Simulation—*R. M. Huey*
A. Use of Modelling in Science and Engineering 328
B. Kinds of Models 329
C. Use of Models 330
D. Conclusion 332

Chapter 27 333
Conclusion: And What of the Future?—*A. E. Karbowiak and R. M. Huey*

Problems 336
Appendices 338
Subject Index 343
Name Index 347

Preface

This book was written originally in response to requests from students in the Faculties of Engineering and Applied Science at the University of New South Wales, and a pilot version of the book was used in first year courses for two years. This version has been extensively modified to make the material more easily accessible to those with interests beyond the sciences.

The book endeavours to provide an overview of methods, powers and limitations peculiar to modern engineering and technology, with some attempt at an estimate of the effects which such developments might have on human societies.

In this present volume an attempt has been made to present the background, the elements, and to give some discussion of systems with minimum usage of scientific and technical jargon. It is hoped that most of the material in Part I is presented at a level that is suitable for all students entering tertiary education.

The more scientific and technical aspects of engineering systems are presented in Part II, while actual systems are discussed in greater detail in Part III. It needs to be stressed, however, that it would be impossible to present in a single volume sufficient detailed material for a technical discussion of systems engineering and the technique of modelling. Instead, various facets of systems engineering are unfolded in stages, but above all, an attempt has been made to help the reader appreciate the methodology and the way of thinking peculiar to systems engineering.

The material for this book is presented in 27 chapters. Readers are advised to study Part I, followed by a selection of chapters from Parts II and III to suit their interest. The next stage is to return to those chapters of Part I, which might require a deeper study before proceeding with Part II, followed by Part III.

Neither the lecturer nor the student should regard the book as lecture material—it is not meant to be—but rather the student should be encouraged to read in his own time, ahead of discussion sessions or study periods. The lecturer should then introduce the topic and the remainder of the period may be spent on discussion of the material studied, and other such material which the student selects of his own interest and which is related to the subject of the study period.

For students with a science and mathematical background it should be possible to cover the whole material in about 40 one-hour discussion periods. Students with non-scientific background might cover Part I and a selection of chapters from Parts II and III (of particular interest to them) in about 20 one-hour sessions.

The modern engineer with the vast technology and financial resources at his disposal can no longer consider a machine or a group of machines in isolation, but rather he must view them in their environment, and include in his estimate the likely effects of the machines on human beings.

The book, among other matters, points to the power of modern technology, notwithstanding that current systems are planned with such a high degree of complexity that no human being can comprehend their detailed *modus operandi* or be capable of using them in an optimum fashion. For example, with large electronic computers, we can put to an effective use only a fraction of their potential power. With many a large system, such as a space project, the costs of a mission accomplished often turns out to be an order of magnitude higher than the original estimates. (It is not beyond the bounds of possibility to find that a project which has been estimated at around $10m finally reaches in excess of $1000m).

The evidence to date is abundantly clear: we do not understand how to design, plan nor manage efficiently a large system. The appalling fact remains that many a multi-million dollar decision is made substantially on an *ad hoc* basis with scanty evidence to support the actions and in the face of a lack of understanding of any possible repercussions on other systems or human societies (or even the very existence of the human race). It is for these reasons that a chapter on human systems has been included.

One must not, of course, conclude that since we do not intuitively know how to plan efficiently for the future that we should demolish the world of today and build on its ruins a better world of tomorrow. For to demolish is relatively easy, but to create is very much harder and requires a disciplined mind. Neither must one conclude, as it is sometimes stated, that since planners and leaders of today are not sure what is best, that all authority should be abolished and everyone should be allowed to do as he thinks is best.

> And when the last law was down, and the devil turned on you—where would you hide, Roper, the law all being flat? This Country's planted thick with laws from coast to coast . . . man's laws, not God's—and if you cut them down—and you are just the man to do it—d'you really think you could stand upright in the winds that would blow then? Yes, I'd give the devil benefit of law, for my own safety's sake.
>
> (Sir Thomas More, in *A Man for All Seasons* by Robert Bolt.)

Nor must one deceive oneself that such "narrow" views on the role of law are characteristic of the older generation.

> A lot of kids complain about their parents. Well, mine couldn't be better. Those of us who are older (I'm 14) are allowed to do things, and go places that the younger ones aren't, but Dad always explains so they won't feel left out and manages to give them a treat too. We often don't like the authority and restrictions, and so on, but our home would be in a shambles if my elder brother got control!
>
> (14-year-old girl, quoted in "Annals' 70" July, 1970)

Evidently, both older and younger generations hold opinions which should be considered, before we try to answer the question: How shall we prepare ourselves for the even larger and more numerous projects of tomorrow? We need foresight and courage: foresight to see the problems ahead and courage to tackle them responsibly. Greater emphasis must be put on co-operative efforts. Hence also the need for more and better education.

Technological problems evolve around patterns and semantics: how to translate semantics into patterns amenable to scientific scrutiny is really the issue behind many a major technological problem. Systems engineering is really an attitude, a way of thinking, a method of putting a major plan into operation. But, in the relatively narrow sense of present day concepts, it is likely, in the long run, to create more problems than it will solve. We can already see the need for a change in attitude.

Thanks to the developments in systems engineering, we can now deal with projects of magnitude hitherto undreamed of, so large in fact that should the project result in failure, a whole nation could be ruined. Furthermore, the operation of a large project could modify the environment fundamentally, and unless this too can be taken into account, the system could lead to disaster. Many of us when faced with prospects of this kind react irrationally and denounce science and technology as basically evil. Such attitudes are dangerous. It was Napoleon who said: "There are two levers for moving men: interest and fear". There is no antidote for fear except education, for with it comes interest and wisdom.

People frequently hold views on various matters with which they have only scanty acquaintance. Such views are often in error and are propagated—sometimes helped by the mass media—to the detriment of the society. Some think that Chinese is a difficult language, yet millions of children in China babble, then soon converse in their native language quite effectively, around about the same age as those in Australia or Germany. Some think that engineering, and electronics in particular, is dull and fundamentally difficult. Yet, the evidence is abundant: to the young, electronics (to take a specific example) is no more difficult a subject than ancient history, mathematics or the classics. It is well-known that the unfamiliar is strange and often appears dull and difficult, even frightening.

There is no such thing as an "inherently difficult discipline". A lively interest is the gateway to abundant knowledge, but the basic knowledge must be acquired in an atmosphere of trust and by playing games with a range of sophisticated toys.

A helpful way to look at modern technology is to regard it as a language or medium in which one can create masterpieces for others to share. But whereas one quickly learns to use a living language from the earliest days of infancy, science and technology remains strange for many years of one's early life and may later on be labelled as something to be feared . . . This is a pity as it stifles creativity. To accept this as a *status quo* is perhaps one of the gravest stains on our educational system that works to the detriment of our societies. Should one run away from reality or, worse still, cry, grumble and despair? Does not the solution lie in one's readiness to labour on the task ahead with a genuine desire to create a masterpiece, however humble?

To be educated means to have a constructive and workable plan for tomorrow with a genuine concern for others.

> "Living is an art and, to practice it well, men need not only to acquire skill, but also tact and taste". (Aldous Huxley)

A. E. K. & R. M. H.

Part I

Systems—Background and Concepts

Foreword

The material in this part of the book is presented in ten chapters dealing with (1) information, (2) engineering materials, (3) building blocks and anatomy of systems, (4) computers (5) systems ranging from communications, through biological systems to human systems, and concluding with a discussion of likely developments.

Discussion on information is fundamental to understanding the system's function: information in various disguises of digital data or electric waveforms is the life-blood of every system without which it could not perform its functions. The information network of a system can be likened to the nervous system of a living organism; both as regards to the functions performed, as well as its complexity.

A system is made up of a variety of building blocks and functional units in accordance with a pre-arranged plan, but ultimately the performance of a building block depends on the fundamental properties of the engineering materials used. Indeed, it would be true to say that the progress with large systems which we have witnessed in the last decade, would not have been possible but for the incredible developments which have taken place in the technology that enables us to produce ultra-pure materials and to control their properties to an extraordinary degree of accuracy. This progress led ultimately to the mass-production of reliable micro-miniature and integrated electronic circuits accompanied by considerable reduction in cost. In fact, in many cases, we have reached the ultimate limit set by the molecular structure of the materials. In a related area, by studying biological systems, the detailed examination of living organisms is likely to suggest new approaches not yet conceived with micro-circuits and systems.

The progress in materials science has been paralleled by an equally spectacular progress in systems (this includes theory, methods and hardware accomplishments). Indeed it is the progress in this area that ultimately led to the development of systems to a degree of complexity peculiar to this age.

It is at this stage that some clarification is necessary. Whatever else technology might be, it is most importantly a medium in which humans can create. Language is to a poet what technology is to an engineer. Much as language can be used for evil (e.g., to incite one to illegal actions) so can technology be misapplied. However, we must not think of technology itself as an evil influence, any more than we should think of language as an evil influence.

The final chapter is devoted to some philosophical issues relating to systems engineering. It attempts to summarise some of the principal issues and terminates on a note of hope for humanity: The key to the solution of the many problems of tomorrow is really in our hands. To a very large extent it is *we* who determine the shape of tomorrow, the ugliness is unlikely to be *their* fault.

Ultimately the only satisfactory solution seems to lie with the introduction of a "global system" in which we have the fundamental obligation to ensure that any new system to be introduced, as well as the progress and the function of monitoring the operation of others already existing, are under the control of a global system performing in the light of a genuine concern for the well being of every nation and every individual.

These objectives can only be achieved through better education. This is a hard road to truth, no one will deny it.

"The roots of education are bitter, but the fruit is sweet."

(Aristotle)

A.E.K.

CHAPTER 1

On Patterns, Systems and Creation

A. E. Karbowiak

> ***"What is this life if, full of care,***
> ***We have no time to stand and stare?***
> ***. . .***
> ***No time to turn in Beauty's glance***
> ***And watch her feet, how they can dance"***
>
> *W. H. Davies.*

A. INTRODUCTION

Applied science—Engineering in particular—represents perhaps the largest and most rapidly growing volume of knowledge and of human skills. At the same time, our progress is almost entirely dependent on developments in technology.

It is now recognised that well-developed technology is not only necessary for the well-being of every country, but that technology is the key to success, certainly a key to power, or a key to destruction, if we so choose. But technology is also a creative way of life, a way to self-fulfilment, to achievement, to greatness, as well as to prosperity.

The challenge here is not so much how to provide for the *wants* of modern societies, but rather, to anticipate and prepare for the *needs* of those around us as well as those of the generation to come.

Engineering will be recognised as a broad discipline with its own tradition and folklore; a rapidly growing profession. But whereas technological knowledge and skills grow in complexity and diversity, the work of an engineer remains essentially creative with intense artistic overtones. It is these overtones with their counterpart of abstract patterns in the minds of those who plan the project that are a source of immense pleasure and impart a sense of professional achievement.

Having said these things, one needs to recognise that in the last few decades engineering has been undergoing great changes. These stem from two directions:

(i) a growing profusion of specialised disciplines, and
(ii) the emergence of very large projects requiring large-scale team work of specialists, with associated human problems.

Indeed, to an increasing extent an engineer, to succeed in his profession, must learn to understand and solve human problems related to his work, otherwise his frustrations will snowball.

The answers to many fundamental problems of a large project are found through applied science, which can only be obtained through extensive research by a variety of specialists. In view of the high degree of specialisation there is, therefore, a need for a large number of specialists to work on separate problems, yet concerning the same goal. Consequently, the organisational problems multiply: partly because of the tendency of many specialists towards a self-centred approach, and partly because of the difficulties associated with communication among an increasing number of specialists who acquire deeper knowledge in narrower fields. In an effort to bridge the gap a greater number of experts are employed; with increasing numbers the communication becomes more difficult to achieve, and so it goes on . . .

How has this come about? And is there a solution? To help answer such questions let us briefly examine the scene. First, we observe that over the years we have come to differentiate between various branches of knowledge and skills involving applied sciences and engineering. Thus, very specialised disciplines have evolved under different headings such as oceanography, communication engineering, management, medical electronics, computers, satellites and many, many others. It should be recognised that this profusion of branches is natural in the development of our knowledge. As the years progress, we learn more and more about the various branches of knowledge, and the compartments grow in size accordingly. As a result, in most professions we tend to specialise, and specialise we must if progress is to continue.

More recently, particularly since the Second World War, there has been another simultaneous development taking place. This is the emergence of very large engineering projects, such as satellite communication systems, particle accelerators for scientific research, complex traffic systems of large cities, large power distribution networks of a country, completely automated factories such as oil refineries, and many others. A peculiar feature of such projects is *not* that an aspect, such as an unusual or a new scientific phenomenon, might be hard to understand, but rather that the problems which might arise in the system as a whole are difficult, if not humanly impossible, to foresee on account of the sheer size and complexity of the project.

A large project brings new problems: various forces which had negligible influence on the performance of a particular component can play important roles when such components become (in great numbers) part of a complete system. Large systems are invariably expensive and the safety factors which one tends to impose are more stringent than those for smaller projects. Finally, large projects contain many interconnected components and sub-assemblies; it thus becomes difficult for a human being to comprehend the functioning of the system as a whole.

These two factors, extreme degree of specialisation and the emergence of large systems of great complexity (which are difficult to comprehend), produce far-reaching human problems. Firstly, there is the problem that no single person can comprehend adequately all matters connected with a large system; no one can, therefore, be said to be wholly in control of it. Team work at all levels, including the uppermost level of management becomes mandatory. This implies a change of attitudes of the specialists and the managers, from a self-centred to a task-oriented motivation.

We thus come to the curious conclusion that further progress, which includes fundamental scientific discovery and which was in the past largely a function of self-centred and self-motivated inventiveness (backroom boy approach), is now recognised to be limited by the rate of progress of technology which we can maintain, and this in turn is primarily limited by human problems. It, therefore, becomes imperative to gain a better understanding not only of the complex system in isolation, but the system in which machines and humans interact on a vast scale in a common environment. If one recognises these problems, one comes to query how it is possible to "design" or manage a large system?

"Systems Engineering" is the science that helps us to go a long way towards the objectives, but before explaining what Systems Engineering is and does, we need first of all to be clear on what we mean by the term "system".

B. WHAT IS A SYSTEM?

Clearly a system is man made.* It is large and complex and made up from numerous parts which perform various functions. The components–of a system are inter-connected to form functional units and these in turn are further connected to form a complete assembly, with suitable input and output units. The various parts of the system may be independent, but are capable of being separated and interconnected in a variety of patterns, but above all the whole assembly is designed to function according to a definite plan and perform as an integral unit.

Usually it is not possible to "design" a system in the ordinary sense of the word. One, therefore, builds various mathematical models and prototype working models before the design stage can be reached and even then the system may be equipped with self-adaptive features to enable it to make decisions in new circumstances not entirely foreseen by its designers.

The self-adaptive and decision-making characteristics make the system's behaviour difficult to predict, but at the same time make the system competitive. It is this last feature which makes systems engineering a fascinating discipline, and in addition the methods which have been developed can have far-reaching effects. At the same time it may not always be possible to foresee at the beginning of the project the particular final outcome of our labours, although we may be reasonably sure of the *likely* outcome.

To clarify our notion the following simple illustrations might help:

A fully automated oil refinery run by a few engineers with the help of a sizable computer is an example of a system. A satellite communication system complete with orbiting satellites, several command stations on the ground, transmitting and receiving units with associated computers is another example. A completely automatic remote-controlled power station system comprising 30 or 50 power stations connected onto a common net, covering a large area, under the command of an engineer with the help of one central and two or three regional computers, would be another example, and so on.

For completeness we need to remark that a more accurate model of a system would need to include the human operator at the relevant points in the system, if a reliable prediction of the behaviour of the system as a whole is to be achieved. Yet modelling a human operator is not the easiest of tasks to undertake.

The question we address ourselves with now is: Where does one start with the design of a complex system, assuming that we have never designed a similar system before and are not even in a position to comprehend it?

It is at this juncture that the new technique of modelling is of immense value. We create conceptual models and mathematical ones; we carry out thought experiments; we model simplified systems on a computer and proceed to more complicated ones. In this way we grow richer in knowledge, understanding and experience, until eventually we are ready to build a prototype model to gain some more experience on the way to reaching the final objective (Fig. 1.1). The technique here is analogous to that of a child playing successively with more and more sophisticated toys, until finally he is ready to venture into the real world.

To give a specific example: if one were asked to design a motor car, then on the first attempt one would make a rather poor job of it. But the second attempt would be much better because one would be that much richer for the experience of having designed a car before. At any rate, even with the first design one could make some sort of a start. However, if the problem were to design a car for an inhabitant on an earth-like planet in the distant world of Andromeda Nebula, no one on this earth could

* See for example: H. Chestnut, *Systems Engineering Methods*. Chapter 1. John Wiley, New York 1965.

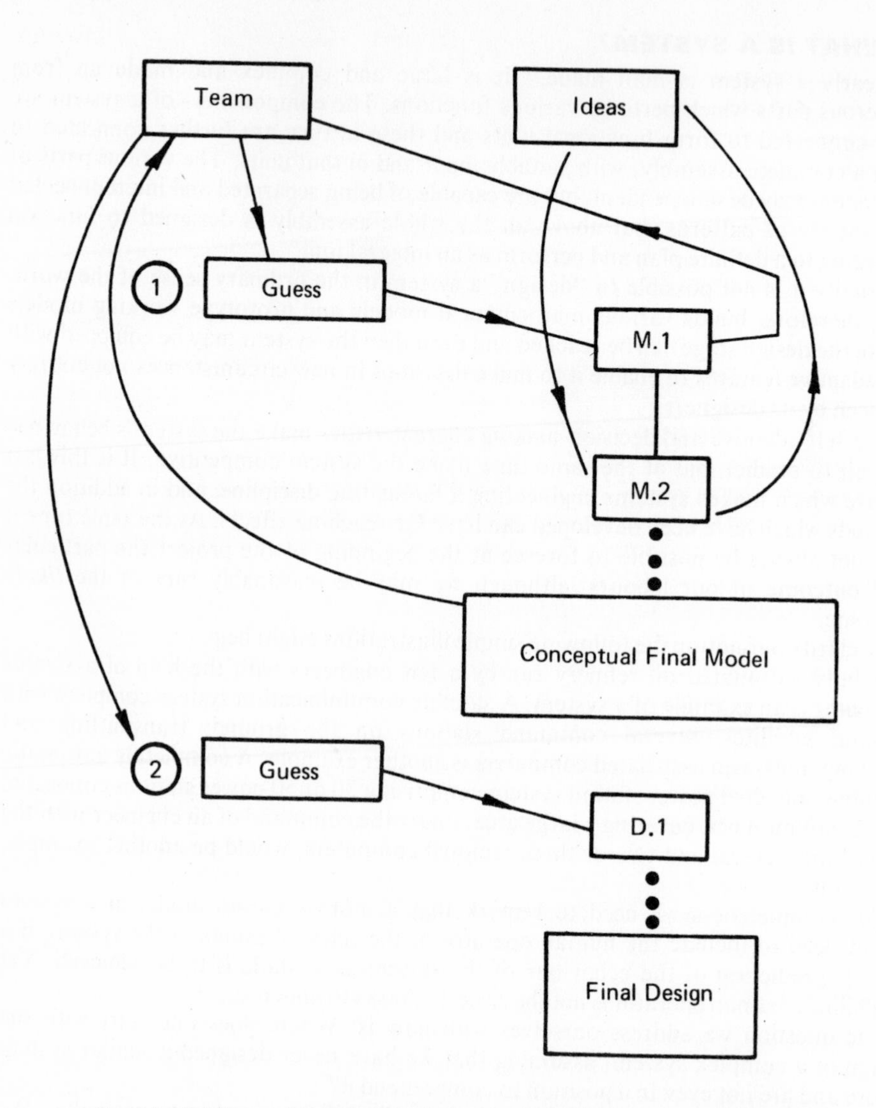

FIG. 1.1 Modelling and Feed-back Technique for Designing a Complicated System.

make any sensible start towards the project. The reason being that we would not know the geometry nor the weight of the driver for whom the car is intended. We would need to know many other factors such as climatic conditions, the local geography, or the state of the roads, not to mention such important factors as likes and dislikes of the intended user. In other words we have no knowledge of the environment in which the machine is to function. Clearly, under such conditions no one could even attempt a vague start at designing a system. The answer is clear: knowledge of the environment is a prerequisite for the design of any system. In practice, study of the environment forms an essential initial phase of the approach to the design of any system.

To meet the needs of tomorrow we must face up to the impending changes and adjust our outlook accordingly.

C. A FRAME WORK

If one were asked to name the most important assets which an engineer should have, then without hesitation most of us would name (apart from professional excellence) imagination and engineering intuition. Yet intuition is something difficult to define and even more difficult to impart, but it usually amounts to an ability to make judicious guesses in the absence of adequate supporting evidence. There is plenty of evidence that a good judicious guess can circumvent years of expensive research.

While in all probability intuition cannot be taught, one can be helped greatly in making inspired guesses by developing a mental framework of measuring rods, with the help of which one can assess a given situation and select out of a maze of alternatives those worthy of detailed exploration. The framework need not be precise but sufficiently accurate to enable one to make, in the first instance, a correct assessment in order of magnitude.

This can be achieved quite effectively by listing the relevant parameters (this can be done in all sorts of areas) in order of their magnitude and committing to memory some of the key points on the lines indicated below. On doing such exercises we soon realise that in engineering we deal with quantities so small, at one end of the scale, and so large at the other that it becomes difficult to imagine the meaning and size of the quantities involved. The numbers themselves are frequently so small or so large that it becomes inconvenient to write them down in full, thus we develop a useful shorthand notation, based on the logarithmic scale. Choosing logarithms to the base of ten, we refer to quantities by order of magnitude with reference to the nearest integral power of ten.

Thus we would say that a man whose height happens to be 186 cm ($= 1.86 \times 10^0$ metres) ranks in height of zero order. At the same time we would say that an insect (approximately 1 mm long $= 10^{-3}$ m) is three orders of magnitude smaller, and that a particular bridge 1 km ($= 10^3$m) long is three orders of magnitude longer than the principal measurement of man.

The various powers of 10 in steps of three orders of magnitude (one thousand) have each been given a name as indicated in Table A.2 in Appendix 1. Using these names we would say that while a man's height is of the order of 1m, the size of a typical bacteria is of the order of 1 μm and that intercontinental distances are of the order of 1 Mm.

D. EXAMPLES

1. Quantity: Length

Take as the first example the physical quantity "length". In the middle of the scale mark a point, I, denoting the height of a human being (of the order of 1 m) and relate on a logarithmic scale the size of other objects to it (Fig. 1.2). In this way we can compress an enormous range of size of objects and compare their relative magnitudes. We note that the total range of physical distances encountered in physics and known to man is of the order of magnitude of 10^{40}, an enormous number beyond human visualisation (it is known under the name of Eddington's number).*

Figure 1.2 also indicates the correct usage of the prefixes shown in Table A.2.

It is a fundamental axiom of modern physics that the higher the accuracy with which a physical quantity is measured the greater the expenditure of energy needed for its determination (see Chapter 2). More precisely, the expenditure of energy increases in direct proportion to the accuracy of measurement. In this way we learn that to determine a distance to an accuracy of the order of 10^{-40}m expenditure of energy equivalent to that liberated when about 1 kg of matter is destroyed would be needed: A quantity of energy liberated when exploding an atomic bomb. Clearly, therefore, such accuracies are inconceivable in any project in applied science.

* Eddington's number is the ratio of the diameter of the whole universe to the diameter of the electron.

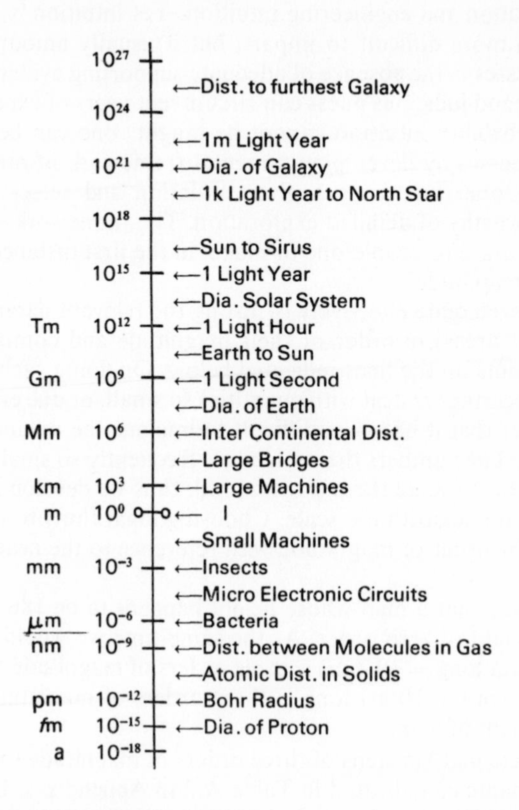

FIG. 1.2 Order of Magnitude of Things Around Us: Length (m = 1 metre).

To illustrate the kind of conclusions which one can draw using data such as those shown in Fig. 1.2, consider the following. Currently manufactured micro-electronic circuits have measurements of the order of 10^{-4} m or less, permitting construction of micro-electronic sub-systems occupying a volume of the order of 1 mm^3. The individual components making up a micro-electronic circuit might be as small as one μm, the size of bacteria. It would be interesting to speculate that technology could be improved to bring about an order of magnitude improvement giving individual dimensions of the order of 100 nm. Yet it is inconceivable—in view of the fact that atomic distances in solids are of the order of 100*pm**— that a further useful reduction in size would be possible if the present day concepts of electronic circuits are to be retained.

2. Quantity: Time

A similar exercise performed with respect to the physical quantity time gives the chart shown in Fig. 1.3. Here too, we find an enormous range of values.

There are a number of most interesting conclusions and speculations which can be made using this chart: the investigation of these is left as an exercise for the reader.

A variety of such charts can be prepared concerning various physical and other quantities. These form a most useful reference framework in which new parameters can be placed in proper perspective.

* 100*pm* = 10^{-10} m is equal to 1 Angstrom unit, written 1 Å.

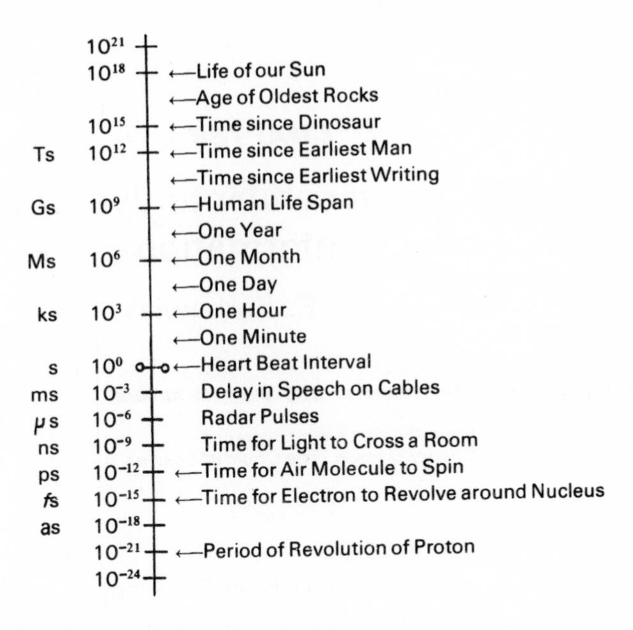

FIG. 1.3 Order of Magnitude of Things Around Us: Time (s = 1 sec).

E. EXERCISES

Prepare an order of magnitude chart based on

(i) weight/mass;
(ii) money and expenditure;
(iii) any other quantity which you feel is important in applied science, or which holds a fascination for you.

In (i) take 100kg, the weight of a human being, as a starting point and work through many orders of magnitude to the mass of the galaxy and the whole of the universe at one end, and down to molecules and electrons at the other.

In (ii) take your weekly income as a starting point and go through many orders of magnitude, including expenses on large scientific projects, and on centres of education. Include consideration of other factors such as costs of the Second World War, or war in Vietnam, or of cancer research. Tabulate the national income of various countries, waste caused by hunger, inefficiency, road accidents, incompetence or ill-will, the value of principal static assets in the country, etc.

SUGGESTED READING MATERIAL

BAILEY, B. and MORGAN, D. *Thinking and Writing*. Rigby, 1966.
MILLER, G. A. *Language and Communication*. McGraw-Hill, New York, 1963.
Physical Science Study Committee *Physics*, D. C. Heath & Co., Boston, 1960.

CHAPTER 2

Energy, Power and Intuitive Notions of Information

A. E. Karbowiak

"There is a grandeur in this view of life, with its several powers . . ."

From the concluding sentence
Charles Darwin "The Origin of Species."

A. THE MEANING OF WORDS

We find in the Oxford English Dictionary the following primary definition of energy: "Force, vigour (of speech, action, person, etc.) active operation, individual powers in exercise of authority and also ability." Energy, therefore, can mean all or any of those things.

The word *power* in turn has the following primary dictionary definitions: "Ability to do or act, a particular faculty of body or mind, vigour, energy, active property, government influence, authority, personal ascendancy, authorisation, delegated authority; influential person, body or thing; Deity, large number or amount etc." We thus see that one word can mean many different things and frequently meaning is inferred in the context.

In every day usage and particularly in literature, a certain degree of vagueness in the meaning of words is sometimes desirable in that it helps to convey the emotional and the artistic tenor of the material presented. In applied science the artistic and creative skills find an expression in the form of the system, the machine or the theory created, but vagueness in expression is an undesirable characteristic and an attempt is therefore made to develop a more precise language than that offered for everyday use.

Thus in science one *defines* energy as: *"Ability to do work"* (potential or kinetic), and power as *"rate of doing work"*. More precise definitions can be formulated with the help of mathematical symbols. Thus we define energy as "force times distance" and power as "the derivative of energy with respect to time", or "the rate of doing work". In addition we understand both quantities to be "scalars".

There are two points which need to be brought to light at this stage. The first one concerns the language of science and its relation to human communication. Thus, with the specific reference to the meaning of the words energy and work, a scientist having understood the physics behind energy (and work) as pertaining to a physical system, *borrows* a word from everyday language—a word which has an *accepted* vague meaning—and ascribes to it a more precise but specific significance which to most people around him is strange. Here lies the first source of confusion: the scientist precisely defines the meaning of a word to signify one thing and his listeners will understand another thing. This of course does not help in the communication. Here an observation is in order.

The re-definition of the meaning of a vague word may help the scientist in his work, but it certainly does not help in the communication of his ideas to the layman. To the scientist, it would have been equally acceptable to have defined a new word and have given to this word the intended precise meaning, thus avoiding all subsequent confusion.

This is a somewhat simplified illustration of the kind of confusion which can arise among members of a team of specialists working together towards a major project, in that they fail to understand each other because of language difficulty rather than the technological complexity of the subject matter. In the first place confusion can arise, and in practice often does arise, from words having an established vague meaning, with new connotations.

The second aspect concerns the opposite effect. Here we have in mind the confusion which may arise from an attempt at a too rigorous (or excessively abstract) association. To give a specific example: we know that energy and power are defined (mathematically) as scalars. By this we mean that the quantity is adequately specified by a number, unlike a vector quantity which needs for its full description a number and a direction. Now imagine an engineer being given a source of power, such as an electric motor, rated 200 h.p. and not being told the direction in which the motor is revolving. Clearly, he would not be in a position to design the actual machine using the data given; he needs to know in addition the direction in which the motor is to revolve when delivering power.

These two examples are simple illustrations, and serve as a warning of the dangers impending in the excessive use of technical jargon. Frequently, it is this factor which is responsible for a technical or scientific discourse degenerating to a second-rate debate on the language itself.

B. ENERGY AND POWER, THE TWO FUNDAMENTAL QUANTITIES OF APPLIED SCIENCE

Every engineering system needs, in order to function, a source of power. There also needs to be a power distribution network to carry the energy to the various parts of the system. But the purpose for which the power flows in the system is twofold.

1. To power the system so as to enable it to perform the specific tasks.
2. To carry the information around the system; here, energy is the vehicle for conveying information.

The problems associated with the first objective are those pertaining to the generation as well as transportation of energy between points which are distant in space (Fig. 2.1). The energy can be transported in a variety of forms such as hydraulic

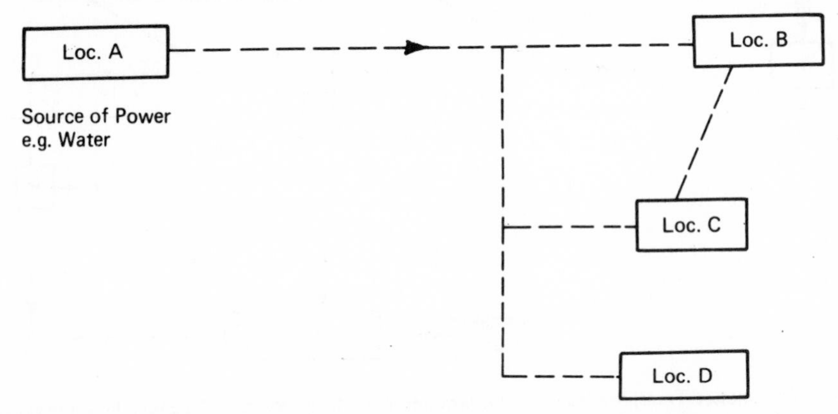

FIG. 2.1 Energy Transportation between a Number of Locations.

or electrical, or, it may be conveyed by means of compressed gas etc. But the actual manner of distributing the energy is of lesser importance in systems engineering, although for convenience, the energy is usually transported by electrical means.

Many years ago in electrical engineering we used to differentiate between power engineering (heavy current engineering) and light current engineering. The former was concerned primarily with generation and transportation as well as utilisation of electrical power, while light current engineering covered such aspects as electronics, telephone, radio and the like, The reason for this subdivision was, in part, due to the fact that light current engineering was concerned principally with aspects of communication and control engineering, where only small amounts of power were involved. However, times have changed, and we now realise that we need energy for communication, and the larger the range and the greater the quantity of information to be conveyed, the greater the demands on power.

To give a specific illustration: if a teacher speaks to a class of some 10 or 20 students, it is then quite sufficient for him to speak at a normal volume but if the class is increased to something like 50 or 100 students, he needs to raise his voice and perhaps, at the end of the lecture he might feel tired. When it comes to even larger audiences, perhaps many hundreds or thousands of listeners, then the speaker in order to make himself heard and to convey the information intended, needs powerful amplifiers to strengthen his voice. Thus we see that the larger the distance and the greater the quantity of information to be conveyed, the greater the demands on power. Nowadays, a typical radio transmitter has a power of 100 kilowatts so that it can broadcast information over a large area of influence. A similar observation would apply to television and other mass media communications.

We have progressed a long way from the early days of electrical engineering. Nowadays we have communication systems where the transmitters have powers of thousands of kilowatts in order to achieve the objectives assigned to such apparatus.

Radar is another example. This is a system designed to search the environment and supply detailed information on the position of various objects, such as aeroplanes approaching a runway, or battleships in formation. In examining the radar problem it needs to be appreciated that a radar system will usually cover a very large volume of space and will be required to supply an enormous amount of accurate information in a relatively short time. It is because of the multitude and the exactitude of such tasks that the power requirements of a radar system may be many megawatts.

In this way the distinction between heavy current electrical engineering and light

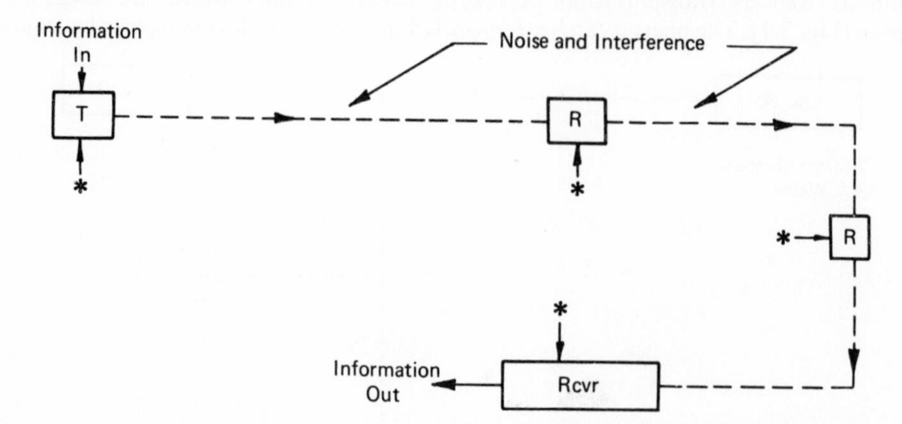

Fig. 2.2 Information Transportation (T = Transmitter, R = Relay, Rcvr. = Receiver). Power in at points marked*.

current electrical engineering can be said to have disappeared, but we still have the conceptual difference in that in power engineering the primary concern is to transport energy between distant points in space (Fig. 2.1) (here, efficient transportation of energy is of paramount importance); while with communications systems the primary objective is to convey, extract and process information, in which process considerable amounts of power may be consumed (Fig. 2.2).

C. ENERGY AND INFORMATION

Having established that energy needs to be expended to convey information, it will also be appreciated that energy must be expended to gain information (radar is a system operating on this principle). If one wishes to acquire a great deal of information, then a very substantial amount of power needs to be expended. By the same token we see that energy must be expended to control energy, and the greater the amount of energy to be controlled and the greater the precision required in controlling the energy, the greater will be the demands on power for performing the task.

From reasoning of the above nature, various corollaries follow. Thus we find that:

1. the higher the accuracy of measurement, the greater the demands on power, and
2. the higher the precision of control or the rate of construction, the greater the demands on power.

Therefore there must be a limit to the accuracy and the precision with which a system can operate.

Thus we find, that there is a limit to the rate with which information can be generated, and also there exist limits to the precision, rate of control and to the rate of construction of devices. The higher the rate, the greater the power needed to achieve the objectives.

The final point which needs to be made, is that every engineering system exists in an environment, and that the environment is a part of the Universe at large. The environment as such, is at a definite temperature, and therefore it is subjected to Brownian motion. At the same time one must accept that the act of measuring any part of the system implies interference with the system or the thing which we set out to measure. Thus the act of communication, in part modifies the communicants as well as the environment; the act of controlling, in part modifies the thing we set out to control, etc.

The reasons behind such characteristics of systems are intimately related to the fundamental concepts of physics. Here we refer to the three fundamental constants of physics:

(i) the electronic charge, e:

$$e = 1.602 \times 10^{-19} \text{ coulombs}$$

(ii) Boltzmann's constant, k:

$$k = 1.38 \times 10^{-23} \text{ joule/degree K, and,}$$

(iii) Planck's constant, h:

$$h = 6.626 \times 10^{-34} \text{ joule sec.}$$

The above constants are important because they determine the minimum discernible quantity which can be measured. In particular $\frac{1}{2}kT$, where T is the absolute temperature, is known as the thermal energy per degree of freedom, and represents the minimum discernible quantity of energy in a system held at the temperature $T°$ absolute. In the same way the quantity $\frac{1}{2}hf$, where f is the frequency, represents the minimum quantity of energy which is meaningful for a system vibrating with frequency f. In quantum mechanics hf is known as the quantum of energy associated with a harmonic oscillator.

Under normal operating conditions the energy corresponding to kT or hf is exceedingly small (see D. 1 below) and might seem unimportant. Yet in many modern systems which handle huge amounts of information, it is these factors that set a limit on what is possible to achieve with a system. For example, with a radio receiver it is the factor kT that sets a limit on the attainable sensitivity and realisable range of the equipment. Frequently the noise which is heard when trying to listen to a distant radio station is the manifestation of the Brownian motion through the all important factor kT.

The above can be called intrinsic forms of noise in the sense that the effects are unavoidable. This can be contrasted with various extrinsic forms of noise (such as aircraft noise, car ignition noise, etc.) which relate to various errors or deficiencies of human or other origin and which are, in principle, avoidable through improved engineering design.

D. EXERCISES

1. Calculate kT and hf for a range of temperature and frequency values and try to appreciate the magnitude of energies involved. Convert the energies obtained into eV.*
2. Sketch a chart (starting with kT and hf at one end) of magnitude to energies and powers around us including those generated by large power stations, lightning discharge, space rockets, motor car, power of a human being, food we eat, atomic explosion, computer elements, biological organisms, neurons in human brain, etc., and try to appreciate the relative magnitudes.

E. A SOLVED EXAMPLE

Assuming that kT represents the minimum discernible amount of energy per degree of freedom. For a system having a bandwidth B hertz (or cycles per second) the total thermal noise power is given by

$$N = BkT$$

Now consider a television signal. Typically for such a signal we would have $B = 4 \times 10^6$ hertz or cycles per second. Therefore, the total noise power admitted into the receiver at room temperature is of the order of

$$N_R = 4 \times 10^6 \times 1.38 \times 10^{-23} \times 300 = 1.5 \times 10^{-16} \text{ watts}$$

(note that 300° K = 17° C = room temperature)

For a good quality television picture we would require signal power to be at least 100 times as much as noise and adding a factor of 10 for receiver imperfection we arrive at 1.5×10^{-13} watts as the minimum power which needs to be received.

This amount of power does not seem a lot until one realises that at a distance of only 50 km from the transmitter the signal intensity is less than one ten-thousandth of a millionth millionth of that obtainable at the transmitter. This means an overall factor of about 10^{-16} which when combined with the quantity of 1.5×10^{-13} watts given above leads to a necessary transmitter power of the order of many kilowatts.

Typically a television station would have output power of 100 kilowatts or more, in order to overcome man-made as well as Brownian noise.

* An electronvolt is the energy equivalent to that needed to move one unit of elementary charge, e, through a potential difference of 1 volt. That is, 1eV = 1.602×10^{-19} joules (see also Appendix 2).

SUGGESTED READING MATERIAL

BRILLOUIN, L. *Science and Information Theory*. 2nd ed. Academic Press, New York, 1954.

PIERCE, J. R. *Symbols, Signals and Noise: The Nature and Process of Communication.* Hutchinson, 1962.

Physical Science Study Committee, *Physics,* D. C. Heath & Co., Boston, 1960.

CHAPTER 3

Patterns, Waveforms and Scientific Measure of Information

A. E. Karbowiak

"... we shall exert all our energies towards the shaping of a plan ..."

New Year message to the Australian people by Prime Minister John Curtin, 29th December, 1941.

A. CONCEPTS AND DEFINITIONS

One readily accepts as self-evident that human voice conveys information. However, speech is not the only means by which humans communicate; facial expression as well as gesticulation also help. We thus see that many aspects of human communications are inherently complex.

Even if we restrict ourselves to communication by human voice, a brief study will soon show that words are not the only carriers of information but that many factors need to be considered, and the problems of measuring information begin to appear formidable.

Intonation, volume and even pitch of voice affect information profoundly. This being the case, how can one measure information? The simple answer is that one cannot really measure information or communication, in the everyday sense, any more than one can measure the suffering of a grief-stricken person. But this need not deter us from an attempt at measuring the seemingly impossible. By trying to measure we improve our techniques, we gain better understanding and in doing so, we gradually appreciate the difficulties involved. Finally, we will find a way of defining something that relates to the quantity of interest, a quantity which we will be in a position of measuring in a precise quantitative manner.

The real purpose in trying to measure the quantity of information and the degree of communication lies in our desire to carry out quantitative comparisons between information carrying systems. In doing so however, the scale of measures must be meaningful in an engineering sense.

When dealing with voice sounds, an obvious choice would be for example, to measure volume of the acoustic disturbance. If we did that, we would obtain results as set out in Table 3.1. We note from the entries in the Table that the average power of human voice is of the order of 24 microwatts, which is an exceedingly small power, when compared with other sources such as a symphony orchestra (of the order of 70 watts). Thus we see that power is one important consideration, but on its own it helps us very little when it comes to speech recognition.

Another possible parameter to study would be the pitch of the voice. An even better description might be to record the actual waveforms of the voice, or the

TABLE 3.1
Power of Acoustic Sources

Source	Power in Watts
Orchestra of seventy-five performers (full volume)	70
Bass drum	25
Pipe organ	13
Cymbals	10
Trombone	6
Piano	0.4
Bass saxophone	0.3
Double bass	0.16
Orchestra of seventy-five performers (average loudness)	0.09
Flute	0.06
Clarinet	0.05
French horn	0.05
Bass voice	0.03
Alto voice pp	0.001
Average speech	0.00002
Violin at softest passage	0.000004

The data given above has been taken from A. Wood *The Physics of Music*. University Paperbacks, Methuen, 1964.

spectrograms* of different sounds, etc. But when we come to examine such evidence, we quickly realise that the parameters which we are studying do not really help us in speech recognition. We therefore conclude that the measures discussed do not necessarily relate to the information bearing capacity of acoustic sounds as generated by humans. Such quite legitimate attempts at scientific study nevertheless produce quantitative data which is of very limited use.

The degree of complexity encountered when trying to measure information, or the degree of communication, can be further appreciated if one realises that information is often conveyed by means of pictures and objects. Here a mere quantitative description is imperfect in that the same number of objects in different juxtapositions will convey different information. Thus, we come to appreciate the tremendous variety of ways, means, and different languages which humans have at their disposal to communicate.

B. ATTEMPTS AT MEASURING INFORMATION

1. Background

"To inform", according to the Oxford English Dictionary, is "to tell" usually an item of news, and "to communicate" is "to impart something". The informer is the source of information but the process of informing does not necessarily ensure that the information has actually been received or understood. Communication, on the other hand, does imply at least a partial two-way flow of information, and some degree of understanding is also involved.

With modern systems, information is usually carried and represented by a variety of electrical signals, and these can be related to the data or patterns transmitted. For example, data on temperature and pressure relate to meteorological patterns; salinity and temperature of sea water give patterns of oceanography, etc. There are patterns of ecology, astronomy, human behaviour, heredity and many others, but this is of lesser concern to communication theory, which deals primarily with translation of patterns

* A spectrogram is a graphical record of the intensity (or loudness) of the various harmonics (frequencies) which are simultaneously present in a given sound.

(or data making up the patterns) rather than with what the patterns stand for.

There is an endless variety of different patterns, but any family of patterns which can be resolved into its constituent data can be analysed in a systematic manner using Information Theory. This is the basis of all scientific analysis and any other sophisticated branch of our knowledge, which can be analysed in this way, can be subjected to the quantitative scrutiny of Information Theory. But no amount of analysis will tell us whether the data which make up the different patterns "make sense" for this would be semantic information, which is outside the domain of Information Theory.

One set of patterns can be translated on a one-to-one correspondence into another set. Such processes are known under different names of "encoding" or "modulation" depending on whether we deal with digital or analogue data. It is also true that digital data can be translated into a corresponding analogue form and vice versa, and the real reason why this is possible is because any analogue pattern realisable in the physical world contains but a finite number of degrees of freedom: the number of independent data needed to describe a pattern is, therefore always finite. (c.f. Chapter 11, F.)

In trying to apply Information Theory to an engineering system, the first step is to find means for translating the various patterns, which we as humans perceive, into some form of quantitative language, and to do that we must devise means of measuring information. One possible way would be to take a poll of opinions and ascribe a scale of values accordingly. This is in fact widely practised in a variety of situations, such as food testing where a group of connoisseurs arrange the produce in their (individual) order of preference, numbering them 1, 2, 3, etc. Then, by taking a weighted average of the results so obtained, we arrive at a quantitative measure which relates to the quality or value of the commodity judged.

In communication engineering a similar approach is frequently used when studying such aspects of human communication as speech intelligibility. However, on the whole, the approach is difficult to generalise, and is used only when other attempts at measuring information fail.

The idea behind the thought leading to the scientific measure of information can be best explained using a simple example. Suppose the objective is to measure the aesthetic value of a lady's dress. The task as stated here, would appear to be impossible to implement. Yet a systematic and scientific approach does enable us to go a long way towards the objective. One simply invents a scale of measure which it is thought relates to the quantity of interest. In this way, the position of the hemline with respect to a convenient reference mark could provide such a scale of measure. If we then say the hemline is 20 inches above the ground, then the statement becomes quantitative and acquires scientific meaning. This is a parallel of what information theory tries to do. Indeed, it is construction of examples of this nature of successively increasing degree of complexity that led to the development of Information Theory in the form in which we know it today.

The subject of communication was outside the domain of science until a sufficiently rigorous definition of "information" and "communication" was established. Undoubtedly, this was a great step forward, but at the same time it opened up a way to endless confusion (in various disguises), by accepting the words "information" and "communication" and giving them a new meaning which was strange*.

Viewing this aspect in retrospect, perhaps much of the confusion could have been avoided had we defined new terms such as "data extraction", "data transmission", "data processing" and the like, because in reality this is what we are concerned with. It does not really help to say that a telegraph message "arrive tomorrow" carries

* Compare a similar dilemma which was encountered with reference to words such as Energy and Power (See Chap. 2, B.).

20 bits of *information*:* it would really be better and less confusing, to say that the given telegraph message takes 20 bits of *data to transmit*, notwithstanding any meaning of information contained in it. The scientific usage of the words "communication" and "information" is now well-established and we shall therefore use these words in this newer sense rather than follow the common usage.

2. Brief History

Man, quite early on in history has been aware not only of the need to communicate but also of the existence of various degrees or levels of communication. Common language was found to be a great asset, while loss of a channel of communication, such as sight, was recognised as unwelcome. In this way "language" and "channel" came to be accepted as necessary for communication. No measure of information was invented until this century, but it was accepted, from experience, that given a language and a channel we also need time to communicate. In general, the longer the available time, the better the chances of conveying more information over existing channels, or of being better understood. These qualitative notions of information and communication were in existence long before communication theory was invented.

The need for a convenient measure of information arose from engineering problems encountered with early communication systems. It was Morse and his co-workers around 1840 who devised a nearly optimum set of symbols for the early telegraph, by assigning simple symbols to the frequently occurring letters of the alphabet and reserving the more complicated symbols for those with lesser probability of occurrence (Fig. 3.1). It is also interesting to note that, the frequency of occurrence of various letters of the alphabet in the English language was deduced from an examination of the quantity of fount kept by printers.

It is here that we can see an early attempt at a successful solution of the problem of optimum coding. Clearly, if we examine the entries in Fig. 3.1 we come to the fundamental conclusion—which at first is not at all obvious—that to minimise the cost of communication (and simultaneously to minimise the time needed for communication) we must assign to frequently occurring letters, such as E, T or A, simple symbols and reserve the complicated symbols for the infrequently occurring letters such as J, X or Z.

Now comes the crucial reasoning. With a simple symbol (a pattern) such as a single dot or a dash, we associate a small amount of information, while with a complex symbol such as — · · — (letter X) we associate a larger amount of information. Therefore, we conclude that symbols having a high probability of occurrence carry a small amount of information, while symbols with low probability of occurrence are deemed to have a larger amount of information. This is not at all like the intuitive notion of information, but our present considerations demand such associations. We shall return to this point later on (see Chapter 11).

Information theory in its modern form started with the work of the communications engineer Nyquist, 1924 and Hartley around 1928, and was developed into a complete theory based on firm mathematical foundation by Shannon in 1949. Around the same time the mathematician Wiener developed a complete theory of optimum detection of signals which differed somewhat from Shannon's approach. The two theories, in a way, could be said to complement one another, and practically all recent developments in this field are related to one or the other of the two approaches.

Much of the formalism of communication theory is submerged in mathematical exposition; this is its strength. But, at the same time, it is its weakness, in that it appears to have little in common with reality. In fact, it has been said that a branch of

* Refer to Sec. C. for the meaning of term "bit"

Symbol	Original Morse Code	Frequency of Occurrence of Printer's Type	p(j)
E	·	12,000	.13105
T	—	9,000	.10468
A	· —	8,000	.08151
I	· ·	8,000	.06345
N	— ·	8,000	.07098
O	· ·	8,000	.07995
S	· · ·	8,000	.06101
H	· · · ·	6,400	.05259
R	· · ·	6,200	.06882
D	— · ·	4,400	.03788
L	——	4,000	.03389
U	· · —	3,400	.02459
C	· · ·	3,000	.02758
M	— —	3,000	.02536
F	· — ·	2,500	.02924
W	· — —	2,000	.01539
Y	· · · ·	2,000	.01982
G	— — ·	1,700	.01994
P	· · · · ·	1,700	.01982
B	— · · ·	1,600	.01440
V	· · · —	1,200	.00919
K	— · —	800	.00420
Q	· · — ·	500	.00121
J	· · · ·	400	.00132
X	· — · ·	400	.00166
Z	· · · ·	200	.00077

FIG. 3.1 The Frequency of Letters in an English Text: Morse Code.*

science, which is in its infancy, needs a great deal of mathematical theory to support it, but once the necessary knowledge has been secured, then the phenomena can be explained without recourse to complicated mathematical analysis. This is partly true of Information Theory in that the formalism is essentially abstract.

C. PATTERN GENERATING CAPACITY OF A SET: A POSSIBLE BASIS FOR A SCIENTIFIC MEASURE OF INFORMATION

1. Patterns and Information

Figure 3.2(a) represents a simple set. The set is defined by saying that there is one site and a convention (or rule of language) that the site can be either vacant or have a black spot in it. Clearly, such a set can be made to represent at most two distinct patterns: (1) a vacant site, (2) a site with a black spot in it. Therefore we say that the pattern generating capacity of the set is 2 patterns.

Figure 3.2(b) represents a different set, in that there is a different convention: whereas previously a black spot was the alternative to a vacant site, in the present case

* The modern day Morse code differs somewhat from the original version above. The letters in question are:

O	— — —	R	· — ·	L	· — · ·
C	— · — ·	F	· · — ·	Y	— · — —
P	· — — ·	Q	— — · —	J	· — — —
X	— · · —	Z	— — · ·		

	Configuration	Information Capacity	
		Patterns N	Capacity Bits
(a)		2	1
(b)		2	1
(c)		$2^2 = 4$	2
(d)		$2^3 = 8$	3
(e)	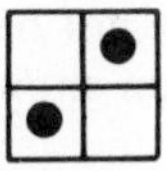	$2^4 = 16$	4
(f)	q sites	2^q	q

FIG. 3.2 Capacity of a Set.

a site coloured black is the alternative to the vacant site. The sets illustrated in Fig. 3.2(a) and (b), therefore, differ in *form,* but are equivalent as far as pattern generating capacity is concerned.

Putting aside, for the moment, the question of form we can extend the above concept to more complicated patterns. Thus with two sites (Fig. 3.2(c)) the pattern generating capacity of the set is $2 \times 2 = 4$ patterns. With 3 sites (Fig. 3.2(d)) the capacity is $2^3 = 8$ and with 4 sites it is $2^4 = 16$ patterns. In general, if there are q sites, the maximum number of distinct patterns which the set (language) can be made to represent is $2^q = N(q)$. It should be noted that the exponential law, giving the number of patterns in the set, is an important property, while the binary base is a direct consequence of the fact that there are two choices per site (either vacant or occupied): if there are 3 attributes per site, then obviously the base would be 3 and the law for number of patterns which the set can generate would then, of course, be 3^q; and correspondingly for other cases.

Knowing the property of the set, it would be adequate to describe the capacity of the set in terms of the exponent q. This has its advantages as will be apparent from the following discussion.

With reference to Fig. 3.2(a), we note that one statement is needed (either Yes the site is vacant, or No the site is occupied) to determine the pattern. Clearly, for a set with 2 sites, 2 decisions are needed, and for a set with q sites, q Yes/No decisions need to be made to determine which particular pattern has been chosen out of the family. Thus q is the number of binary decisions (Yes/No, Black/White, etc.) needed to identify the particular pattern unequivocally. Therefore, q is the binary information, measured by the number of binary decisions (or bits for brevity) needed to reconstruct any one pattern of the set. In fact, q bits of information is a characteristic of the set.

There are, therefore, two ways of describing the pattern generating capacity of a set:

(a) by the actual number of distinct patterns which the set can generate

$$N = 2^q \text{ for binary patterns} \tag{3.1}$$

or

(b) in bits

$$C = q \text{ bits.} \tag{3.2}$$

The utility and simplicity of the concept of measuring information in bits is one of the principal attractions of the method.

2. Patterns and Form

Let us now return to the question of form. We observed in connection with the patterns illustrated in Fig. 3.2(a) and (b), that the two sets, although identical as regards the pattern generating capacity, differ in form. Obviously, if form is important, then it can also be taken into account by a suitable digital description. Thus, with reference to (a) and (b) in Fig. 3.2, if we say that there are 3 attributes per site (vacant, black spot, completely black) then (a) and (b) will be sub-sets of a one-site set with three attributes. The total number of patterns which the set can be made to represent is 3 (not 4, since the vacant site repeats and must not be counted twice). This conclusion also follows from consideration of the diagram shown in Fig. 3.3. The diagram is known under the name "Venn diagram".

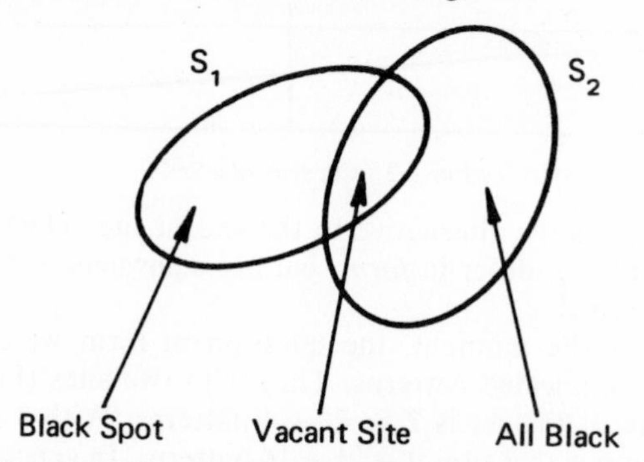

FIG. 3.3 Venn Diagram Representation.

Returning now to Equation 3.1 we see that on taking the logarithm to the base of two ($\log_2$ will be denoted by lgg for brevity) we get:

$$\text{lgg } N = C \text{ in bits.} \tag{3.3}$$

Hence follows the general rule given the total number of patterns which the set (or the language) can generate within a given framework (grammar), the total number of Yes/No binary decisions needed to determine any one number unequivocally is given by taking the logarithm to the base of two of the total number of patterns which the set can generate.

We can use Equation 3.3 to derive a formula for the capacity of a multi-attribute set. Clearly if there are n attributes per site the pattern generating capacity per site is n. If there are q sites, the total number of patterns which the set can generate is clearly

$$N(n,q) = n^q \tag{3.4}$$

taking the logarithm to the base of two of both sides of (3.4) we get for the capacity of the set

$$C = \text{lgg } N = q \text{ lgg } n, \text{ bits.} \tag{3.5}$$

Therefore, the information conveyed by any one pattern of the set is q lgg n bits. Consequently, it follows that the information can be "translated" on a one-to-one basis and be represented by a binary set of q lgg n sites.

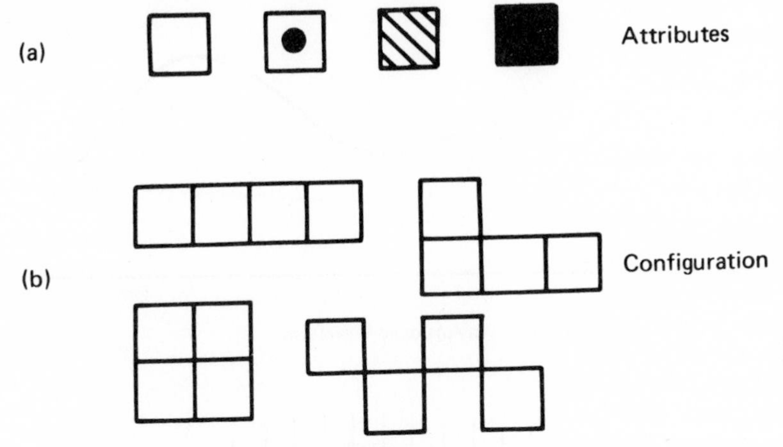

Fig. 3.4 Patterns and Form.

Varying the attributes of the sites is one way of changing the form of a set. There are, however, many other alternatives. Figure 3.4(b) shows a number of sets of identical capacity but differing in configuration. No doubt, we could consider such configurations as being sub-sets of a much larger set, and take configurational changes into account quantitatively (in much the same way as we did take a number of attributes), but if we do not put any restrictions on the configurational changes, then the mother set will grow in size accordingly. In the end, to describe a particular pattern and the configuration may require an inordinate quantity of data, so that the description in such terms becomes impractical. Such are the difficulties associated with detailed description of form. The fact however remains that, in principle, forms of any nature can be described quantitatively, and the scientific measure of information as described above is of immediate and direct use.

D. WAVEFORMS: A CLASS OF PATTERNS

In an earlier chapter we indicated that with systems, information is commonly carried by electric currents in the form of a waveform. Such waveforms represent data essential to the proper functioning of the system.

Figure 3.5 represents a typical waveform which can be approximated by a series of steps as indicated for a fragment of the waveform in Fig. 3.6.

To calculate the pattern generating capacity (or the information carrying capacity) we consider a waveform approximated by a series of steps. Clearly, if the system can effectively resolve n steps in amplitude and k steps in unit time, then we can regard the waveform as a pattern with k sites, each with n attributes (the steps in amplitude), per unit of time. Thus the information carrying capacity is obtained by adapting Equations (3.4) and (3.5). Thus the capacity of a typical waveform is

$$\begin{aligned} C(n,k) &= \text{lgg } N(n,k) \\ &= k \text{ lgg } n \text{ bits/sec.} \end{aligned} \tag{3.6}$$

To give a numerical example, consider a waveform with 8 resolvable steps in amplitude and 10 resolvable steps in every one second interval. Here $n = 8$ and since lgg $8 = 3$ therefore

$$C(1) = 10 \times 3 = 30 \text{ bits/sec.}$$

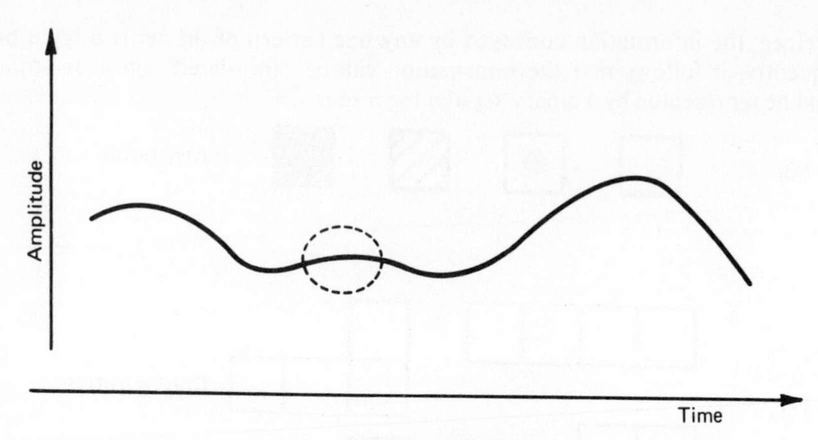

FIG. 3.5 An Analogue Waveform.

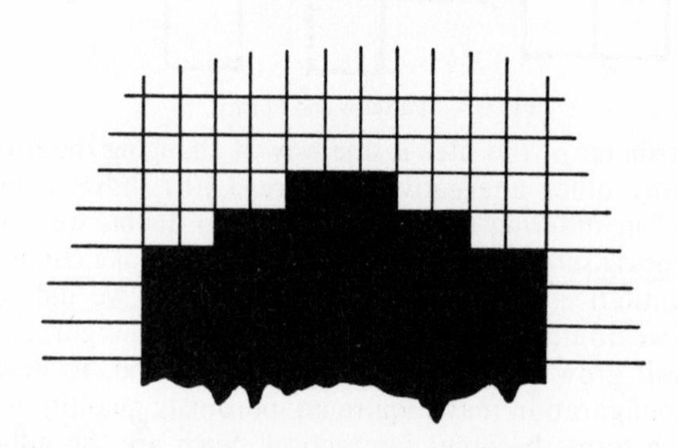

FIG. 3.6 Fragment of Waveform Shown in Fig. 3.5.

We thus conclude that the system can generate $N = 2^{30}$ different waveforms (patterns) in a one-second interval of time.

Note that a voltage waveform of 2 sec duration contains $C(2) = 60$ bits of information and in general a waveform of p seconds duration contains $C(p) = 30\,p$ bits of information. Thus it appears that the chosen scale of measuring information in bits is particularly convenient when dealing with time sequences, in that the information is simply proportional to time. The total number of distinct sequences which the waveform can represent is an exponential function of time. In our numerical example

$$N(n,t) = 2^{C(n,t)} = 2^{30t} \tag{3.7}$$

when t is measured in seconds.

If any one pattern were selected at *random* then, clearly the chance of making a particular selection in any one of the one-second intervals is equal to the reciprocal of the total number of patterns, i.e., 2^{-30}, assuming the patterns to be equi-probable. Therefore, the figure of 30 bits describes the uncertainty as to which pattern might have been transmitted in the given one-second interval, and at the same time gives us a quantitative information as to how many bits (Yes/No decisions) are needed to determine which particular selection has been made. The fact that we might conceivably have guessed the right pattern after only a few guesses is irrelevant, because on other occasions we would not be so fortunate, and *on average*, we could not be sure that a correct choice has been made unless we are given 30 bits of information in every second. Thus 30 bits of information are necessary and sufficient to determine any one pattern.

To summarise: We have thus established, both mathematically and conceptually, that for a waveform which can take on any one of n discrete steps in amplitude in every interval $1/k$ seconds long, the potential capacity is given by

$$C(n,k) = k \text{ lgg } n \text{ bits/sec.} \tag{3.8}$$

The total number of distinct patterns which a waveform could be made to represent increases exponentially with time, and is given by

$$N(n,kt) = n^{kt} = 2^{C(n,kt)} = 2^{kt \text{ lgg } n} \tag{3.9}$$

The exponential law for the information capacity is an important property of waveforms as well as sequences, and we shall return to this point later on.

The use of logarithms to base 2 for the description of system capacity is not essential although it is a convenient choice when dealing with patterns with binary attributes per site as we have discussed above. In other applications, logarithms to the base 10 or to the base e are used. There are three units of information in current use. These are

1 bit being a choice between 2 equi-probable events
1 nit ,, ,, ,, e ,, ,, (equals 1·44 bits)
1 hartley ,, ,, 10 ,, ,, (equals 3·32 bits)

These units involve logarithms to the base of 2, e, and 10 respectively. Short tables of powers of 2 and powers of e are given in Appendix 3.

E. EXERCISES

1. Calculate the pattern generating capacity of a (true) 6-sided die in bits per trial. (Ans: 2·58 bits).
2. A source of signal in the form of a waveform can assume 16 distinct values in successive microsecond intervals of time independently of all the other values. Calculate the capacity of the source in bits/sec and estimate the total number of distinct patterns which could be generated in 1 second. (Ans: roughly a number with one million significant digits which corresponds to 4 million bits.)
3. A chess board consists of 8×8 sites, calculate the capacity of the set in bits and then estimate the capacity of the set in terms of the total number of binary patterns N. (Ans: $N \approx 2 \times 10^{19}$.)
   Assuming that a human brain contains 10^{10} effective neurons and that each one could store one pattern, compare this number with N.
   If a human being could learn to memorise the pattern (N) at a rate of one per second, compare the time needed to memorise all N with human life span.
   Postulate an ultra-high speed machine that could classify patterns at the rate of 10^9 per second and estimate the time needed for the machine to accomplish the task.

SUGGESTED READING MATERIAL

PIERCE, J. R. *Symbols, Signals and Noise: The Nature and Process of Communication*. Hutchinson, 1962.

SINGH, J. *Great Ideas in Information Theory, Language and Cybernetics*. Dover, 1966.

MILLER, G. A. *Language and Communication*. McGraw-Hill, New York, 1963.

KARBOWIAK, A. E. *Theory of Communication*. Oliver & Boyd, 1969.

WOOD, A. *The Physics of Music*. University Paperbacks, Methuen, 1964.

CHAPTER 4

The Semiconductor Revolution

L. W. Davies

"There was reason for fear that, like Saturn, the Revolution might devour each of its children in turn."

Pierre Vergniaud.

A. INTRODUCTION

In this chapter we are concerned with real systems, rather than with models. We trace the way in which electronic systems, and particularly the building up of very complex electronic systems such as computers, have been profoundly influenced by the advent of semiconductor electronics.

It turns out that many complex electronic systems in fact owe their existence to the invention of the transistor, and to the later development of semiconductor integrated circuits. Satellite communication systems fall in this class. Other systems are being developed to an increasing extent which in part involve electronics in process control, or in other applications; the attractive costs of integrated circuits, or "micro-electronics", have played their part in accelerating to a quite remarkable extent the rate at which such developments are being made.

Thus a story of complete and fundamental reconstruction in the world of electronics is unfolded, and we use the term "semiconductor revolution" to describe it. The reader may feel that use of such a radical descriptor is a little inappropriate in a technological context, and it is true that no actual violence has been done other than to the outlook of engineers and manufacturers concerned with the design and construction of electronic equipment. But so far-reaching have been the cost-savings, improvements in reliability and in functional capabilities of integrated circuits that we shall persist with the use of the term.

As will be seen in the following sections, the revolution involved the overthrow of vacuum tubes (valves) and other discrete circuit components by integrated circuits. In an integrated circuit, a number of transistors, diodes, resistors and capacitors are fabricated together, and inter-connected, on a single small crystal of silicon. In order that the important engineering consequences of this new micro-electronics may be more fully appreciated, some of the essential background understanding of semiconductor and device physics is elaborated in more detail in Chapters 15-17 of Part II.

B. COMPONENTS AND DEVICES IN SYSTEMS

In previous chapters of this book we have been concerned with information, and with communication, with a view to the application of these concepts to systems, and to computers in particular. Systems are large, complex, man-made equipments which are designed to perform specific functions. Examples of systems include electrical power generation and distribution systems, communication systems, co-ordinated traffic lights, automated rolling mills in steel fabrication, and so on. A system is made up of a number of parts, each of which performs a specific function. We are particularly concerned in this book with those of the sub-systems which perform an electronic function such as sensing, amplification, computation, transmission, etc. Carrying this generic subdivision a stage further, each electronic sensor, amplifier etc., is composed of electronic components which have specifically designed electrical characteristics.

As an example of this more detailed subdivision, consider the case of a television receiver. A receiver is one of many similar elements of a television broadcast system, shown schematically in Fig. 4.1; other parts of the system are the studio in which the programme originates, the television cameras and microphones, the transmitter, and the transmitting and receiving antennas. Each receiver, as presently manufactured, is made up of a number of components contained within its cabinet: valves or transistors, capacitors, resistors and inductors, a loudspeaker, and a picture tube which acts as a transducer to convert the amplified output of the receiver into a visible pattern on its screen.

Corresponding to the subdivision of a complete television broadcast system outlined above, creative engineering activity has been required at three distinct levels. In the first place there was the invention and development of electronic techniques for the conversion of video and sound information into a form compatible with the known methods of transmission and reception of electromagnetic waves; in other words, the communication system was devised. This involved the discovery of suitable transducers (television camera tubes, and microphones), of modulation techniques which allow the signal information generated by the transducers to be impressed on the transmitted electromagnetic wave, of antennas which can be made sufficiently cheaply for use with a receiver in the home, and so on.

In the second place, there has been the continuing task of designing each sub-system, of which the receiver is one example, so as to incorporate improved performance and reliability. An additional requirement is, of course, that such improved designs shall always be economically competitive.

Finally, we must not forget that the electronic components of which the television receiver and other sub-systems are constructed are themselves man-made: engineered from materials such as conductors, semiconductors, insulators, phosphors, and so on. The application of new physical phenomena and processes to the invention and development of useful electronic devices or components is a continuing process, and it is with this aspect that we shall be concerned in Chapters 15-17. The development of an improved electronic component in general has some influence on the second of the levels of engineering referred to above, but we shall see that in the case of the new semiconductor components this influence may indeed be revolutionary. Semiconductor devices have opened up possibilities of completely new systems, as will be seen in the next section.

C. VACUUM TUBES AND TRANSISTORS

Of the electronic components referred to above, let us first focus our attention on the active devices, those that are capable of amplification and switching. The earliest active electronic devices, the vacuum diode and triode, depend for their operation on two basic phenomena: the thermionic emission of electrons from a heated cathode, and the laws of motion of electrons in a vacuum when subjected to electric fields. A vacuum diode has the properties of a nearly ideal rectifier because the electrons emitted from the cathode will flow through the vacuum space to the anode, and thence into the external circuit, only when the anode is made positive with respect to the cathode. If a control grid is placed in the space between cathode and anode, as in a vacuum triode, the application of a negative voltage to the grid controls the flow of electrons to the positive anode, and may even reduce it to zero. Again the ability to

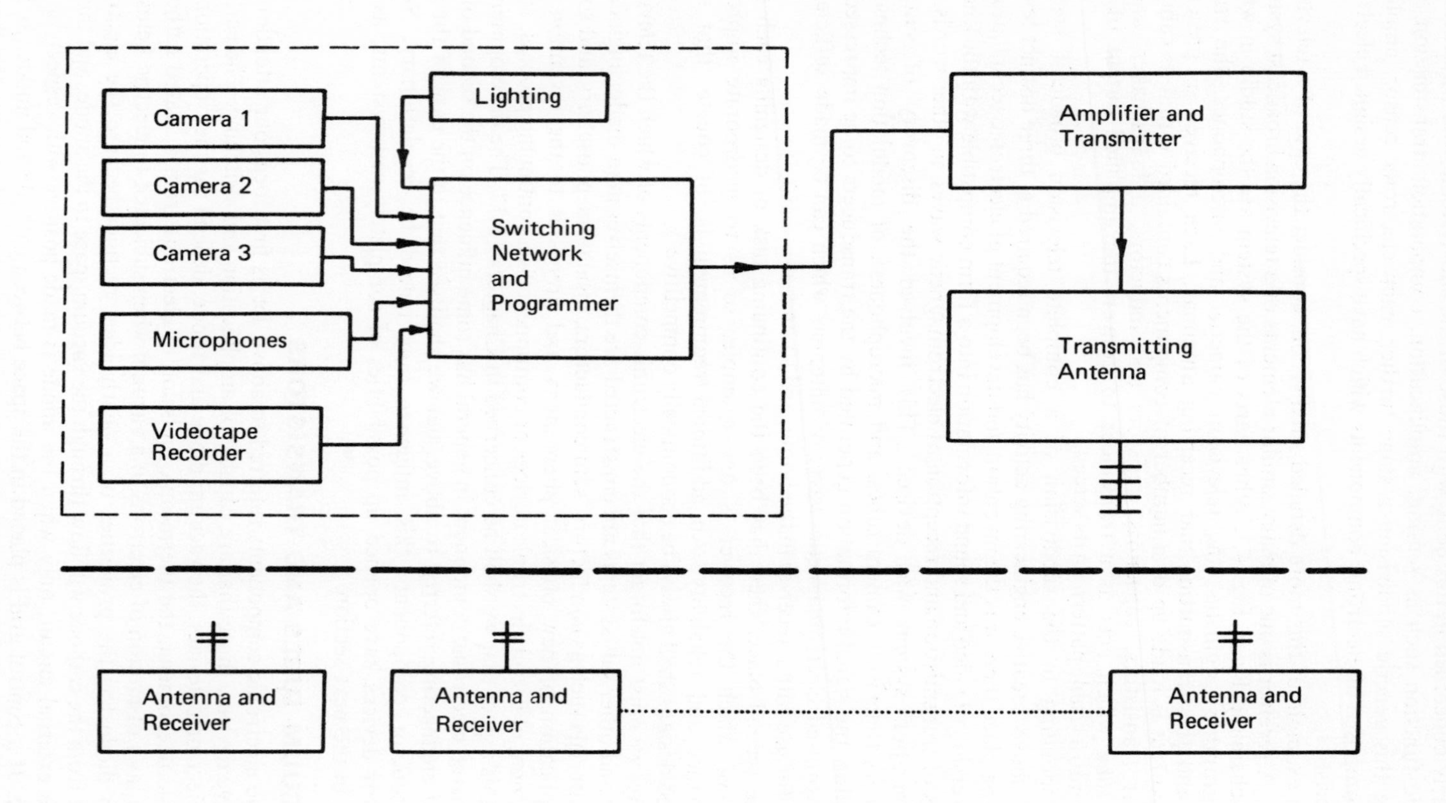

Fig. 4.1 Television Broadcast System.

control the magnitude of the anode current arises from the effect on electrons of the electric field set up in the vacuum space. Grid control of electron flow to the anode is so strong that one can obtain in the anode circuit an amplified version of a signal voltage applied to the grid.

The physical phenomena which form the basis of operation of transistors, and of semiconductor $p - n$ junction rectifiers, will be discussed in more detail in Chapter 16. It is sufficient to say at this stage that a transistor owes its amplifying and switching properties to the motion of electrons in a semiconductor crystal, into which impurity atoms have been introduced in a controlled way. Both the vacuum triode and the transistor are three-electrode devices, but in the case of the transistor it is not necessary to supply heat to the electrode corresponding to the cathode in order to produce a supply of electrons whose motion in the crystal can subsequently be regulated by the control electrode. In the transistor, the electrons are already present at ambient temperatures.

The transistor possesses a number of distinct advantages when compared with a vacuum triode, from an engineering point of view, and we may enumerate them as follows:

1. Transistors may be operated at much lower power levels than vacuum triodes. In the first place there is no requirement for cathode heater power; secondly, it turns out that the bias voltages required are very much lower for a transistor than for a triode. A further incidental advantage in some applications arises from the finite time required for the cathode of a vacuum tube to reach operating temperature after switching the equipment on; there is no such delay in transistorised equipment.
2. Vacuum tubes or valves are inherently less reliable than transistors, principally on account of the heated cathode in the former. Whenever a solid, such as a thermionic cathode, is heated to temperatures of the order of 700°C or above, it becomes possible for atoms to move around to some extent by a process of solid-state diffusion, and eventually to lower the thermionic emission efficiency of the cathode. There are no such effects in transistors, which are generally operated within a range of temperature up to 150°C above room temperature, depending on the amount of power being dissipated in the transistor under amplifying or switching conditions.
3. Transistors have considerable advantages of size and weight over vacuum tubes. It is difficult to manufacture a vacuum tube and enclose it in a cylindrical vacuum envelope which is much under 1 cm diameter, and 2 cm long. Transistors however can be fabricated on a silicon crystal as small as 350 microns (μ) square and 100μ thick, on the other hand.*
4. Since they are fabricated from a solid crystal, transistors have greater resistance to mechanical shock and vibration than valves, for in the latter there are at least three metal electrodes which must be constrained in position in a vacuum enclosure.

It is a natural consequence of these many advantages that transistors should be used to a greater and greater extent in the design of electronic equipment. There was a gradual transition in this direction in the fifteen years following the invention of the transistor announced in 1948. The trend to transistorised equipment was not complete, however, because not all the capabilities of vacuum tubes can be emulated by transistors; this is particularly the case in respect of high power output, and amplification at high frequencies. Vacuum tubes are likely to remain superior in these areas for some time to come. Furthermore, economic considerations have sometimes

* 1 micron = 10^{-6} metre = 10^{-4} cm.

precluded the replacement of valves by transistors in many applications. For these reasons, and also because it is necessary to maintain a supply of vacuum tubes to replace those failing from time to time in equipment already in use, there are today (1971) approximately equal numbers of transistors and vacuum tubes manufactured each year.

One important consequence of the greater reliability of transistors has been the ability of engineers to consider the design of more and more complex equipment, incorporating an ever increasing number of active devices. The way in which this consequence comes about is not too hard to see. Suppose that vacuum tubes have an average life expectancy equal to 5000 hours of operation. A piece of equipment which incorporates 5000 vacuum tubes will then be expected to cease operation, due to failure of one of the vacuum tubes, at intervals of approximately one hour, on the average. If we wish to design a complex computer incorporating 50,000 vacuum tubes we see immediately that the likely period of operation between failures is only six minutes! Since such a level of reliability is totally unacceptable, we can say that the complex computers of today owe their existence for the most part to the invention of the transistor, with its greatly enhanced reliability. As will be seen in Section D, an even greater improvement in reliability can be achieved with the use of integrated circuits, by which additional modes of failure can be circumvented. It is estimated that in the present decade (1970s) failure rates for integrated circuits will be reduced to values as low as 0.001 per cent per 1000 hours of operation.

Reference was made in Section A to the possibility of devising completely new systems as a consequence of the invention of a new device. Satellite communication systems were specified as an example, the new device involved in this instance being the "solar cell", a by-product of the invention of the transistor.

One of the interesting properties of a silicon $p-n$ junction, to be described in detail in Chapter 16, is its ability to convert incident light into electrical power with a conversion efficiency of approximately 10 per cent. Solar energy, in the form of sunlight, arrives at the earth's surface at a rate of approximately 1 joule per m sec. (1kW) per square metre normal to the direction of the sun. A one-square-metre array of silicon $p-n$ junctions, so-called solar cells, is therefore capable of providing useful electrical power amounting to 100W when sunlight falls on it.

Present satellite communication systems would be quite impracticable but for the power available from solar cells. Such systems were first contemplated early in the last decade. Initially a satellite consisting of a large aluminised sphere, Echo 1, was used to reflect signals from a transmitter on earth and thus to provide intercontinental line-of-sight communication at microwave frequencies. Severe technical difficulties were encountered, however, arising in part from the low level of the signal power reflected from the satellite. Today's commercial systems are able to receive, amplify and re-transmit on board the satellite, the necessary electrical power being supplied by solar cells. The Intelsat IV satellite is to be launched in 1971. The greater part of its cylindrical surface is covered with more than 30 square yards of solar cells, supplying the electronic equipment on board the satellite with more than 1kW of electrical power. The development of transistors, with their low power requirements, and of solar cells, has therefore led to a completely new system for world-wide communications.

The reader may well enquire why it is that solar cells, using "fuel" of zero cost, have not replaced the more conventional methods of electrical power generation on the earth's surface itself. The reasons why this replacement has not taken place are, of course, entirely economic. Silicon solar cells are at present rather expensive to manufacture, but even if the initial cost of solar cells were substantially reduced one is still faced with the fact that electrical power is by no means consumed only when the sun is shining. An essential element of such a power generation system would therefore be a cheap means of storing electrical energy. If such a means is devised in the future, this may well be the genesis of another completely new system.

D. THE REVOLUTION IN ELECTRONICS

We have seen that the invention of the transistor, and its subsequent development, has produced far-reaching changes in the design of electronic equipment. Because of the increased reliability of the active devices, engineers are able to contemplate the satisfactory construction of equipment of a much higher degree of complexity than could be undertaken with the use of vacuum tubes. The lower consumption of electrical power by transistors has similar consequences also. But changes in design concepts of even greater magnitude than these have arisen in the past ten years, with the development of silicon integrated circuits. It is with the revolution constituted by these changes that we shall be concerned in this section.

In Section C it was mentioned that a silicon transistor could be fabricated in, or on, the surface of a silicon crystal of square cross-section, 350μ on edge. This dimension was actually specified for the reason that it is the minimum size of crystal that can conveniently be handled in manufacturing processes. (Such processes are naturally carried out under a microscope, in view of the small dimensions of the crystal). As will be shown in Chapter 17, it is actually possible to fabricate many transistors on a crystal of silicon of these dimensions, and to interconnect the transistors subsequently by an appropriate pattern of metal placed in the form of a thin film on the surface of the crystal. Since we may also at the same time fabricate resistors and capacitors as an integral part of the small silicon crystal, we have available a technique for manufacturing a complete integrated electronic circuit, containing up to some 20 components, on a crystal whose dimensions are the smallest that can be handled with ease.

These developments have indeed proved to be revolutionary. As we should expect, the possibility of fabricating entire circuits, thousands at a time on a single slice of silicon crystal, leads to substantial economies in cost—provided that there is a market to absorb the particular circuits produced. It also turns out to be the case that integrated circuits are inherently more reliable than circuits constructed of discrete components which are soldered together in the conventional fashion. Integrated circuits thus enable engineers to undertake the design, with satisfactory reliability, of even more complex equipments than were possible with transistors and other discrete components.

From the individual design engineer's point of view, some of the most exciting and interesting aspects are the creative opportunities opened up by integrated circuits. No longer is the design of a circuit necessarily a process of minimising the number of transistors to be used; one is instead faced with an area of silicon crystal on which resistors, capacitors, diodes and transistors can be fabricated in any chosen ratio, the area of silicon used being minimised in an optimum design. In general, transistors and diodes are found to be more economical of space than resistors and capacitors, leading to further changes in design considerations. Furthermore, because it has not proved to be possible to devise satisfactory inductors as integral elements of a silicon crystal, new ways of carrying out particular circuit operations (e.g., frequency-selective amplification) have had to be devised. This is indeed a revolution in electronics, one which is still in progress and to which graduating electrical engineers can look forward to contributing.

The revolution in electronics since the invention of the transistor in 1948 points up the rapidity with which new inventions and processes are incorporated in modern manufacture. This rapidity is a characteristic of the electronics industry. The period since 1948 has for example seen the rise and fall first of the point-contact transistor, and then of the germanium junction transistor. There have been other exciting developments as well in solid-state electronics, also marked by extreme rapidity of the transition from research laboratory to factory.

E. PROBLEMS

1. (i) If a given integrated circuit occupies a square of silicon 350μ on edge, approximately how many ICs may be obtained from a silicon wafer 5 cm in diameter?
   (ii) Suppose that each of the above circuits contains ten circuit elements, and that it costs ten dollars to process a wafer. If the yield of good devices is 50 per cent, what is the cost of each useful circuit element?
2. (i) Consider a house in the tropics, with the sun shining vertically on its flat roof. If the roof area is 1000 square feet, at what rate is solar energy incident on the house?
   (ii) If the roof of the house were completely covered with a cheap solar cell of efficiency only 1 per cent, what electrical power would be available to the householder?
3. Referring to the Intelsat IV satellite discussed earlier the height and diameter of the "solar drum, are respectively 111 inches and 93.5 inches.
   (i) Calculate the area of solar cell.
   (ii) What is the electrical power supplied to the equipment on the satellite when in orbit, assuming a solar cell efficiency of 12 per cent?
   (iii) Is a storage battery necessary on a communication satellite in synchronous orbit?

SUGGESTED READING MATERIAL

HITTINGER, W. C. and SPARKS, M. "Microelectronics", *Scientific American*, **213**, No. 5, p.56, 1965.

HEATH, F. G. "Large-scale integration in electronics", *Scientific American*, **222**, No. 2, p.22, 1970.

CHAPTER 5

Building Blocks for Systems

R. M. Huey

"The world should know by this time that one cannot reach Parnassus except by flying thither"

From the notebook of G. M. Hopkins, Oxford, April, 1864.

A. INTRODUCTION

In describing the nature of the building blocks from which descriptions of systems both large and small may be built up, it will often be convenient to describe building blocks as if they were electrical or electromagnetic in nature. There is a fundamental reason for this choice, which is really a desire on our part to follow a pattern of development (i.e., of the ideas of system, system analysis and system design) which is similar to the pattern by which, historically, such ideas were formulated and put to use for mankind.

Examples of building blocks and systems will also be used which are not of an electrical nature. There are certain general principles which are true no matter what sort of system we may consider. Both sorts of example (i.e., electrical and non-electrical) have been chosen with the aim of fixing some of these general principles in the mind of the reader.

It is our intention to point out some of the mathematical descriptions which are used by the group of professional people nowadays known as "systems engineers". Originally developed mainly within the disciplines of electrical engineering and operations research,* these very general formulations are being found useful in many fields of engineering and applied science including the life sciences. They are finding widespread application too in the social, political and military sciences. Already extensively used in administration and modern business management, they may also secure a place in the professional armoury of the educator and the humanist.

It is easy to see therefore, that we must strive towards an extremely generalised description (i.e., a generally applicable mathematical model) for our building blocks in order that the same sort of description may be used profitably and easily for the extremely varied building blocks that we may encounter in all these different kinds of "systems". The way in which this can best be done is to seek out the simplest mathematical model capable of yielding the sort of information which we require.

This generalised but simple description was achieved in the electrical engineering field by three main steps,

* The name *operations research* was given to the use of certain fairly simple mathematical techniques in order to obtain the *optimum* (or "best") answer to a defined problem. The technique is not really research in the sense that a scientist or engineer uses that word. *Operations research* is the study of the optimum use of resources, and we will be saying more about it under the heading of *optimisation*.

(i) a wise choise of variables,
(ii) by linearisation of the mathematical models chosen to describe the building blocks (i.e., by discarding the complexities of non-linear behaviour),
(iii) by choosing to use what is known as the "black box" concept to describe the behaviour of a number of building blocks which have been connected together.

The same steps were the basis of the success of the operations research technique, together with a fourth one,

(iv) The choice of a suitable criterion so that *optimum* results can be clearly and unambiguously recognised.

Non-linear problems[1] may be dealt with by graphical methods, by numerical computation, by more complicated mathematical analysis or by the artifice of choosing and solving the nearest equivalent linear problem—a much-used trick known as "linearising".

We will say more about each of these wise choices in what follows.

B. HOW CAN A SYSTEM BE DESCRIBED?

In some of the suggested references at the end of this and other chapters, are listed some of the attributes of a system. These are expressed in a way that is particular to each writer. If one looks at other authors one will see other expressions of what a system is and also of what a system is not. Probably not one of these catalogues, by itself, is a complete one. It is only by wide reading, by considering many different mathematical models and by working out a variety of examples that one acquires a firm notion of what is meant by the deceptively simple word: "system".

Quite often when one is reading or talking on this subject it is not even clearly stated whether the "system" under discussion is the actual physical collection of apparatus, devices, instruments, machinery and control knobs, levers, etc., or whether the discussion refers to the mathematical model which has been built up in our minds and by using symbols written on paper in order to represent the actual physical system.

When one speaks of the behaviour of the system, one may mean either the actual behaviour of the system (i.e., speed, energy, position, etc., at known times) or one may mean the algebraic or arithmetical behaviour (i.e., the manner of variation of the relevant variables concerned) of the mathematical model which has been conjured up or chosen to represent the actual system. Borrowing from the everyday jargon of computer engineers we might speak of the actual physical system as the "hardware" and the mathematical model as the "software".

In the jargon of the computer world the word "software" is commonly used to mean a set of instructions for the computer, i.e., a program or a compiler. Rather than being written as a set of symbols on a piece of paper, these might be "stored" as alphanumeric symbols on punched cards, in magnetic-core memory stores, on magnetic tape or even in some of the newer computer-graphic systems in visual or photographic or holographic form. A mathematical model could be recorded in this way too. See also Chapters 6 and 18-20.

We will not attempt to give a canonical* definition of a system, but will proceed by way of examples and discussion to build up ideas as to what it is and what it is not.

C. SOME KINDS OF SYSTEMS

Let us list a few kinds of systems:

Electrical systems, e.g., a large electric power supply system, a widespread telephone system, an electronic digital computer.

Mechanical systems, e.g., a steel rolling mill.

* The word canonical occurs in mathematical terminology. Roughly speaking, it indicates the simplest possible form or description which is nevertheless complete or unambiguous.

Chemical systems, e.g., an oil refinery.

Economic systems, e.g., government finance, industrial and commercial firms, personal monetary arrangements.

Biological systems, e.g., a mink farm, the tuna fishing grounds, an afforestation plantation.

Irrigation systems, e.g., the Aswan High Dam, the Mekong Delta scheme, the Murrumbidgee Irrigation Area.

Transportation system, e.g., a state-wide railway system, the escalators in a large department store.

Biochemical systems, e.g., the life support system in a space vehicle.

The list is a long way from being exhaustive; it should be easy for the reader to add other examples.

Not all of these systems may best be analysed in exactly the same way; however, there are certain basic principles which may be applied in all cases to the models which we set up to describe their behaviour. Certainly, too, the methods developed in the last few decades for the analysis and design of electrical systems can be applied to many other kinds of dynamic systems.

The first common attribute of all the systems mentioned is that they are dynamic. By the word dynamic, is meant a system whose state is varying with time.

The state of a system may be described by one or more variables: for example, in the telephone system we might be interested in the number of conversations being carried on at a given moment and to describe this we need to symbolise the number of conversations as a variable function of time; thus we might choose the symbol $n_1(t)$ to represent this quantity at each specified instant of time t. However, we might also be interested in the number of trunk line channels occupied at a given moment and we would then need another symbol to represent this quantity, say $n_2(t)$. Again for example, in the life support system of a space capsule we would certainly be interested in the pressure, temperature, percentages of oxygen, carbon dioxide and other compounds and we could write each of these variables as a function of time in order that we might study the dynamics of the system, i.e., the way in which these variables change as time progresses.

We have taken our first step towards a mathematical model by recognising that if we are concerned with system dynamics then it is desirable to describe the state of the system by variables which are functions of time. Let us note, at this stage, that some of these variables may be independent variables and others dependent variables.

In addition to system dynamics, we may also be interested in problems concerning the state of a system in relation to some property of the path by which the state is reached. In the real world the later sections of a sequential path would necessarily be traversed at a later time than earlier sections, but our attention is focussed on the sequence rather than the dynamics. An example of such a problem would be the optimisation problem faced by a building contractor in trying to decide how many times and in what sequence each sub-contractor should be called onto the job, in order to minimise his total expenditure. Such a problem would involve a state resulting from possible sequences of decisions and actions, rather than a state which could be represented mathematically as a function of time. Problems of this sort could be described as "system statics", although this is not a generally accepted name among systems engineers. In relation to these problems the term *network* is often used (e.g., the so-called critical path technique used in the construction and manufacturing industries is really a network problem) and the term *optimisation* will also be seen frequently.

Another example of such a path is a computer flow diagram. In this case we are usually concerned with the total time for execution of the program, since this is the basis for the cost of the job but we are unlikely to be concerned about the exact state of the computer during each microsecond of the computer run.

Let us list a few kinds of systems where we may encounter problems of "system statistics":

Information processing systems, e.g., a communications system, an electronic digital computer.

Economic systems, e.g., a banking system.

Transportation systems, e.g., a railway system, an urban public transport system.

Constructional systems, e.g., the completion of a civil engineering work.

Again it should be easy for the reader to add other examples. Notice that a particular system may appear in the lists for both statics and dynamics. This would depend not so much on the system, as on the nature of our interest in it.

D. CONSERVATION LAWS OR PRINCIPLES

Common to the problem of describing all sorts of systems is the notion of choosing a suitable conservation law or principle as a basis for our mathematical model. Such conservation laws enable us to build mathematical models which can be used not only to calculate the dynamics of a system, but also to enable us to study its static properties and optimisation problems within a system.

Examples of these are:

In a chemical system—conservation of mass.

In an electrical system—conservation of charge, or conservation of energy.

In a combined system (e.g., electro-mechanical and thermal)—conservation of energy.

In a closed economic system—conservation of money.

In a constructional system—continuity of (i.e., conservation of) time, or conservation of money.

We must realise that any one of these conservation laws will hold only in a suitably closed and defined system. Whenever we have to extend beyond our previously closed system, we may have to modify the conservation laws.

For example, in an atomic explosion the two everyday laws of conservation of energy and conservation of mass are both violated—some mass disappears and a large amount of energy appears. In this case the modification is to take account of $E = mc^2$ and to say that (Mass + Energy) is conserved.

In addition we must also know all imports and exports to and from our closed system and include these in the balance or conservation equation. Let us commence with a homely example.

We all know the phrase to balance one's budget. Most of us have been faced with this problem in the past and no doubt most of us will continue to be faced with it for the whole of our lives. In essence the law of conservation of money* may be expressed by the equation.

$$\text{INCOME} - \text{EXPENDITURE} - \text{CHANGE IN SAVINGS (STORED MONEY)} = 0$$

When expenditure exceeds the sum of income and stored money we have to go outside our closed system and borrow (assuming we do not wish to steal) from friends, family or banker. This will mean that we need an extra term in the equation, i.e., Borrowings.

Unfortunately the act of borrowing (unlike stealing or begging alms) usually means that we must include the further term Repayments and very often too, Interest. The

* This is not a generally recognised name for this relationship. It is used here to draw the analogy with other systems.

equation of conservation of money now looks like

INCOME + BORROWINGS − EXPENDITURE − REPAYMENTS −
INTEREST − CHANGE IN STORED MONEY = 0.

It is this equation (possibly with extra terms to allow for gifts, acts of God such as fire and flood, etc.) that accountants have used to set up daily, weekly, monthly or annual accounting systems.

It should be observed that all these terms would normally be functions of time.

It should also be noted that there could be some difference of opinion about the correct way to specify what is meant by stored money—if we are considering a time interval such as a week or a month we should really write down the change in stored money over this interval of time. The example below, dealing with stored energy may make this clearer. Also to be noticed is the adoption of a negative sign for this term and the reader should make sure he understands why this is so written.

Another way of expressing a conservation law is to say (as in the conservation of mass in ordinary chemical reactions)

$$(\text{TOTAL MASS INVOLVED}) = \text{CONSTANT}$$

There are two ways in which the last equation may be rewritten to get a zero on the R.H.S. and both these forms are useful.*

(1) $(\text{TOTAL MASS AT START}) - (\text{TOTAL MASS AT FINISH}) = 0$

OR,

(2) $\frac{d}{dt}(\text{TOTAL MASS INVOLVED}) = 0$

Form (1) holds for a finite time interval.
Form (2) is an instantaneous balance.†

Other laws may be written down as examples, e.g., conservation of energy.

$$\begin{pmatrix}\text{TOTAL ENERGY}\\\text{INPUT}\end{pmatrix} - \begin{pmatrix}\text{TOTAL ENERGY}\\\text{OUTPUT}\end{pmatrix} - \begin{pmatrix}\text{INCREASE IN}\\\text{STORED ENERGY}\end{pmatrix} = 0$$

This equation would refer to a closed system with inputs, outputs and capability of storage. It refers also to a defined time interval.

It is well known that $\frac{d}{dt}(\text{ENERGY}) = \text{POWER}$.

So by differentiating the above conservation of energy equation we would get

$$\begin{pmatrix}\text{INPUT}\\\text{POWER}\end{pmatrix} - \begin{pmatrix}\text{OUTPUT}\\\text{POWER}\end{pmatrix} - \begin{pmatrix}\text{TIME RATE OF}\\\text{INCREASE OF}\\\text{STORED ENERGY}\end{pmatrix} = 0.$$

This is also a conservation equation but it refers to instantaneous conditions in the closed system with specified input and output arrangements. It is therefore a more convenient formulation when we wish to express the three quantities INPUT, OUTPUT and STORAGE as functions of the variable, time.

In a biological system (it may be more correct to call this example ecological) we might have for the total population of a closed system (such as, e.g., the human population of Australia, or the tuna population of certain fishing grounds off the coast) a conservation equation such as

$$(\text{IMMIGRANTS}) - (\text{EMIGRANTS}) + (\text{BIRTHS}) - (\text{DEATHS}) - (\text{INCREASE IN POPULATION}) = 0$$

written for a specified time period.

* This is a convenient but not essential way in which to write down these equations.

† The expression $\frac{d}{dt}$ is termed the differential with respect to time t, and represents the *rate* at which the quantity concerned is varying per unit time.

Again by differentiating with respect to time would arise a similar equation but with each term representing a rate (hence the terms birth rate, death rate, etc.) with the sum of these terms equal to zero.

From these seemingly obvious statements may be developed the dynamics of systems.

E. INEQUALITIES AND CONSTRAINTS

In addition to the basic conservation or continuity laws, certain physical necessities or laws can be expressed as inequalities. For example, hydrostatic pressure may never be negative and this can be expressed by an inequality*

$$\begin{pmatrix}\text{SUM OF TERMS CONTRIBUTING} \\ \text{TO HYDROSTATIC PRESSURE}\end{pmatrix} \geq 0$$

Particular problems may have constraints which may be also expressed as inequalities. An example of this is the statement which might be considered by an electronics equipment designer while he is designing a transistor circuit, that a voltage greater than 30 volts between collector and emitter is liable to cause breakdown of the

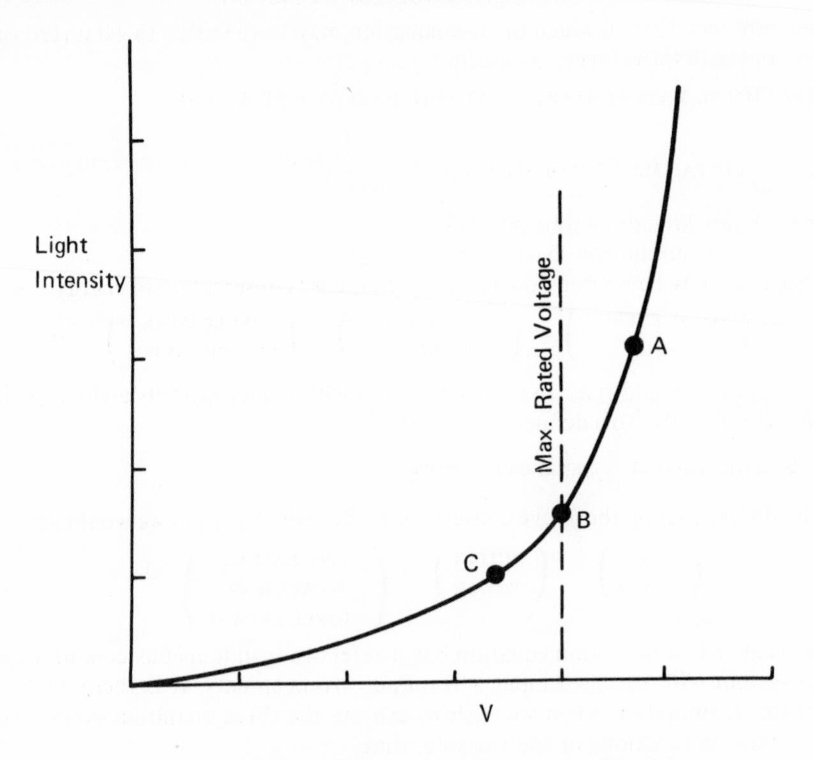

FIG. 5.1 Choice of Operating Voltage for a Filament-type Electric Lamp.

Operating Point	*Probable Life*
A	50 hours
B	1000 hours
C	3000 hours

* The symbol $>$ means greater than.
The symbol $\geq$ means greater than or equal to.
The symbol $<$ means less than.
The symbol $\leq$ means less than or equal to.

transistor. This could be expressed as an inequality

$$\begin{pmatrix}\text{SUM OF VOLTAGES CONTRIBUTING} \\ \text{TO } V_{ce}\end{pmatrix} \leq 30.$$

This inequality is one of the constraints chosen or imposed in order to make the circuit reliable in operation.

As another example, in a system designed to keep control of ordering replacement stocks in the material store which services an engineering workshop two constraints may exist

(1) $$\begin{pmatrix}\text{NO. OF UNITS} \\ \text{IN STOCK}\end{pmatrix} \geq 0$$

due to the physical impossibility of a negative number of units in the stock bin

AND

(2) $$\begin{pmatrix}\text{NO. OF UNITS} \\ \text{IN STOCK}\end{pmatrix} \geq n$$

an artificial constraint which may have been established by the stores supervisor and which indicates that once the stock falls below the number n then action must be taken immediately to place an order for replacement stocks.

Inequalities may exist as well as equations in the mathematical model of a system, and *both* equations and inequalities will need to be satisfied simultaneously. More about this will be discussed in later chapters, but for the moment let us consider as a homely example, the incandescent electric bulb.

Referring to Fig. 5.1, the lamp will give much more light if operated at point A on its characteristic. The constraint "maximum rated voltage" is chosen so that, on the average, lamps will have a life in excess of 1000 hours if operated at point B. An increased lifetime may be achieved if the lamp is operated at point C, at the expense of lower brilliance and less efficient utilisation of energy. Operation at point A would involve a considerable decrease in life expectancy.

F. SYSTEM STATICS

In using the word *statics* we do not mean that there is, necessarily, a complete lack of motion in the system. Rather the use of the word *statics* implies that we are dealing with a steady state of the system so that any transient disturbances will have died out. Velocities or rates of flow (e.g., flow of fluid, heat, electricity or information) will generally be steady, although of course zero rates of flow would also be included in the definition.

Whenever a knowledge of the steady state behaviour of a system is sufficient for our purposes, it is just common sense to investigate only the statics of the problem and to avoid the greater complication and cost of solving the dynamic problem.

If, however, we know that we will eventually need to know something about the transient behaviour (i.e., the behaviour of a system during its settling down time after a disturbance or after initial switching-on) then we might as well tackle the more complicated dynamics problem from the start. The static solution can then usually be obtained (as a bonus) very easily from the dynamic solution by putting all the time-rate of change terms equal to zero and/or by looking at the magnitude of the dynamic solution after a long time has elapsed.[2]

G. SYSTEM DYNAMICS—BLOCK DIAGRAMS

The next four sections of this chapter will be concerned with some of the concepts which will be needed in order to understand how to calculate the dynamics of a system. To assist in visualising the role of each part of a complete system, we will make use of the pictorial scheme known as the block diagram. Each block in a complete block diagram (see Fig. 12.10) represents a mathematical operation or equation. Each block will receive an input (representing the independent variable) and will deliver an output (representing the dependent variable). An assembly of such blocks and directed (i.e., arrowed) lines forms a block diagram and represents the group of simultaneous equations which form the mathematical model of the whole system.

A second kind of pictorial representation is known as the signal-flow graph or network chart where the arrows represent the operations (or equations) while the joins (or nodes) represent the magnitudes of the quantities in which we are interested (i.e., the variables in the equations).

From what has been said in the previous section it is easy to understand that both dynamic and static problems may be represented pictorially either as block diagrams or as network graphs (or flow-charts).

A third kind of pictorial representation is the circuit diagram which can cope with equations based on two separate conservation laws each of which is applicable simultaneously throughout the whole circuit or system.

More will be said about these alternative schemes in later chapters. In the next four sections we will develop the idea of blocks being used to represent parts of a dynamic system.

H. INPUT AND OUTPUT

Let us now turn to a new use of the words *Input* and *Output*. Suppose we have a dynamic system G, as is shown in Fig. 5.2.

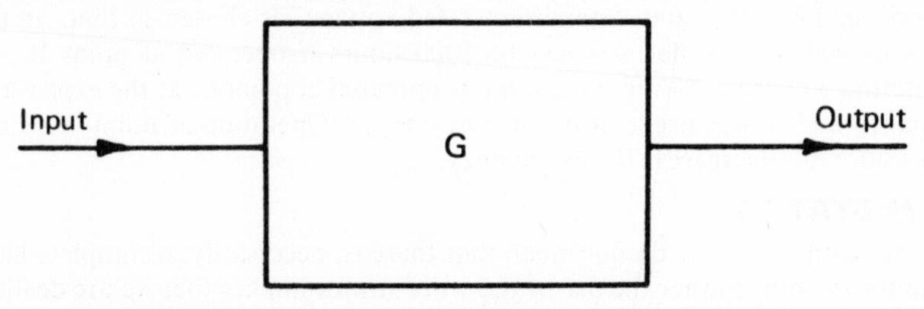

FIG. 5.2 A Dynamic System G represented as a Closed Box linked to its Environment solely by Input and Output.

We can consider the system receiving from the outside world a certain input which is a function of time (e.g., consider the system comprising a motor car plus driver on a straight open road, whose input is the position of the accelerator pedal as a function of time).

We can also consider the system as giving to the outside world a certain output which is also a function of time (e.g., the distance of the car from some specified starting point).

Note 1: we could have specified other inputs and outputs. This is quite arbitrary and our choice simply indicates what we are interested in.

Note 2: we could have specified two inputs (e.g., accelerator pedal position and gear change)

Note 3: we could have specified two or even three outputs in which we are interested (e.g., distance covered, speed, acceleration).

In the latter case we need additional equations or, we can if we wish, use arrays of numbers and the rules of matrix algebra in order to calculate all the outputs simultaneously. For those readers who have not yet studied matrices it may be sufficient to say that matrices look like determinants, and are treated in more advanced texts. An array of numbers (or symbols) is used to designate the inputs or outputs. These are single column or single row matrices and are often termed "vectors". They should not be confused with directional vectors in real space. They are however vectors in the "space" formed by the two (or more) variables we are dealing with. Most of the readers of this book will have already encountered determinant arrays as a means of solving simultaneous equations. In that case most of the numbers in the array represent coefficients of terms in the equations. The solution of simultaneous equations by the use of these arrays of numbers (or symbols) is possible because there are certain formal rules for manipulating determinants (and matrices).

Now let us return to single input and single output systems.

I. SIGNALS

It is part of the terminology of systems theory that we speak of input and output signals to a system. Basically, these are simply variable quantities which are either known or else have to be calculated, as functions of time. Electrical engineering circuit theory and systems theory as we know it today is based on the response (i.e., output) of a linear system to a specified excitation (i.e., input). Both response and excitation are functions of time.

The theory may be extended also to non-linear systems but these are considerably more complicated and also very much less well understood.

The linear theory is based on the mathematical analysis of linear ordinary differential equations. Most of you have dealt with some examples of ordinary differential equations in your Maths courses at school. Some of you may also have realised that differential equations in general are divided into two classes:

(i) ordinary differential equations, and (ii) partial differential equations.

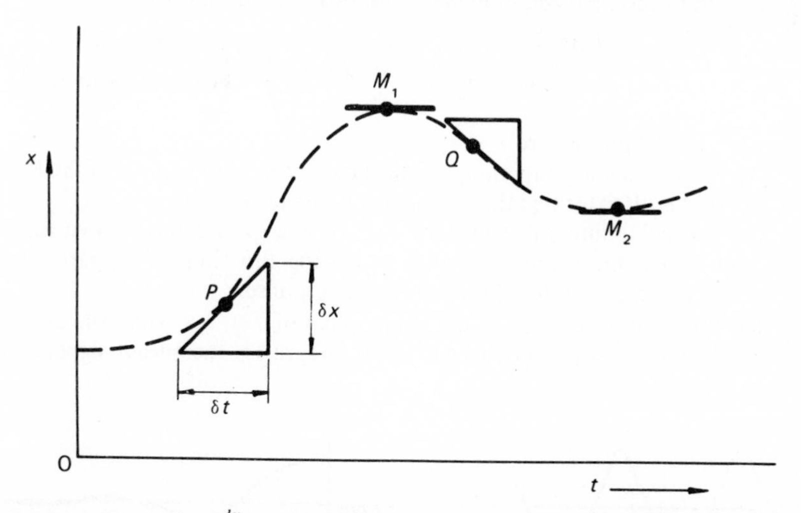

FIG. 5.3 Illustration of $\frac{dx}{dt}$ the First Differential of x with respect to t, as the Slope of a Graph.

For others who may not yet have encountered differential calculus, the following line of summary may be of help. An algebraic equation will contain the variables, say x and y which in a dynamic problem will be functions of time. A differential equation will also contain differentials or derivatives of these variables with respect to time. The first derivative of x with respect to time t is written as $\frac{dx}{dt}$. The same process of differentiation with respect to t may be applied to $\frac{dx}{dt}$ thus forming the second derivative $\frac{d^2x}{dt^2}$ and so on. Readers not yet familiar with the differential calculus, may be helped to visualise the process of differentiation by drawing a graph to represent how a variable x, behaves with the passage of time t. The value of $\frac{dx}{dt}$ at any moment is simply the slope of the curve of x plotted against t.

In Fig. 5.3, to find the value of $\frac{dx}{dt}$ at point P, draw a tangent to the curve and make δt equal to unity. Then $\frac{dx}{dt} = \frac{\delta x}{\delta t} = \delta x$. Note that the slope at P is positive, at Q is negative. At a maximum (M_1) or minimum (M_2) point, the slope is zero.

In each of these classes we may have either linear or non-linear cases. We will say a little more about partial differential equations when we come to distributed systems in a later chapter. For the moment let us consider only the systems describable by ordinary differential equations. These may be recognised by the fact that the differential operators are written in the familiar form $\frac{d}{dt}, \frac{d}{dx}, \frac{d^2}{dt^2}$, etc., while partial differential operators are written as $\frac{\partial}{\partial t}, \frac{\partial}{\partial x}, \frac{\partial^2}{\partial t^2}$ etc. or by making use of the symbol ∇ (pronounced del or, by a few people, nabla) of the branch of algebra known as vector analysis.

A complicated system may be represented by a set of several simultaneous differential equations. We will not use examples of this complexity—the necessary concepts may be conveyed more easily by considering examples where the system is representable by a single differential equation. The solution of such a single equation is usually found in two parts. To be specific, let y be an unknown variable which we wish to find as a function of the known variable t.

Let us write the differential equation in the following standard format

$$\left(\begin{array}{c}\text{L.H.S.}\\ \text{Collect terms in } y \text{ and its derivatives } y\ \frac{dy}{dt}\ \frac{d^2y}{dt_2}\ \text{etc}\end{array}\right) = \left(\begin{array}{c}\text{R.H.S.}\\ \text{Collect terms in } t\end{array}\right)$$

The two parts of the solution are

(1) The complementary function, obtained by putting zero in place of the R.H.S. This part is often termed the *transient* solution or free response.

(2) The particular integral, when the R.H.S. is a (particular) known function (often called the forcing function or driving function). This part is often termed the *steady state* solution or forced response.

Notice that the forcing function is the input signal of Fig. 5.2 while the complete solution (complementary function plus particular integral) is the output signal of Fig. 5.2.

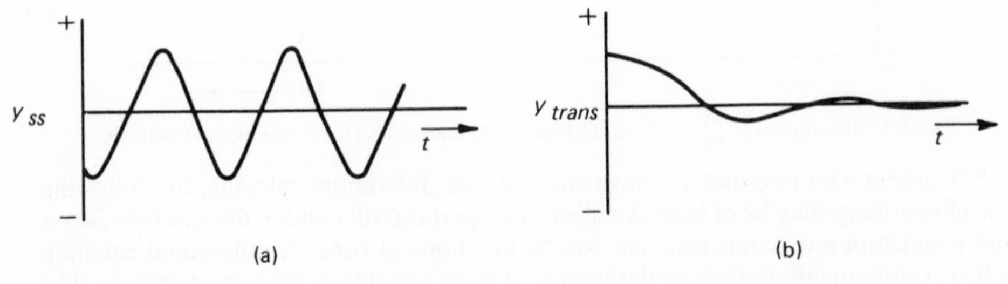

FIG. 5.4 The Two Components of Response to a Sinusoidal Input Signal switched on at time $t = 0$.
(a) Steady State Component.
(b) Transient Component.
The Total Response is the Sum of (a) and (b).

A typical response to a sinusoidal input signal switched on at time $t = 0$ is shown in Fig. 5.4

The complete solution is $y = y_{trans} + y_{ss}$ and each of these three quantities is a function of time.

In a well behaved (i.e., stable) system the transient dies out while the steady state response keeps on as long as the input signal is present.

J. METHODS AVAILABLE TO CALCULATE THE RESPONSE

We will return in a later chapter to look a little more closely at the methods available to calculate the response of a dynamic system to a specified input. Other texts used in later years of undergraduate courses study each of these methods in detail.

The same methods or simplified versions of them, may be used to calculate the steady state behaviour (i.e., the static problem) of a dynamic system. In addition, the whole class of problems which go under the general name of *optimisation*, is of great practical importance. Because we have chosen to be interested in either the statics or dynamics of a system, it follows that optimisation problems may be either static or dynamic, the latter being more complicated. The essence of an optimisation problem is to choose a satisfactory mathematical model for the system (i.e., one which will provide all the necessary information, preferably with the least effort) and then to choose a satisfactory criterion (in particular, one which is unambiguous or single-valued in mathematical terms) for what we mean by *optimum*.

Readers should learn more of the advantages and disadvantages of each scheme (and the variations of each scheme) by undertaking a good deal of problem solving as they progress through their education and their careers. Some parts of this task can be quite tedious, but paradoxically it is useless just knowing how to catch fish unless we practise the occupation of fishing and succeed at it. It is a waste of time if we come up with discarded rubbish.

To conclude this chapter let us say something about a number of actual input/output devices. That is, let us look briefly at some hardware.

K. AMPLIFIERS, MACHINES AND TRANSDUCERS

It would have been possible to add a great many more nouns to those chosen for the above subheading. Notice that each of the three nouns describes an input/output system. The reader should try to make up his own list of nouns each describing a particular kind of input/output system. Even if we confine the list to man-made hardware systems by excluding biological, social, sexual, political, managerial, intellectual, emotional, educational, systems, etc., we will still quickly compile a very long list.

It is quite natural to ask—how is it that we can describe these many different things by the one word "system"? Is it just an omnibus word, a dust bin of sorts? Or can we find attributes which are common to all "systems"? The answer to the last question will turn out to be "Yes", but first of all let us look at a few particular devices, remembering that while each one is a small system just by itself, and therefore worthy of attention, nevertheless our real aim is to gain understanding of how to connect together a number of devices in order to form larger systems.

In Fig. 5.5 (which is almost but not quite a repeat of Fig. 5.2) is represented an amplifier, by which we mean (usually) a device which when provided with an input signal will produce a similar but larger (i.e., amplified) output signal. In an ideal amplifier the output would be a replica of the input, produced instantaneously and with increased amplitude. We may write for the response of such an ideal amplifier, using subscripts "*o*" for output, "*i*" for input

$$f_o(t) = Gf_i(t)$$

where G is a constant real number greater than unity. In practical amplifiers there will be some difference in the shape of $f_o(t)$ and $f_i(t)$ in addition to the difference in amplitude described by a constant real G. This difference represents distortion and is described in a variety of ways. Which way we choose to describe distortion depends on the manner in which the input signal itself is described. Thus G may be complex, rather than purely real; G could be a function of the input signal; or we could recognise additional classes or modes[3] of signal as being present in $f_o(t)$ over and above those which are present in the input $f_i(t)$.

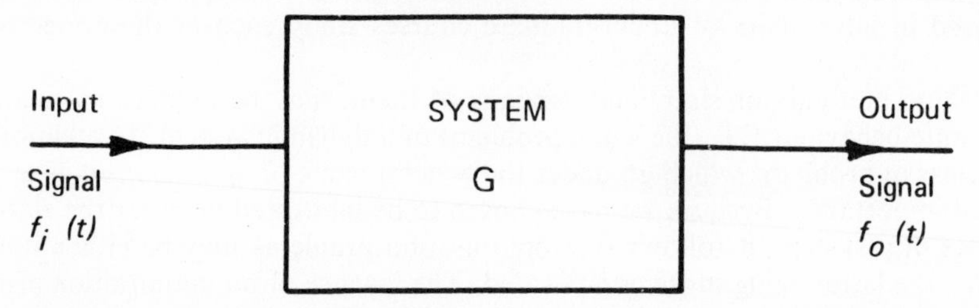

Fig. 5.5 Illustrating the Concept that both Input to and Output from a System, will be Signals which are Functions of Time.

Amplifiers in which the input and output are electrical signals are usually made up with electronic active devices (i.e., controllable amplifying devices as distinct from the passive electrical components such as resistance, inductance and capacitance). However, in addition to the various kinds of vacuum tube and transistor, active devices may be made up by utilising the non-linear properties of ferro-magnetic cores and of some dielectric materials. In addition it is possible to utilise the non-linear properties of electric conduction through ionised gases. A device which behaves very much like a transistor (but is much slower) can be made up using ionic conduction in liquids and semi-permeable membranes. No technological use has been found for the latter device but a study of its behaviour could serve to give extra insight into the mechanisms of the peripheral nervous system in animals and humans. Rotating electrical machines may also be arranged to act as amplifiers.

Other forms of amplifier are the controlled switch either in the form of the electro mechanical relay,or the controlled gas-discharge tubes known as thyratrons and ignitrons, or in more modern guise the silicon-controlled-rectifier (SCR) or thyristor. The relay contacts are controlled (i.e., opened or closed) by action of current in an electromagnet. Conduction in the gas-discharge tubes is initiated by an electrical impulse applied to a grid or igniter electrode. In the SCR conduction is initiated by an impulse applied to a small gate electrode (actually a subsidiary *p-n* junction) built into or onto the main *p-n* junction. In the latter two cases conduction is quenched (or commutated, if it is a repetitive on-off operation) by allowing the current to fall to zero for a sufficient time or by superimposing a larger current pulse in the reverse direction.

Such switched-mode amplifiers are capable of extremely high power gains. The power gain may be defined as the ratio of "power controlled" to "power needed to operate the switch". Power gains between 10^2 and 10^5 are readily achieved.

Switched-mode amplifiers may also be built up using transistors or vacuum tubes. For example the so-called Class "C" amplifier used as the final or power stage of a broadcast or T.V. transmitter operates in a switched mode and may have a power gain approaching 10^2. As another example the so-called flip-flop circuit using transistors is really a two-way switch which changes state in response to an impulse input. This device finds very wide application in digital logic circuits (see Chapters 6 and 18).

In switched-mode devices we usually exploit the high energy efficiency made possible by having the device either fully On or completely Off.

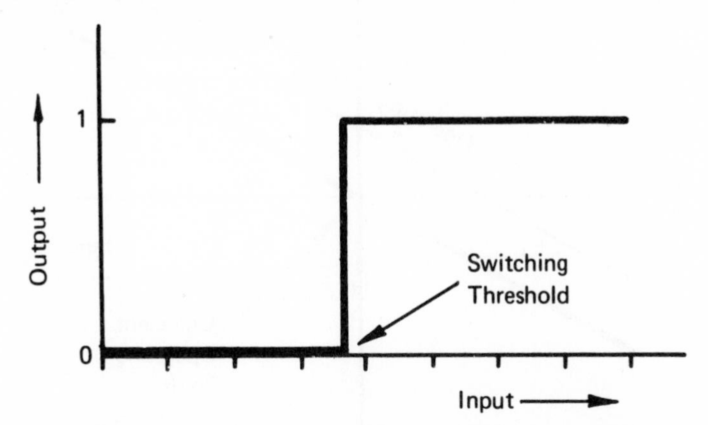

FIG. 5.6 The Input/Output behaviour or "Characteristic" of a Simple Switched Device.

Figure 5.6 illustrates the "characteristic" or input/output behaviour of a simple switched device. It is easy to see that G in the expression

$$f_o(t) = Gf_i(t)$$

is a non-linear function which is zero below the threshold and such that the product $Gf_i(t)$ is unity above the threshold.

In distinction to the switched-mode amplifiers is the class of linear amplifiers. In these devices we endeavour to make the change in output proportional to the input. A characteristic for this class of amplifier is shown in Fig. 5.7. The relation between output and input is a linear function, in an ideal case.

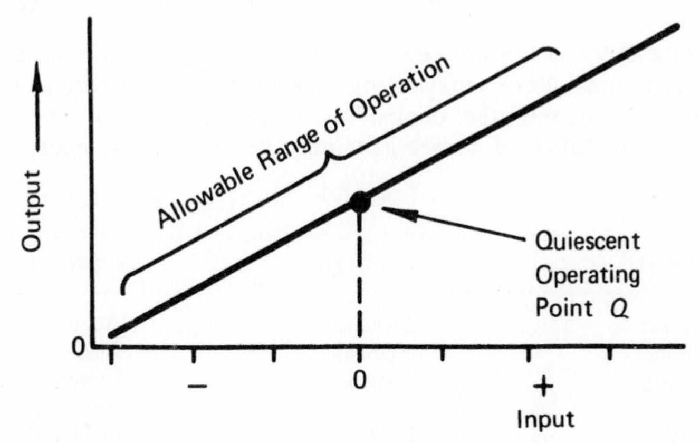

FIG. 5.7 The Input/Output behaviour or "Characteristic" of a Linear Amplifier operated about a Quiescent Input value Q (so called single-ended or unbalanced mode).

By connecting two amplifying devices in a so-called "push-pull" connection the operating point may be established at the origin, as is suggested in Fig. 5.8. This artifice reduces distortion and increases the allowable operating range. The word quiescent is used to indicate the condition of the amplifier at zero-input.

Clearly switched-mode and linear-mode amplification are two extreme cases and in that sense are ideal. Practical amplifiers will have characteristics which depart to some extent from those shown.

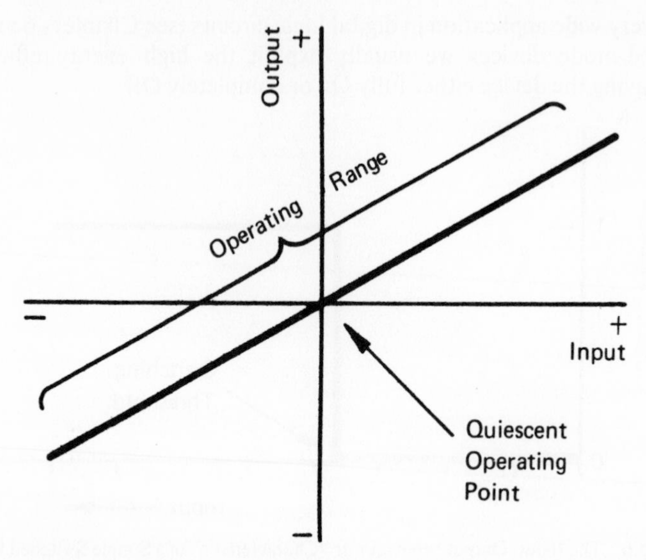

Fig. 5.8 The Input/Output behaviour or "Characteristic" of a Linear Amplifier operated about a Quiescent Input Value Zero (so called push-pull or balanced mode).

In addition to electronic or electrical amplifiers, other forms of switched-mode and linear amplifier are possible. Among these are fluidic devices in which both the streamline and turbulent flow properties of a fluid are exploited—by the expenditure of a small amount of input flow or pressure energy a much larger amount of output flow or pressure may be controlled. The devices commercially available are particularly suitable for switching-mode operation and fluidic logic systems are increasingly coming into use. Air if used as the working fluid has the advantage of much lower inertia than a liquid with consequent reasonably rapid response even from a multi-device system.

Hydraulic amplifiers have been used since the last century. The mechanical forces developed by quite small devices may be made very large and, by careful design of the control valves to minimise the controlling force needed (e.g., by a balanced arrangement), large ratios of power-gain may be achieved. Typical applications

Table 5.1

Transducer	Input Energy	Output Energy
Loudspeaker	Electrical	Acoustic
Microphone	Acoustic	Electrical
Electric lamp	Electrical	Luminous and Infra red
Photo-cell	Light	Electrical
Air compressor	Mechanical	Pneumatic
Electric generator	Mechanical	Electrical
Steam engine	Chemical	Mechanical
Petrol engine	Chemical	Mechanical
Thermostat	Thermal	Electrical or Mechanical
Storage battery		
(discharging)	Chemical	Electrical
(charging)	Electrical	Chemical

include the raising and lowering of aircraft landing wheels, the controlled movement of large masses such as guns or gun turrets, the operation of certain powered auxiliaries in motor vehicles, etc.

Without having specifically defined what we mean by the terms, we have already cited some examples of machines and transducers. It would not be easy to give a satisfactory definition of what we mean by a machine. Of the various likely attributes of a machine, which ones should we emphasise? Some of these likely attributes include—it is man-made, it involves mechanical forces and movements, it consumes or delivers or stores mechanical energy.

A transducer is more readily defined—its meaning has not become diffuse since it has not yet passed into everyday non-technical use nor has it (yet) come into the hands of the advertising copywriters. A transducer is simply a device for the conversion of energy from one form into another. The input and output can both be expressed in energy terms but different kinds or forms of energy. Examples are given in Table 5.1.

It will be noticed that in every case the transducer may be represented by the block diagram of Fig. 5.2.

Individual devices may be arranged so that an output of one serves as an input to another device. We will learn more about the ways in which this may be done in Chapter 12.

L. SUMMARY OF HIGHLIGHTS IN CHAPTER 5

(i) A system possesses inputs and outputs, but is otherwise isolated.

(ii) The outputs depend on the inputs and the initial state of the system.

(iii) The behaviour of a system may be described by a set of equations. These are known as the mathematical model of the system.

(iv) The choice of variables is very important and will decide whether the mathematical model is a useful one.

(v) In a dynamic system inputs and outputs vary with time (i.e., are functions of time) and are often termed signals.

(vi) The set of equations is usually derived by applying suitable conservation laws to the system.

(vii) The closed system is represented conventionally by a closed box with one or more inputs and one or more outputs, e.g., Fig. 5.2.

(viii) Amplifiers, machines and transducers are important devices because they serve as building blocks for dynamic systems.

(ix) Amplifiers, machines and transducers may operate either in a switched-mode or in a linear mode.

(x) In the representation or model of a system, each device may be represented as a block.

(xi) Individual blocks may be connected together to form a larger system, which could in turn be represented by a single block if we so wished.

(xii) A few simple rules exist for the construction and manipulation of block diagrams. These should *not* be disobeyed.

(xiii) We will learn more in Chapters 12 and 13 about the ways in which devices (or the blocks representing them) may be connected together to form larger systems. We will also learn something of the way in which a mathematical model for a system may be formed provided we know the properties of and the connections between the component blocks. In addition to discovering more about the block diagram we will also learn something about how other kinds of diagram can assist our understanding of systems theory. In particular we will be introduced to circuit diagrams, and network graphs which can be used to plan and control sequences.

M. EXERCISES

1. Make a list of further examples to add to Table 5.1.
2. Draw a block diagram for a small record-player. Which blocks represent transducers?
3. Draw a block diagram for a thermostat controlling the temperature of an electric oven. (The thermostat is a switch.) Draw a block diagram for a thermostat controlling the temperature of a gas oven. (The thermostat regulates the flow of gas, but it is not a "switch").

N. END NOTES

(1) A linear system or problem is one with a linear mathematical equation and variable terms, raised to the first power. Squares, cubes, etc., of the dependent variable terms must *not* be present, nor special functions such as sin, cos, exponential, etc., of the dependent variable, nor terms involving the product of two separate variables. The equation will usually contain also a term, or terms, involving an independent variable, but note that in one very important case the restrictions may be relaxed as far as the independent variable is concerned. This is the case where both the dependent and independent variables are written as being functions of a third quantity (e.g. functions of time); in this case the independent variable should be termed a "forcing function" and its value is always known prior to solving the problem.

(2) In any stable system the terms which are changing with time will tend to zero as the time t gets large. An example of this would be any term containing the factor e^{-at} where e is the base of natural logarithms and a is a positive real number. The presence of a factor e^{+at} would indicate a tendency for the term to diverge to an infinite value after a long time. Such behaviour is termed instability and is usually considered an undesirable property. *Knowledge* of whether our system is stable or unstable is of course very desirable. The phrase "long time" means a period much longer than the settling down time of the system.

(3) A simple example of this occurs when a sinusoidal input generates harmonic frequencies which occur in the output, i.e., signals at two, three, four, etc., times the frequency of the original input or so-called fundamental frequency. A slightly more complicated case is when the input is the sum of two (or more) sinusoids of differing frequency. Distortion modes or signals occur at the sum or difference frequencies of the various fundamental and harmonic frequencies. For a square step wave input the distortion is usually described by referring to the finite rise-time of the output and its degree of "overshoot" before settling down to its new steady value. For certain pulsed waveforms as may occur in a television signal it is convenient to describe the distortion in terms of paired echoes (Reference S. Goldman, *Signals, Modulation and Noise,* McGraw Hill, 1948.) or in modern television practice in terms dictated by convenient and accurate measurement techniques.

SUGGESTED READING MATERIAL

LYNCH, W. A. and TRUXAL, J. G. *Introductory System Analysis.* Preface, Chap. 1 and Chap. 3 up to and including Section 3.7. McGraw-Hill, New York, 1961.

CHESTNUT, H. *Systems Engineering Tools.* Chapter 1. John Wiley and Sons, New York, 1965.

CHAPTER 6

History and Applications of Computers

M. W. Allen

Dear Reader, this notice will serve to inform you that I submit to the public a small machine of my invention, by means of which you alone may, without any effort, perform all the operations of arithmetic, and may be relieved of the work which has often times fatigued your spirit, when you have worked with the counters or the pen. As for simplicity of movement of the operations, I have so devised it that, although the operations of arithmetic are in a way opposed the one to the other—as addition to subtraction, and multiplication to division—nevertheless they are all performed on this machine by a single unique movement. The facility of this movement of operations is very evident since it is just as easy to move one thousand or ten thousand dials, all at one time, if one desires to make a single dial move, although all accomplish the movement perfectly. The most ignorant find as many advantages as the most experienced. The instrument makes up for ignorance and for lack of practice, and even without any effort of the operator, it makes possible short-cuts by itself, whenever the numbers are set down.

Blaise Pascal
On His Calculating Machine 1619.

A. ANALOGUE AND DIGITAL COMPUTERS

Automatic computers may be classified broadly as analogue or digital. An analogue computer works with continuous variables and makes use of the analogy between the values assumed by some physical quantity such as an electrical voltage or current, distance, or shaft rotation. The slide rule is an example of an analogue computer. On the other hand the digital machine works with numbers directly, although it must of course represent such numbers by a physical quantity. However, the analogy only requires that the variable represented should have restricted discrete values. Figure 6.1 is intended to show a broad classification of computers.

The two kinds of computer have their characteristic advantages and limitations. The analogue machine is restricted to a rather narrow range of calculations; for example, a slide rule to multiplication, division and the evaluation of powers and roots. Further the accuracy is limited by the mechanical or electrical accuracy of physical measurement of the result. For example, with a slide rule it is not possible to multiply exactly 2 by exactly 2.5; all one can do is to multiply a number in the range of $2.000 \pm .002$, say by one in the range of $2.500 \pm .002$ and get a result perhaps in the range of $5.000 \pm .005$. On the other hand analogue machines are comparatively simple to build, are able to deal with continuous variables and can work at high speed in real-time. These can be significant advantages in certain applications.

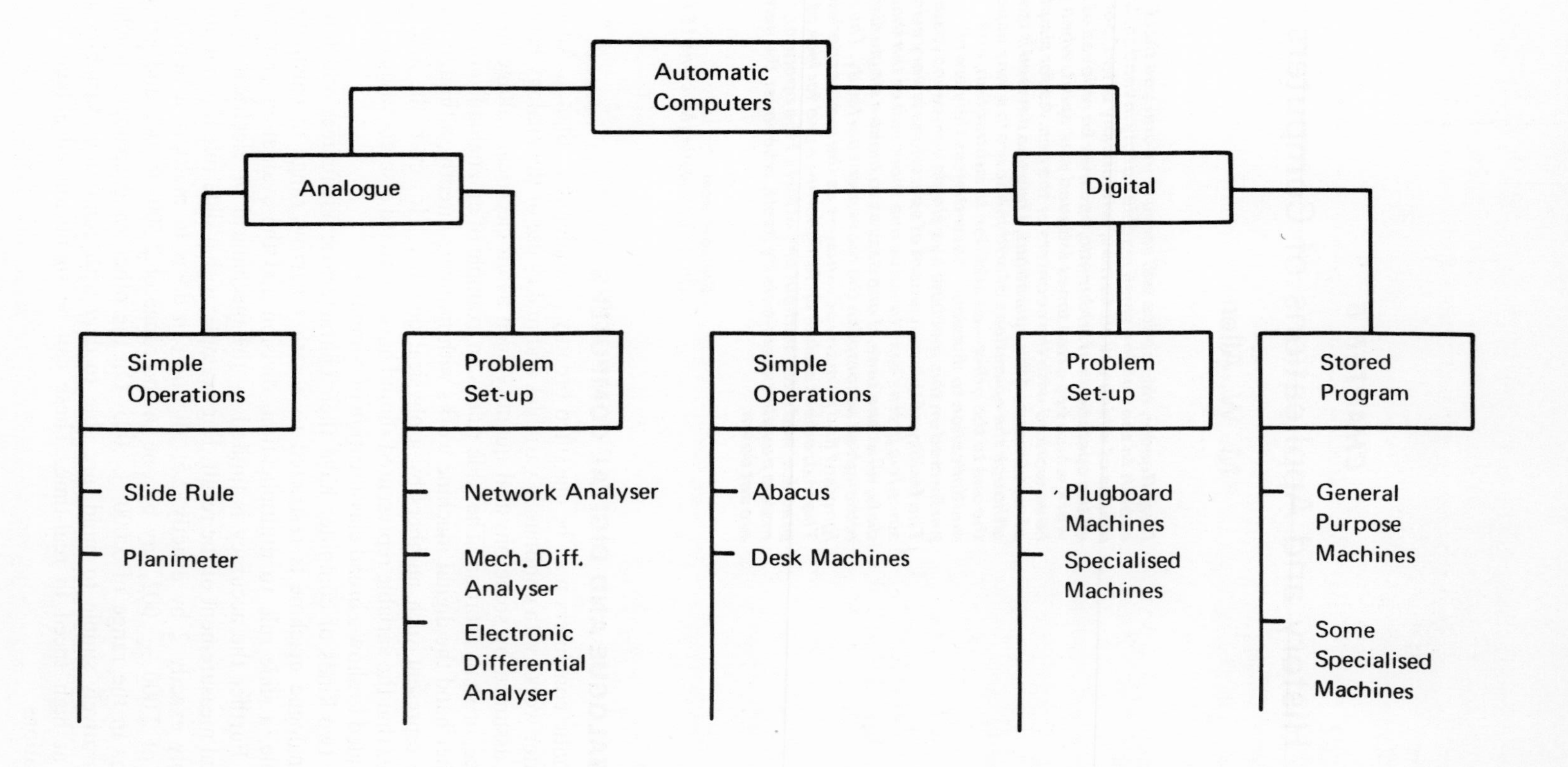

FIG. 6.1 Classification of Computers.

The development of analogue and digital machines has been almost independent and their characteristics are so different that they can be considered quite separately. However, the term computer is now generally understood to mean digital computer since the rate of progress with digital machines has been so outstanding. In the chapters which follow we will consider only the digital machine.

B. HISTORY OF DIGITAL COMPUTERS

1. Early Concepts

Mechanical aids to counting and calculating have been known and used for centuries. The earliest form of digital calculation is the abacus which is still in use today and was known to be in use in China as early as the sixth century (B.C.). The early form consisted of a grooved, clay board with round pebbles in the grooves—later forms used a wood frame with wires or rods on which beads were strung. Each group of beads represents a "place" in the decimal system such as tens, hundreds, thousands, etc. A skilled operator with an abacus can calculate at speeds comparable with that of a simple mechanical desk calculator.

The first mechanical device capable of addition and subtraction in a digital manner has been generally credited to Blaise Pascal (1642) the well-known French scientist and philosopher. The incentive is said to have arisen from the fact that, in his youth Pascal was required to spend his vacations doing calculations needed by his father in his tax-collecting activities. The device employed number wheels the positions of which could be observed through windows on the cover. The stylus operated pocket adding machines, which are still widely used, are descendants of Pascal's machine.

A method of extending the principle to include multiplication and division was proposed in 1671 by Leibnitz. Unfortunately, this extension was regarded as a curiosity and it was not until 1820 that a successful commercial machine was made which could reliably carry out the four arithmetic operations of addition, subtraction, multiplication and division. This development was called the Arithometer and is credited to C. X. Thomas. The commercial development of such machines then followed, but did not really reach wide acceptance until the latter part of the nineteenth century. A related invention which had a significant effect on computing was the Jacquard loom punched-card system. This method of weaving allowed a completely automated loom by connecting a series of punched cards into an endless belt. The code holes in the cards operated the weft mechanism and the loom could thereby automatically weave complex patterns. Thus it was the first successful application of the principles of punched card and tape control. It was demonstrated between 1725 and 1745 by Bauchen, Falcon and Jacques, and came into widespread use in the early nineteenth century. It is interesting to note that "on-line" computers are now being used to design textiles, and to automatically control textile production equipment.

Automatic calculating machinery probably saw its inception with the invention in 1786 by J. H. Muller of the "difference engine". This machine enabled automatic calculation of tables of functions whose higher order differences are constant. It required a register for each order of difference and a means of adding the contents of each register to those of the next lower order register. The operation consisted of setting the constants and then turning a handle.

Muller did not build such a machine and the construction was first attempted by Charles Babbage. Parts of his original machine are still in various scientific museums. Babbage is generally regarded as the inventor of many of the concepts employed in modern computers. He was a British mathematician and engineer, also Lucasian Professor of Mathematics at the University of Cambridge (1829-39). He devoted a great deal of his life to the concepts, design and construction of computers. His interest in the "difference engine" led to the concept of a much more advanced and comprehensive machine which he called the "analytical engine". The analytical engine was to consist of three main parts: one which Babbage called a "store" in which numerical information could be recorded on a bank of counters; another, which he called the "mill" in which numerical operations could be carried out on numbers taken from the store; and a unit to which he did not give a name but which corresponds to the "control unit". This unit controlled the sequence of operations, the selection of numbers, and the disposal of the result. Babbage envisaged a store of 1000 numbers of 50 decimal digits and an operation speed of one addition per second and one multiplication or division per minute. Control was to be carried out through a set of punched cards like those of the Jacquard loom. Normally operations continued in sequence but means were also provided for causing control to skip forwards or backwards to any extent according to a criterion evaluated in the calculation. Provision was made for access to function tables stored on cards. Babbage prepared thousands of drawings and devoted his life and money to the project which was also supported by the Royal Society. However, the technology of the time was not able to meet the demands of the design and he was only able to complete a few parts of the machine, some of which are in the British Museum.

The idea of the machine readable unit record card of the type still in use today is due to H. Hollerith. His proposal was for an electro-mechanical means of sensing holes in punched cards and for machines based on this principle to be used for entering, classifying, distributing and recording data used in compilation of the 1890 census reports of the U.S.A. The punched card equipment in use today is a direct descendant of Hollerith's machines.

Another series of developments were based on the work of G. Stibitz (1938) and involved the construction of several relay computers at Bell Telephone Laboratories. These machines used paper tape input, allowed branching of control and introduced a number of self-checking features.

2. Transitional Machines

The Harvard Mark I Calculator was planned before the war and completed during the war years at Harvard University. In principle and in its functions it is broadly similar to Babbage's conception of the analytical engine. Although it was the first machine to operate successfully on those principles it was of course structurally different from anything Babbage could have foreseen. The mechanical counting units were driven through electromagnetic clutches controlled by relays which could themselves be controlled by contacts in other relays on counting units, or from switches or punched cards. There were 72 counting units each of 23 decimal digits. Storage for tabular functions was held on punched tape or set on hand switches. The sequence of operations to be performed was coded onto paper tape. It could automatically perform the four arithmetic operations, and required about 3 seconds for multiplication. The machine operated successfully for about 15 years and produced many of the tables in current use. A Mark II model with added flexibility and using relays was developed for the Naval Proving Ground at Dalgren, Virginia.

The ENIAC (Electronic Numerical Integrator and Calculator) was the first electronic machine. It was planned by J. W. Mauchley and J. P. Eckert and developed at the Moore School of Electrical Engineering of the University of Pennsylvania for the Ballistic Research Laboratory at Aberdeen Proving Ground. It was not designed as a general purpose machine but was intended to produce firing and ballistic tables for the U. S. Army by solving the corresponding system of differential equations. It was most impressive in physical size and occupied a space of 30 ft by 50 ft, contained

18,000 vacuum tubes and required 130 kW of power. It was a decimal machine with a word length of 10 digits plus sign and in operating principle the arithmetic and storage was an electronic equivalent of the Harvard Mark I. The outstanding feature was that addition time was reduced to 0.2 milliseconds and multiplication to 2.8 milliseconds, and that it did perform with adequate reliability despite the large number of vacuum tubes used in its construction. Thus the advantages of electronics in computing were established. However, the ideas generated during the development of ENIAC led to new concepts in organisation and a new generation of machines which became the true basis of the industry as it is now established. ENIAC operated successfully as a production computer from the day of its dedication (15th February, 1946) until October, 1955.

3. Electronic General Purpose Computers

The concept of the general purpose stored program computer was first published in 1945 in a draft of a report that proposed a new computer the EDVAC (Electronic Discrete Variable Computer). This draft was prepared by John von Neumann, a consultant to the ENIAC project.

Construction of EDVAC commenced in 1946. A similar but rather less ambitious machine was started in early 1947 at the University of Cambridge (England) under the direction of Maurice Wilkes, who had spent the previous summer at the Moore School. EDSAC performed its first calculations in May, 1949 and was the world's first stored program computer. Both EDSAC and EDVAC were serial machines using acoustic delay lines of mercury for storage. A number of similar machines were built in various places; SEAC (1950 Standard Eastern Automatic Computer), SWAC, ACE (England), CSIRAC (Australia). One of the important early research documents was a series of reports produced at the Institute for Advanced Studies at Princeton, New Jersey by A. W. Burks, H. H. Goldstine and J. von Neumann. These tutorials led to the I.A.S. series of machines which were parallel binary using cathode-ray-tube storage. The first of these machines operated in 1952; they included ORDVAC and ILLIAC (University of Illinois), JOHNIAC (Rand Corporation), MANIAC (Los Alamos), WEIZAC (Weizman Institute, Israel), SILLIAC (University of Sydney, Australia).

Another very early but somewhat different machine was built at the University of Manchester (England) in 1947-49. This machine was the first to use electrostatic cathode-ray-tube storage and the first to include index registers (called B-lines). It also used a magnetic drum memory organised in pages. The M.I.T. Servomechanisms Laboratory was also active in early computer development and Whirlwind I was probably the first computer designed with eventual real-time applications in mind. It could multiply two 16 bit numbers in 16 microseconds. This group was also responsible for the development of the coincident current core memory.

In 1947 Eckert and Mauchley left the University of Pennsylvania to form the Eckert-Mauchley Computer Corporation. Their first successful commercial product was UNIVAC I and was delivered in June, 1951. The group was subsequently absorbed into the Remington Rand Corporation in 1952 together with another pioneer group (E.R.A., Engineering Research Associates). A number of other companies also entered the commercial field in the period 1948-50. These included the Raytheon Corporation (RAYDAC), Radio Corporation of America (BIZMAC), Ferranti Limited (England) and a number of others. IBM continued their main interest in punched card equipment and did not enter the large machine field in a commercial sense until later. (The IBM 701 was first delivered in 1953.) However, having entered the field IBM fairly soon became the dominant commercial company and has retained this position ever since.

The development of the magnetic core memory and of high speed switching transistors led to a new generation of machines commencing in 1957. It also led to the formation of a number of new companies and the merging and absorption of some of the earlier ones. The additional speed and reliability encouraged the development of "super computers" which were intended to achieve speed gains of the order of 100 times those of the early machines. The best known of these were STRETCH (IBM) and LARC (Univac). Although these early ventures were not a commercial success they undoubtedly accelerated the rate of progress and led to some very successful products such as the IBM 7090. The Ferranti Group and Manchester University (England) also co-operated on the development of the ATLAS system which pioneered some novel concepts in the organisation of memory hierarchy, interrupts, read-only memory and extra-codes. One of the most commercially successful super machines was first delivered in 1964, the Control Data 6600. This company was formed in 1957 by a group which broke away from Univac (including key members of the original E.R.A. group).

The hardware realisation of early machines differed markedly from each other; however, the basic concepts of organisation were essentially similar, and have remained so. Thus the subsequent development of computers has been evolutionary rather than revolutionary. The major breakthroughs have been in technology and have led to dramatic improvements in storage, speed, and reliability. The improvements in speed alone have been in excess of 1000 to one. At the same time a great deal has been achieved in reducing the amount of human effort necessary to express problems in a form suitable for computer solution. We will be discussing all of these topics in greater detail in later chapters. Before doing so we will discuss some of the more outstanding uses of computers in modern science, engineering, education, and business.

C. APPLICATIONS OF COMPUTERS

Computers were first used for scientific research and indeed a great deal of the incentive to build machines came from this need. The reason was that several important lines of research were making very little progress because of the requirement for very large amounts of computing. Nuclear physics and astronomy were probably the main areas of need. Astronomers were unable to test new theories on the evolution of stars and to correlate these with observations until computers became available. Nuclear research creates computing problems in almost all of its aspects and nuclear research establishments have some of the biggest installations in the world. One of the more interesting problems is in the analysis of tracks in bubble chambers which are produced by charged particles resulting from nuclear collisions. It is not uncommon to require analysis of several million photographs per year each carrying tracks of a number of particles. These tracks have to be analysed to determine the possible existence of a new particle. The structure of crystals can also be determined by patterns resulting from X-rays which are directed at a sample. Analysis and refinement of this data can yield a structure for the crystal, knowledge which is of considerable value to chemists. However, these calculations make very heavy demands on computers, particularly now that small computers are also being used to control the apparatus and collect the data.

Engineers have made a great deal of use of computers. Many engineering projects involve huge expenditure so that design refinement can yield substantial savings. Perhaps the most outstanding example of this is in the aircraft industry where we find some of the biggest computer installations in the world, outside of the nuclear field. A new aircraft design requires the most thorough analysis possible; the price of error can be costly in every way. Computer analysis of frames for buildings, bridges and other structures is also an established practice. Study of the motion of fluids is important to many aspects of engineering; e.g. aircraft, ships, water flow. These problems can involve a great deal of computing.

Electrical engineers have been using computers for a very wide variety of problems; from studies of electric power distribution to the automatic production of masks for fabrication of integrated circuits. The computer is also becoming an important unit in communication systems; in fact the new electronic telephone exchanges are based on computer techniques. This is an interesting change since computers relied for their initial development on the components and techniques of the communication industry. Computer control is also becoming established in applications ranging from control of huge power stations to control of aircraft, radio telescopes, industrial processes and space vehicles. As might be expected computers are also used to assemble computers; the production of logic cards, and the connection of their base wiring is now largely done by machines controlled by computers. The rôle of computers in education is also developing rapidly. The more advanced systems work through personal consoles allowing interaction between machine and student. As more suitable hardware becomes available and the learning process better understood, this type of use can be expected to increase very rapidly. The concept and value of personal consoles has already found widespread acceptance and need in many other applications.

Business applications of computers have also found wide acceptance. The traditional role of computers in business is in areas such as payroll, stock and inventory, and general accounting. However, business is now making use of computer methods to improve plant scheduling, to simulate or model business situations, and in other ways designed to optimise current operations and evaluate future plans.

The future of the computer is hard to predict. In less than twenty years it has grown from zero to hundreds of thousands of installations and has become an indispensable part of science, engineering and business. The computer industry now ranks as one of the largest in the world. The range and diversity of products continues to grow so that small machines are available for less than $10,000 and large systems for $10,000,000. The speed of machines has improved by the order of 1,000 to one in a period of less than twenty years, a rate of progress which is probably unmatched in any other field. At present the industry is utilising the gains made in micro-electronics and this alone will ensure continued progress for years to come. These gains in hardware have been matched by corresponding progress in computer software which amongst other things seeks to provide a more acceptable product by allowing problem specification in more "natural" languages. There does not seem to be any major reason to suppose that progress will not continue; the main limitation could be in learning how best to use computers, and how to make them even easier to use.

D. EXERCISES

1. Classify the following as being analogue or digital:
   (a) Cash register
   (b) Car speedometer
   (c) Car mileage meter
   (d) Wrist watch
   (e) Car fuel gauge

2. Suggest some new areas of application of computers which could result from reductions in the cost of machines.

SUGGESTED READING MATERIAL

MORRISON, P. and MORRISON, E. (ed.). *Charles Babbage and his Calculating Engines.* Dover, 1961.

BOWDEN, B. V. *Faster Than Thought.* Pitman, 1953.

GOLDSTINE, H. H. and GOLDSTINE, A. *The Electronic Numerical Integrator and Calculator (ENIAC).* "Math Tables Aids Comput. 2" pp.97-110, July, 1946.

SERRELL, R., *et al. The Evolution of Computing Machines and Systems.* Proc. I.R.E. May, 1962 p.1039. (This includes an extensive bibliography and also covers early history.)

ROSEN, S. "Electronic Computers: A Historical Survey". *Computing Surveys* Vol. 1, No. 1, p.7 March, 1969. (Concentrates on the period 1945-69 and has an extensive bibliography.)

WILKES, M. V. *Automatic Digital Computers.* Methuen, 1956. (An interesting account, including early history by one of the pioneers.)

CHAPTER 7

Man-Made Systems and their Description

R. M. Huey

"Education with inert ideas is not only useless; it is, above all things, harmful."

A. N. Whitehead, 1962; quoted with useful comment by George Pickering "The Challenge to Education" Pelican edition, 1969, Chapter 6.

A. CONTROL SYSTEMS

When the phrase *control systems** is used, one usually means a system by which a desired state may be achieved. Control systems may include a human operator, or they may be entirely automatic. An example of a control system in which a human operator is an essential link in the chain of events, is a motor car and a human driver. Another such example would be a crane for lifting material onto a construction site, controlled by a crane driver.

An example of an automatic control system is a passenger lift or elevator in which the control sequences for going to the desired destination floor level are completely automatic, once the destination has been indicated by the passenger pressing a button. Another example of an automatic control system is the thermostat controller of the oven in a domestic cooking range. The temperature inside the oven is automatically maintained within a fairly small variation from the desired temperature which is preset or chosen by the cook who had adjusted the thermostate dial.

Control systems are of many different kinds. We may think of many examples of systems dealing with the movement of material or people. We may think of examples of systems which control the conversion (i.e., so-called "generation") of energy, its distribution or its application to a specific task. Economic and financial systems may be added to our list, where the flow of money is controlled by legislation and regulations in order to achieve (or perhaps to *try* to achieve) a desired result. The set of laws and procedures known as patent law could be termed the patent system.

A chemical system could be set up in which the rate of production of the end-product is governed by the rate of input of raw materials and the temperature of the reaction vessel. The latter quantity will depend on the amount of heat generated or absorbed by the reaction itself and the amount of heat supplied or extracted (i.e., the rate of heat input to the reaction vessel) together with other factors such as heat loss to the surroundings. The boiler (or steam generator) in a thermal power station is an example of a chemical system from which heat is extracted and used to boil water and to superheat (i.e., to heat above the temperature of boiling water) the steam so generated.

* For more detailed descriptions of Control Systems, refer to Chapter 22.

Elaborate automatic controllers are built into such boilers to ensure that the temperature and pressure of the steam are maintained at desired values, while the flow of steam is automatically adjusted to meet the demands of the steam turbine driving the electric generators (alternators[(1)]).*

Other automatic regulators control the voltage and power output from each alternator in accordance with a pre-arranged pattern, while a master controller is arranged to keep constant the frequency of the power supply system as a whole. *All* the alternators feeding power into the one system must be running at the same frequency in order for all machines to keep in step and thus avoid the catastrophic surging of power which occurs if electrically connected machines get out of synchronism.

Summing up, we have given some examples of systems which deal with energy, with the regulation of human activities, with transportation, with the conversion of matter into different chemical forms. The systems were controlled either manually or automatically, or by some combination of manual and automatic control.

Both with manual and automatic control, the aim is to make the system perform its function in some desired fashion. This is sometimes expressed by saying that we wish the system to meet specified performance criteria.

All the examples, and many others which readers may care to add to the list, have in common that they are systems *designed, established and used by human beings.*

B. COMMUNICATIONS SYSTEMS

Communications systems differ in one essential point from the examples quoted in the preceding section, namely that they are systems designed or established to deal with *information.* They may discover, transform or transport information (see also Chapter 23).

However, it should be borne in mind that *all* systems of whatever kind need to acquire and utilise information in some form or manner, in order to carry out their intended functions. The phrase communication system is thus usually reserved for a system which is designed to acquire or to convey information. The telephone system is an elaborate world-wide communication system for the transportation of information in the form of human speech, from one place to another.

A computer is an expensive and elaborate information system whose purpose is to transform information or data in accordance with a scheme which will re-pattern the information into a (usually) more intelligible or useful form.

A radio telescope is an expensive and elaborate system intended to acquire information of interest to astronomers, to astro-physicists, to engineers concerned with space travel and to many other classes of people. During the acquisition of this information (which is often almost blanketed by unwanted signals, or "noise") it may be transformed into a variety of forms. These transformations of the signal gathered by the aerials of a radio-telescope are very carefully *thought out and constructed* so that they can be *used* to emphasise the desired information and to suppress, as much as possible, the undesired noise. Looking at the three words which have been accented in the last sentence, they will be seen to be identical meaning with the three words accented in the last sentence of Section A.

It seems that all systems, control and communications and any other categories we care to nominate, will have in common the fact that they are *designed, established and used by human beings.* This is indeed so. It is tempting to add to the phrase, the extra words *for the benefit of mankind*, particularly since this is the expressed ethic of the engineering profession. Unfortunately, the purposes for which systems are established

*Endnotes are signified by (1), (2), (3) etc.

are not always beneficial, or even when the purpose is so conceived the end-result of the system may not always be in accord with its purpose. A responsibility therefore rests on technical people who are acquainted with systems theory, to encourage the spread of such knowledge and, to be at some pains to explain it to the community at large. They carry the added responsibility of urging that the methods of systems science and systems engineering should be utilised to integrate or to mesh together technical systems with human systems and with everyday activities in ways which will be beneficial to mankind. Over the whole global system this is a very large responsibility indeed and it is one of the hopes of the authors of this book that many of their readers will consciously assume a share of this burden.

There are also many "systems" which are not man-made, and Chapter 8 has something to say about one such class.

C. MEASUREMENTS ON SYSTEMS

It is not always easy to be sure that reliable and accurate measurements can be made, which will display to us the *significant* behaviour of the system in which we are interested. Let us consider some of the reasons why measurements on a system are important, and try to develop a working rule which can help us to say whether the ability to make one particular measurement is important or unimportant.

Firstly, let us remember that a system is *designed, constructed and operated* by humans and that these three different stages in the life of each system* will normally be the concern of three different groups of people. It is understandable therefore that a technical description of the system which will enable the technologists and managers in these three groups to communicate meaningfully with each other, must be able to highlight the sort of matters in which *each* of these three groups is interested. A technical description must therefore be written in terms of the variables and parameters (or, if the reader wishes, for variables he may read the word *signals* and for parameters he may read the phrase *system function*, as described in Chapters 5 and 12) which have significance to each of these three groups. These quantities of interest will not be identical for each of the three groups nor even for different persons within the same group. However at each stage, measurements must be made which will enable us to say something, *in numerical terms*, about both the signals and the system.

In the same way that designers will tend to use *easy* design and computation techniques (provided of course, that they are *effective*), so also will the experimentalists, installers and operational people tend to use easy and cheap measurement techniques, provided that the resulting measurements are *significant*.

There are thus two questions always to be asked about a proposed measurement:

(1) What will our experiment actually succeed in measuring, and

(2) is that what we really want to know?

Sometimes the measurement will only tell us something about the measuring instrument itself. Suppose for example, that we wish to measure the time in which a new circuit (intended for use as a computer logic element) can change over from On to Off. Let us suppose that our first attempt to measure this is with a cathode-ray oscilloscope which indicates that the change over time is 50 nanoseconds, a figure which is many times slower than the hoped for result. A check on the specification* for the particular oscilloscope reveals that it is a "slow" instrument in the sense that it cannot display changes in input which occur more rapidly than this figure. Our next step is to make agreeable conversation with the custodian of a faster

* The fourth stage and the final one is the death of the system, usually due to a human decision to effect replacement with another system for economic, technical or social reasons.

† That is to say for the rise time of the vertical deflection amplifier in the oscilloscope. Remember 1 nanosecond is 10^{-9} seconds about equal to the time for light to travel 30 centimetres in air.

oscilloscope in a neighbouring laboratory in order to borrow the instrument, which is known to have a response time of a few nanoseconds. The next measurement indicates a switching time of 15 nanoseconds for the circuit under test, still about three times the hoped for result. This is still puzzling until we investigate more closely and discover that due to a careless choice of the way we have connected the oscilloscope to the circuit under test the resistance and capacitance of the circuit under test, plus the oscilloscope and the connection leads are such that the resulting time constant(2) accounts almost exactly for the observed rise time of 15 nanoseconds.

It has now become obvious that the answers to our first question have been: on our first try we have measured the response time of the "slow" oscilloscope, on our second try we have measured the response time of our connection circuit.

A third try, with the fast oscilloscope connected by shorter, properly disposed leads to a more suitable circuit point on the new logic element, finally yields a value of 4 nanoseconds. We may be fairly confident at this stage that the measurement is meaningful. It is no longer the case as McLuhan* says, that "The message is the medium". In our third try the message really did originate with the device under test.

It is now time to come to the second question "Is that what we really want to know?" In the case of our hypothetical logic circuit design task let us suppose now that we announce to the person responsible for the overall computer sub-system that the new logic element can achieve the desired speed of 5 nanoseconds switching time. Let us now imagine that this announcement is greeted with "That's fine, but don't forget it has to be 5 nanoseconds *when* it is feeding pulses into the group of three logic gates next in line in the assembly".

The next obvious step is to try the new device either feeding into an actual group of three logic gates, or into a *simulated* load (3) which will behave in a similar way to the actual group of three logic gates. If the design specification for the new circuit had been sensibly conceived and executed we may find that the desired speed can still be achieved, together with pulses of sufficient power to operate reliably the three logic gates.

This example in which we talked about designing a single element of a system has shown us that we must always consider the behaviour of each element, not in isolation but *embedded* in the system for which it is intended.

It has also shown us that the results of measurements may not always have meaning unless we consider the whole system—device under test plus connection arrangements, plus measuring instrument.

D. SIMULATION

In the last section we have also introduced the very powerful idea that parts of a system (or even whole systems) may be simulated and that the success of such a simulation will depend on whether we can provide an adequate description which will enable the simulated portion of the system to be constructed.

Such descriptions may be verbal or numerical, but if they are going to be useful (i.e., so that the simulation could actually be constructed) they will almost certainly enable us to set up a mathematical model. The idea of the mathematical model to represent a system will crop up again and again in this book. Some of its necessary attributes are described in the next section of this chapter.

To return to simulation, it should be observed that simulations may be:

(1) actual physical simulations of the same kind as the system being simulated (sometimes termed homologues)

* H. M. McLuhan, *Understanding Media: The Extensions of Man.* A chapter so entitled. Sphere Books, London, 1967.

(2) actual physical simulations of a kind different to the system being simulated (usually termed analogues)
(3) abstract simulations which may be verbal, graphical or mathematical.

Let us give some examples which can be fitted into the above categories. Readers who live in Sydney, Australia will probably recognise the hydraulic model described, as one which is located in a large factory-type building erected on waste land between the Mascot airport grounds and Botany Bay. The hydraulic model concerned is an accurate small scale model of the bed of Botany Bay which is filled with water and in which the effects of tides, waves and winds outside the heads of the bay, may be simulated by raising and lowering the water level, by wave-making machines and by large fans. The model will be used to ascertain the effects of dredging for port development, to accommodate both cargo vessels and larger oil-tankers, of works such as the extension of the airport north-south runway into the bay to cope with the next generation of jet aircraft and the effects of natural forces of sea and air which may tend to destroy or to silt-up these developments. In addition, the model can serve a political-social purpose since the results of experiments can be used to re-assure the local residents that the sands of the popular bathing beach stretching for several miles from Brighton-le-Sands to Sans Souci will not be swept away by seasonal storms, due to new siltation and erosion patterns created by the new works. Technically and economically the hydraulic model will enable the design of the new works to rest on assurances that the processes of siltation and erosion can be controlled in an intelligent manner. The model falls in the class of homologues since it is of the same kind (i.e., the actions of moving water and air) as the system which it simulates.

Other homologue simulations which may be mentioned are the laboratory simulation of kilometre-size radio astronomy aerials by metre-size small scale models, the wind-tunnel tests on small-scale models of aircraft structures and tall buildings to determine aerodynamic forces, the laboratory simulation of lightning strikes on electric transmission lines using model towers of 1-2 metres in height, and the attempt to model the propagation of light in an insect's eye structure using a large-scale model (several thousand times larger) and radio waves in place of light waves (see Chapter 21). These models are all homologues since they utilise the same physical actions in the simulation as in the original system, although on a different size-scale and time-scale.

An analogue simulation differs from a homologue in that different physical actions are utilised in the simulation. For example electric voltage and current may be used to simulate temperature and heat flow in a thermal situation. Such an analogue was actually used in the design of the buildings housing the repeater amplifiers needed for the coaxial cables used to convey television, telephone, telegraph and data signals between the capital cities of Australia. The problem in this case was to ensure that the amplifying equipment would not be subjected to more than its allowable range of temperature, even on the hottest inland day. Electric circuit analogues are convenient to represent many other problems, such as the the movement and acceleration of mechanical devices, the diffusion of oil or water through sand beds, the acoustic properties of an enclosure which needs to be sound-proofed, the dynamic behaviour of the human living system, and many others. Nevertheless it is sometimes convenient (e.g., as a teaching aid so that we may render visible in an analogue the invisible actions of electricity or magnetism) to use a mechanical simulation of an electrical situation; a series of masses on springs coupled one to the next can simulate the propagation of an electrical impulse along an electric transmission line, rendering the action both visible and slow enough for us to comprehend, thus adding meaning to our visualisation of the somewhat abstract picture of one or more pulses of electric charge propagating along an electric transmission line or across the junction of a transistor or other semi-conductor devices.

Analogue computers are constructed from electric circuits to simulate systems which can be described by differential equations (a widely used form of mathematical model). The simulation is exact in the sense that the same differential equation is the mathematical model for the actions of voltage and current in the analogue computer and for the actions in the original system. This leads us on to the question of what is meant by the term mathematical model.

E. MATHEMATICAL MODELS

The mathematical models for the various systems mentioned in the preceding section of this chapter would usually be equations dealing with the dynamics of the system involved—for example the behaviour of the hydraulic or electrical models mentioned could be described by differential equations written with respect to time, or by simpler algebraic equations which represent* these differential equations. It is easy to understand that these equations plus a specification of the commencing state of the system will enable us to calculate its state at subsequent times. The mathematical model is thus simply a set of equations plus a description of the initial state, which enable us to predict the future state of the system at any time. This is the central problem of dynamics and very often we are interested in discovering this sort of answer.

Sometimes, however, it may answer our purpose if we can calculate the ultimate state of the system—this may be termed the steady state and in this case we need not concern ourselves with the behaviour while the system is settling down to its steady state. Usually the problem of calculating steady state behaviour may be classified as a problem in statics, and is as a rule a much easier problem to solve than the equivalent dynamic one.

This usually means that a simpler mathematical model will serve our purpose if we need to calculate only a static solution. By a "simpler" mathematical model one means that some of the terms in the equations can be discarded (i.e., all those terms which imply change with time—for example the acceleration terms in a mechanical system—will have become zero when the steady state has been reached).

Dynamic problems will usually involve differential or integral equations with time as an independent variable (or the somewhat similar difference equations) while static problems can be written as or turned into equations in which the variable time does not appear. Some of the models will be "deterministic" in the sense that the inputs are known functions and the outputs can thus be calculated or determined. Other models will be statistical or "probabilistic" in the sense that we do not know in advance what the inputs and hence the outputs will be, but nevertheless we can assign probabilities to the occurrence of certain values and hence can talk about the likelihood of certain events occurring. Mathematical models for system reliability (i.e., expected time between failures, expected time till "death" of a component, expected percentage of rejects from a manufacturing process, etc.) are usually based on probabilistic ideas.

No matter whether a mathematical model is deterministic or statistical in nature, to be of service it must be simple enough to be understandable, and computable by the group of people who will be obliged to use it. It is sound policy to choose the simplest model which is capable of serving the intended purpose.

This shows us that the choice of a mathematical model to represent a system is not a unique one. It shows us also that the choice must be made to fit in with the reason

* See Chapters 5 and 12 for a mention of phasor algebra and Laplace transform methods where the solution of differential equations is transformed into simple algebraic routines.

for using a model and to fit in with the people who will be required to use it.

A model which produces too much information can sometimes be as bad a choice as a model which produces too little information. In choosing which model to use, a considerable degree of insight and judgment is required.

F. EXERCISES

1. Write down, in note form, what information you possess concerning a man-made control system with which you have some familiarity. This could be the group of elevators and escalators in a tall city building; the traffic control arrangements at a busy intersection; the hydraulic system comprising a heart and the blood circulation; a pop-up toaster; the automatic transmission system in an automobile; the automatic gain control in a radio receiver, etc. Devise two or more *models* of your chosen system. Write down one or more choices for variables to represent an input and an output in each case. Make a list of what parameters of the system would need to be known in order to enable computation of the output for a given input. List which quantities are sensitive or insensitive to each other. Discuss your model with other students. Did such discussion help you to improve the model? If not, why not?

G. ENDNOTES

(1) The term alternator is used because the electric current produced, alternates in direction many times per second, as the alternate north and south magnetic poles in the rotor of the machine pass in front of and so induce a voltage in a particular set of coils or windings. In Australia a typical alternator would utilise four poles on the rotor (two north and two south polarity) and would run at 1500 revolutions per minute, generating an alternating voltage of about 11,000 volts at a frequency of 50 hertz (i.e., 50 cycles per second, each cycle comprising successive positive and negative voltages). In Europe the frequency of 50 hertz is commonly used, while in the U.S.A. the power supply frequency is 60 hertz (see also Chapter 25).

(2) The time constant of a single resistor R ohms in value connected to a single capacitor C farads in value, will be RC seconds. The time constant may be reduced by splitting the total resistance and capacitance into separate sections or by making proper use of the inductance property of the connections in order to speed up the response of the connection circuit.

(3) In this case we might be able to simulate the load by a resistor connected in shunt with a capacitor; depending on the values involved, connection of such a load could either slow down or speed up the switching time. An improvement in speed of response can usually be gained only at the expense of other quantities, e.g., a decrease in the amplitude of the available pulse. Carried out with a knowledge of the quantities involved this process is termed "trading-off" between the various performance capabilities of the unit. Trading-off in an intelligent manner involves both skill and judgment, and is used at all stages in the life of a system—design, construction and operation.

SUGGESTED READING MATERIAL

Chestnut, H. *Systems Engineering Tools*. John Wiley and Sons, New York, 1965. Part III of this book particularly Chapters 22, 23 and 26.

CHAPTER 8

Biological Systems

G. A. Horridge

***"Wha wadna pity thae sweet dames
He fumbled at, an' a' that,
An' raised their bluid up into flames
He couldna drown, for a' that."***

*Robert Burns
The Merry Muses of Caledonia*

A. A MIXED DIET

If electrical engineers were grandmothers and computers were simple things like eggs, then it would be presumption indeed for a biologist to offer advice. But a biologist need have no fear, for he that is low fears no fall: electrical engineers clearly prefer a mixed diet, and computers are quite incapable of turning into chickens.

One continuing source of inspiration for us all is the natural world, and it is certain that lessons for engineers will flow from biological examples for those that are willing to make the effort. The unrolling of the information stored in the sequence of building blocks of the DNA* molecule in response to the progressively developing situation in every part of a growing embryo, and the control of the chemical turnover in any living body, provide examples of information flow that are amenable to mathematical approach if they can be written down as a series of suitable relationships. A more attractive candidate for analogy with man-made mechanisms is the brain, or, in general terms, the nervous system. Here it is more obvious that communication between parts of the body occurs, and the activity has a more striking time scale.

The control of development and the control by nervous systems differ from computers in the following ways.

(i) Many thousands of parallel channels are active, i.e., they are running numerous partially interacting programs simultaneously. An embryo may have 10^8 differentiating cells, each separately unrolling its own DNA. A human eye has 10^7 nerve cells, an ear 10^4 nerve fibres to the brain, a small region of the brain or spinal cord has 10^5 nerve cells, all more or less active simultaneously. For n independent subsystems which can be either active or not active at any given moment, the number of combinations of activity in different pathways is 2^n. Because the combinations must change continuously the possible number of patterns of activity soon becomes prodigious.

(ii) Computers perform their operations many times over and such "in depth" or iterative methods make use of long waveforms as a means of building up information content; e.g., they use differences between large numbers. They must therefore be

* DNA, short for deoxyribose nucleic acid, is the chemical stuff out of which the building blocks of inheritance, genes, are made. Basically, the controlled differentiation of living structures is all coded by appropriate combinations of a few units in the long DNA molecules.

accurate to better than 1 part in 10^{12}. By contrast, nerve cells are rarely accurate to 1 part in 100, and a figure of 1 part in 10 is more realistic. Therefore, on this account alone, nervous systems must perform only "shallow" or non-iterative operations.

(iii) In computers the timing of signals is supremely important in the carrying of information: in biological systems the significant feature is the choice of labelled line or pathway.

(iv) The *state* of a biological system or of its components is never fully definable. In particular, biological systems are not digital and, therefore, every measure has a range of error. The state of a biological system is also indefinable because it contains a memory of unknown extent.

(v) Computers are systems that are built in a particular way; on the other hand biological systems are ones about which we are still *finding* things out; although a great deal is known about tracts and centres in nervous systems, only the most trivial circuit diagrams are known in terms of the single nerve cells by which the nervous system operates. Even if the computer components are thrown together at random, and include random number generators, these features are potentially *all known to be there*. This much cannot be said for any biological system.

Notwithstanding, computers are valued by biologists for:

a. doing calculations. In the past most calculations could not be done because they were too hard. Successive approximations were too time-consuming. Use of computers makes it possible to do a larger proportion of all possible calculations;
b. storing and sorting information into categories that are given. In addition, the new types of program search out likely categories first and then sort for optimum separation of previously undefined categories. In short, computers organise data to produce sophisticated summaries.
c. making mathematical working models. These are usually conceptual models within the program of the computer and they illustrate how a set of interacting parts work out in practice. But this approach is always to be viewed with suspicion. A mathematical equation summarising the relation between input and output of a black box is a contribution that biologists will not accept as an *explanation*. They may accept it as a description if they see that they are dealing with the transfer function of a single, simple interaction which is really there. The mechanism must be understood in terms of its actual components, which are not relevant in the pure mathematical model approach. One good reason for this is that the biologist always wants to say how the information is distributed among his many parallel channels of flow at any one time. Without the visualisation of the mechanism, as a collection of real biological components arranged in space (called Anatomy) the biologist will always say that he does not see how his system really works. The most meaningful models in biology are always structures, and sometimes a computer "work-out" of them will show that they are models which fit the facts quantitatively as well as in composition.

The future value of biology to computer designers has no foreseeable limit. One obvious animal mechanism to copy is pattern perception. The eye, even in simple animals, is much better designed for pattern perception by having many lines in parallel than is the one-channel scanning system of a TV camera. The abstraction of the relevant features of visual patterns is a very diverse, and interesting product of independent natural selection in many animal systems. To copy these would be fun; to make effective artificial systems would be useful for sorting mail, recognising approaching accident situations in traffic, sorting goods in a warehouse or in lieu of visual inspection on quality control. Partly in response to this need, a new type of computer is being explored. A large number of independent sensors each feed into a network of coincidence and anti-coincidence circuits which are able to pick out particular spatial patterns. But the number of possible patterns rapidly gets out of hand as the number of inputs and possible selections increases. Most lower animals have conquered this problem by restricting their attention to certain classes of pattern which occur commonly round them, and to a very limited number of choices in their behaviour. This is their *repertoire*. Therefore, it may follow that computers built with many parallel inputs must be limited to a particular repertoire. But without a memory such a system gives only a rapid answer to an immediately presented pattern. To achieve more the problem is to replace a time-sequence memory on a tape by numerous interacting memories in many locations, laid out in as many dimensions as there are input lines in parallel, and all simultaneously available. This is difficult but not impossible, for apparently many types of animal brains, e.g., bee, frog, octopus, rat and spider, manage to do it.

B. EXERCISES

1. A pattern perceiving device has n input channels which may be either On or Off at a given moment. The device has m possible outputs that it can bring into action in different combinations as a result of x different classes of patterns in the inputs. What is the relation between m, n and x? (A system with $n = 2$, $m = 4$ and $x = 4$ is easy.) Draw in terms of hardware, as relays, a system with $n = 4$, $m = 4$ and $x = 4$. (Clue: you now find that some patterns are not distinguished.)
2. You lead off with a random probe the electrical activity from *an unknown point* inside a digital computer, and on another occasion from inside a brain (both having a memory). In each instance how would you distinguish signal from noise?

SUGGESTED READING MATERIAL

BARLOW, H. B. "The Coding of Sensory Messages" and GREGORY, R. L. "The Brain as an Engineering Problem" in a book entitled *Current Problems in Animal Behaviour*. W. H. Thorpe and O. L. Zangwill (eds) pp.424. Cambridge University Press, 1961.

CRAIK, K. J. W. *The Nature of Explanation*. Cambridge University Press, 1943.

DEUTSCH, J. A. *The Structural Basis of Behaviour*. Cambridge University Press, 1960.

HORRIDGE, G. A. *Interneurons*. Final chapters. Freeman, 1968.

CHAPTER 9

Human Systems

(With reference to modern concepts of management)

R. C. Olsson

"I would not trust the management of the nation's affairs to men who had no first-hand knowledge of what was happening in this quickly changing world, and who did not understand even the language of technology. But, I do not want the world to be run by scientists and technologists, because science and technology are of value only as servants of society. I want to see at the head of affairs basically educated men, science-oriented humanists. They must understand the values of mankind, they must have a view of history and a view of the future. They should have a very strong flavour of science about them, but not enough to turn them into scientists."

Sir Leon Bagrit, The Age of Automation,
Penguin Books, Middlesex,
England, 1965.

A. INTRODUCTION

Science has already become a most pervasive force in our modern society, having a widespread influence over all of man's activities. It has had an impact on man's work environment, standard of living, health, social institutions, and on his recreational activities.

Scientific and technological achievements have had a similar impact upon business organisations. Almost every industry has seen major advancements not only in its products but also in its means of production.

Scientific and technological advancements have greatly influenced decisions made by governments on such matters as health, education, foreign policy, welfare, and national defence.

There are many other areas where science and technology will undoubtedly make spectacular strides in the years ahead—in biological and medical research, nuclear technology, computers, transportation and communication systems, and in space technology; and it would appear that this advancement will continue at an ever-increasing rate.

One of the major factors underlying these achievements has been man's ability to develop social organisations for the achievement of his purposes. But as major scientific and technological projects become more complex the problems of organising and managing these projects becomes more difficult. In addition, there is a growing awareness that the development of very large and highly complex systems are now producing problems of far reaching consequences for the human race. Some of these problems are as follows:

1. There is the need for someone to translate the scientific achievements into

beneficial and usable products or services; and the achievement of human, moral and ethical controls over the utilisation of many of these developments (e.g., nuclear energy) is as important to mankind as the discoveries themselves.

2. There is the problem of optimum allocation of resources. Many of these developments will present simultaneous claims on the limited material, human, natural, financial and time resources. The means (resources) available at any point in time are never sufficient to meet man's wants (objectives), and choices have to be made. Criteria for determining priorities and for evaluating the highly complex technological proposals will present a formidable task.

3. The amount of money and time required to design and develop new complex projects has increased (for example the long design periods of modern aircraft). At the same time accelerating technology leads to shorter and shorter life-spans for each new generation of products.This trend of longer development cycles and shorter life spans requires more attention to long-run planning than hitherto, if the risk of expending valuable resources on products with little chance of success is to be avoided.

4. Large and complex programs require for their solution greater specialisation in narrower fields. As a consequence the management of such programs requires the co-ordination of an enormous number of people—many of whom have a variety of specialised training. With greater specialisation, integrating the activities of people into effective organisational units will be increasingly difficult.

5. The problem of conflict between managers and scientists. The managers seek to obtain a reasonable return on the company's investment within a reasonable time. The scientists, on the other hand, seeks to expand his knowledge and to apply this knowledge through an organisational team. Unlike the manager, financial, time, and other resources are not always viewed by the scientist as operational constraints. Conflicts arise when the scientist wants tools and equipment which the project cannot afford. Thus a great deal of attention will have to be given, in the future, to the study of organisational goals and to the factors motivating scientists, managers, and other employees, if optimum results are to be achieved.

6. The emergence of very large and complex projects will need, for their production or manufacture, the facilities of more than one corporate entity. This will require the development of new and more effective approaches to the management of these programs where inter-company and inter-industry relationships are present.

7. The development of large and complex projects also introduces uncertainty into the decision making structure. Uncertainty may arise in respect to the selection of a particular project and this will be met by the development of better techniques for evaluation. Another area of uncertainty relates to costs. Large projects require large amounts of funds but it is often not possible to define in advance, and within reasonable limits, the amount of money likely to be involved. This area of uncertainty poses larger economic problems when the difference between the estimated and the actual cost is substantial.(For example, the cost of developing the atom bomb in 1942 greatly exceeded the original estimate.) Such problems create difficult problems of economic management at the national level with consequent effects on national income distribution and therefore on human welfare.

These are some of the problems now emerging. The rate at which the technological and scientific developments will continue in the future, and whether or not these developments will benefit mankind, as it is hoped, depends on man's ability to organise and manage human systems in this rapidly changing scientific and technological age. What is required is a better understanding of human systems—systems in which men and machines interact on a vast scale within a common environment.

This chapter is about organisations and their management. Its object is to provide an introduction to problems of organisations and of management, and to indicate the difference between physical or mechanical systems and human systems. Indirectly it attempts to draw attention to the growing need for men in the community possessing a genuine breadth of education and a sense of social and human responsibilities.

B. THE NATURE OF HUMAN SYSTEMS

The basis of human systems is man; and it is said that man is a social animal. The term social implies a tendency to develop co-operative and interdependent relationships. It is also said that man is socially inventive. He expresses his social tendencies in a number of ways by developing many differing patterns of relationships with his fellows. Society is therefore a system, a human system made up of myriads of sub-systems—groups or organisations—all formed to provide outlets for man's social tendencies.

1. The Rôle of the Group or Organisation

Groups or organisations play a large part in the life of man. He is born into a family. His family forms part of a larger group of relatives or kinship groups, and at the same time part of a larger group of friends. He receives his education and pursues his recreational interests in groups. Later on he joins a particular work group and forms associations with other work associates. For his material needs he relies on further groups of people.

These groups or organisations form the environment, various aspects of which are inter-related and in which the whole complex of the inter-related parts is supposed to serve his fundamental purposes.

What are the essential features of a group or organisation? If a particular social group is examined a number of features can be identified. Firstly, its members are bound together in a pattern of relationships; secondly, they appear to share a common set of values; thirdly, their activities are often differentiated; and fourthly, they appear to be striving to achieve certain common goals. These factors suggest that the relationships and activities of the group are organised or planned.

This tendency to organise or co-operate in interdependent relationships is inherent in man's nature. Equally evident in society is the tendency for men to compete. Many examples of this dual behaviour pattern can be witnessed. For example, members of a family compete with one another for limited family resources and yet co-operate in many other endeavours. Companies within a given industry compete with one another for markets, yet co-operate in Associations for particular purposes. Individuals compete with each other in an athletic team, but co-operate with team members against opposing teams. This suggests that although organisation implies co-ordination and integration, it also implies competition or conflict. The nature and intensity of co-operation and conflict will have a bearing on how well the organisation performs. It is the task of management to integrate these diverse—sometimes co-operative, sometimes conflictive—elements into a total organisation endeavour.

2. The Rôle of Management

Essentially, management is the process whereby human and material resources are co-ordinated towards objective accomplishment. Ever since man got together to work in groups for a common purpose such as survival there has been management. However, the study of management is relatively new in our society, stemming primarily from the growth in size and complexity of organisations. It is not surprising that there is, as yet, no universally acceptable definition of management. Typical definitions suggest that management is a process of planning, co-ordinating and controlling activities. Others define management to cover the entire concept of decision making, suggesting that decisions are made in establishing objectives, in the planning process, in task allocation and structural relations, and in controlling the activity in question. Whatever approach is used, management must be considered as the primary force that co-ordinates the activities within the organisation and relates them to the environment.

The study of management is still evolving. Its subject matter is extremely complex as it deals with the nature of man and his intricate relationships with his fellow men. The development of a body of knowledge relevant to organisations and management is an extremely important endeavour in society. Because of the importance of organisations even slight improvements in their effectiveness and performance can have far reaching consequences. Further, if society at large is to benefit fully from the scientific achievements of our age it is imperative that each group in society perfect its functioning. In business, government and other spheres, growth in size, complexity and diversity of operations has made managerial functions more critical to the success of the organisation. However, the question of improving organisational performance (i.e., systems improvement) is one that poses broader issues. For example, if it is supposed that organisations can be improved, what is meant by improvement? Can improvements be made to systems without understanding the whole system? Is it possible for human beings to understand the properties of whole systems? In order to determine whether or not a system can be improved it is first necessary to arrive at a definition of "improvement". This definition must be sufficiently precise in order to develop empirical measures of improvement and thus translate the data collected into criteria of better or worse with respect to that system. If it is assumed that some reasonable measures of better or worse with respect to systems can be developed, the next task is to translate a physical description of a system into a measure of systems performance. Scientists have a remarkable ability to describe the physical world and, by means of mathematics, to translate the physical description into measures of effective performance of the whole system. It is very exciting to contemplate the possibilities of using science to study improvements in large-scale systems. Such investigations suggest to the young creative mind the possibilities of using his intellectual attainments in the improvement of human systems.

The idea that scientific descriptions may be able to cast light on methods of systems improvement was realised in 1940 when military managers began using scientific resources to study various critical problems of defence systems. The scientist using his own training and background, studied the nature of military systems and translated the physical description of a defence system into concepts of improvement. This early study which was called operations research laid the foundation to what is now called "management science".

The advent of the computer has also opened up new avenues of approach to systems improvement. In many cases, the physical description of even a fairly simple system involves thousands of variables and the mathematical relationship between these variables may be very complicated. Today many scientists and engineers devote their lives to the study of systems and their improvement. They try to formulate the physical descriptions of systems and then translate these physical descriptions into mathematical "models" which enable them to predict what will happen if certain changes are initiated in the systems. This trend has introduced a new relationship between scientist and manager, because now the scientist is working much closer to those who are directly concerned with the responsibility for systems performance. There is also an implication that the scientist in his endeavours to improve systems may undertake to plan them. Why should the scientist have a servant-master relationship? Is not the scientist obliged to go beyond the manager's needs and ask whether these needs are morally correct ones? Thus there is not only a need to develop techniques for improving performance given the goals that certain groups wish to attain. There is also a need to determine whether the goals are proper ones. This is only one of the issues that must be examined in order to improve the efficiency of group operations whatever the nature of those operations.

C. FACTORS LEADING TO COMPLEXITY IN ORGANISATIONS

To return to the problem of social relationships within the organisation, many factors which are now emerging are adding to the complexity of relationships between the organisational participants. Some of these factors result from natural evolution, others are induced.

1. Natural Evolution

In the evolutionary development of all living matter there is an unmistakable trend from the simple to the complex. The essential element in this trend is specialisation. It allows organisms a means of dividing up the work performed in each sub-part for more effective performance. (The development of the human embryo is an excellent example of this process.) However, as in human development, specialisation requires the integration of activities, if the organism is to pursue identifiable goals. These same trends are evident in organisations. More specialisation requires increasingly sophisticated methods of co-ordination and integration. These factors have led to increasing complexity within the organisation and have made the task of management more difficult.

2. Size of Organisations

There is a trend towards increased organisational size. Increasing rate of population growth; larger amd more complex plant and machinery requiring capital outlays, and the opening up of national and international markets, are some of the dynamic environmental factors underlying this basic trend. In small face-to-face organisations, management was relatively straightforward. As groups grow in size, simple face-to-face relationships are no longer possible. The number of inter-relationships between participants increase dramatically and personal contact by managers is lost.

3. Science and Technology

One of the most important influences leading to complexity in organisations is the accelerating trend of scientific knowledge and its implications for society in the form of technological developments. Developments in science and technology have magnified the trend towards specialisation. Scientists and technicians are becoming more prevalent in organisations. Integration of their efforts towards organisational accomplishments can be difficult. These trends have definite impacts on organisations and hence are important managerial considerations.

4. Increasing Educational Levels

The increase in the general level of education of the population provides a more sophisticated atmosphere in organisations. People are no longer prepared to accept, without question, the dictates of the organisational leaders.

5. Increasing Government Influence

Another important influence relates to the growing influence of government at all levels. Trends in all areas of government concern, including economic defence and social spheres, provide the framework and constraints within which managerial decisions are now made.

6. Growth of Specialised Disciplines

Developments in the social sciences are now beginning to appear at increasing rates.

More and more attention is being devoted to disciplines such as anthropology, psychology and sociology—all topics which have particular concern to organisation theory and management.

All of these factors are adding to the complexity of organisations and the management of them. Clearly there is a need not only to understand the nature of organisations themselves but also the factors affecting their form and development and in addition there should be concern about new directions of enquiry.

D. TRADITIONAL THEORIES OF ORGANISATION AND MANAGEMENT

Throughout recorded history there is evidence that men have pondered the problems of human organisations and the management of governments, armies, churches and other complex social groups. However, it is only since the end of the nineteenth century that a systematic body of knowledge concerning organisations and their management, has emerged. Even after the start of large scale commercial and industrial enterprises in the latter part of the eighteenth century, the development of management thought was relatively slow. Problems of organisation and the use of labour were solved *ad hoc* for each establishment. Knowledge about the solutions was transmitted by observation but in most cases had to be rediscovered by most new firms. What was needed was the development of some theory.

1. The Value of Theory

The value of theory is its capacity to explain and predict. Theories are developed from certain basic statements which are assumptions. These assumptions are based upon experience or tests or "hunches". But "hunches" will not help unless they are derived from knowledge. For example, the Romans built beautiful bridges but they had no theory of strength of materials or friction, or co-efficient of expansion. They employed no theoretical knowledge but were guided only by observation and experience. The young Roman builder learned how to build bridges by following the example of the experienced builders.

Today, however, we require more than these simple, though effective, methods of bridge building. We have to build longer, higher and stronger bridges. We have to take account of costs. The Romans were not constrained to think in terms of costs or labour or the effects on the human and institutional system in which they operated. The fact that Roman bridges have lasted over the ages may indicate that they were overdesigned. It points out the need for concern on the part of designers of today and tomorrow to design and construct projects so that the quality and precision is related to the purpose for which the projects are designed. More quality or precision, than is required to meet the needs, is costly and wasteful of economic resources. Less quality and precision is equally bad as it results in inefficiency and loss. When engineering and other technological projects involve costs which represent a substantial part of a country's national expenditure, the consequences of irrational or inadequately managed programmes can have damaging effects of great magnitude.

It was not until the Scientific Management Movement initiated by Frederick W. Taylor (1856-1915), the writings of Max Weber on Bureaucracy and the early Management Theorists such as Henri Fayol, that there developed a systematic body of knowledge related to management of complex business and other organisations. These concepts are now generally referred to as the Traditional Theory.

At the risk of over simplification a brief review of some of the major concepts follows.

2. The Scientific Management Movement

The followers of the scientific management movement considered the organisations as a system which could be improved by the application of scientific principles. The concept was based on the idea that each man had a rôle to play in society. It was oriented to an engineering and mechanistic emphasis that focused primarily upon increasing worker efficiency. It was thought that work could be analysed scientifically in order to determine the best way of doing each task. Once this "best method" was determined it provided the basis of standardisation. Workers could then be selected on the basis of their ability to perform the specific tasks. This procedure was expected to lead to increased productivity which would benefit both employer and employee. The assumption was that workmen would be motivated by greater economic rewards which would come from greater productivity. The rôle of management was to develop a "scientific" approach to the method of performing work tasks, to scientifically select and train each man for specific tasks; to co-operate with men to see that the tasks were performed according to "scientific" principles and to divide the responsibilities between management and workmen. The concept led to the development within organisations of specialist departments including industrial engineering, personnel, and quality control.

This concept of organisation and management was not without its opponents. The workers resented the fact that they were being treated like cogs in a machine and claimed that it destroyed humanistic practices in industry.

The followers of the scientific management concept were basically concerned with efficient task performance. However, they provided many of the ideas for the conceptual framework later adopted by other management theorists. These included a clear delineation of authority and responsibility; the development of functional organisation—that is, the division of the organisation into functions such as production, marketing, finance, and engineering; the use of standards in control; task specialisation; and the development of incentive payments.

Despite its critics the scientific management concept operated during the latter part of the eighteenth and beginning of the nineteenth century.

3. Administrative Management Theorists

In contrast there developed in the first half of the twentieth century another approach to the study of organisations, which has been called administrative management theory. It also forms part of what is called classical or traditional theory of management. The primary emphasis of this approach was on establishing broad principles of management with particular emphasis on organisation structure and the delineation of the basic processes of general management. Henri Fayol, a leading French industrialist was one of the earliest exponents of a general theory of management. The general management approach sought to improve the efficiency and effectiveness of organisations by the establishment of a formal structure. Formal structures were to be designed on the basis of a number of principles. The principle of co-ordination, the pursuit of a common objective by means of unity of action; the scalar principle which emphasised a hierarchical structure; the functional principle, organising tasks into departmental units; the staff principle, which recognised the differentiation of line management (to exercise authority) and staff management (to provide advice and information). The basic ideas of the administrative management theorists are currently applied in the development of many organisations. Criticism of this concept relate primarily to the rigidity of the approach and to the failure to recognise human and sociological factors as important determinants of organisational behaviour and management practice.

4. Bureaucratic Model

Another important influence in the development of modern organisational theory is attributed to Max Weber and his bureaucratic model. The term "bureaucracy" as developed by Weber refers not to the popular notion of red tape and inefficiency but to certain characteristics of organisational design. He suggested that the bureaucratic structure was based on the principle that the right to exercise authority is based on position. Within the bureaucratic structure each member of the staff occupies a position with specific delineated powers, the various positions are organised in a hierarchy of authority, compensation is in the form of a fixed salary, selection for office is based upon technical competence and the organisation is governed by rules and regulations.

Critics of the bureaucratic model are many. Some say that the bureaucratic form encourages its members to adhere to rules and regulations for their own sake. Others suggest that bureaucratic mechanisms develop certain forms of autocratic leadership and control. Many modern writers stress that the bureaucratic view might be appropriate for routine organisational activities where productivity is the major objective but is most inappropriate for highly flexible organisations involved in many non-routine activities and where creativity and innovation are important.

5. Summary of the Traditional Concepts

Although the traditional theory represented by the preceding review has been developed from a wide variety of observations and experience there are some common assumptions underlying the various concepts.

1. Man was assumed to have complete knowledge of opportunities and act in a rational way.
2. The goals of an organisation could be accomplished most efficiently through specialisation in a well-defined hierarchical structure.
3. The organisation could be controlled in a mechanistic way by the legitimate authority of management.
4. Workers were primarily motivated by economic incentives.
5. In order to ensure co-operation leading to the attainment of organisational goals it was necessary to establish tasks, provide detailed instructions and closely supervise participants.

The traditional management theory thus evolved into the development of concepts such as formal structure, specialisation of functions, line and staff relationships and other principles which would be appropriate for all organisations.

6. Criticism of the Traditional View

One of the criticisms of traditional theory is concerned with the unrealistic assumption concerning human behaviour. It is now recognised that economic incentives are not the sole motivating force of workers.

Secondly, traditional theory assumes that the organisation operates as a closed system. Under such assumptions the model fails to take into account environmental influences on the organisations.

Thirdly, the concept of hierarchical structure with its emphasis on authority according to position is no longer a useful concept in dealing with modern complex organisations. The development of technological specialisation with specialised knowledge and therefore authority, is leading to conflict between the two types of authority.

A fourth criticism now emerging is the failure of traditional theory to recognise man in the organisation. Traditional theory was concerned primarily on the organisation and its structure and functioning.

E. DEVELOPMENT OF MODERN VIEWS

Many forces have modified traditional organisation and management theory. New knowledge has been acquired following studies and research in a number of fields including psychology, sociology, anthropology, mathematics, statistics and industrial engineering. These developments may be classified into two major forces: the behavioural sciences and the management sciences.

Behavioural science refers to those aspects of anthropology, psychology and sociology concerned with establishing generalisations about human behaviour that are supported by empirical evidence collected in an objective way.

Management science refers to the contribution by mathematicians, statisticians, economists and engineers to the economic-technical aspects of organisations. Management science seeks to determine systematic approaches to problem solving with emphasis on the scientific method.

1. Contributions by the Behavioural Scientists

The behavioural scientists developed a number of concepts about human behaviour in organisations. They viewed the organisation not as a technical, economic system but as a social system. They found that human behaviour is affected by feelings, sentiments and attitudes; that economic incentives are only one of many of the social and psychological facts that motivate men, and emphasised the importance of communications between the various levels of the hierarchical structure.

The writings of many of the behavioural scientists emphasise the value of a more democratic, less authoritarian structured organisation. They emphasise human values and often tend to depreciate economic and technical considerations.

2. Contribution of the Management Scientists

Management science is an extension of the earlier scientific management concepts and the later Second World War development of operations research. The early philosophy of operations research was directed towards total systems and optimal solutions. As such it had a significant influence on organisation theory. In order to achieve an optimal solution it was necessary to ensure that all significant factors were given consideration and this called for a total interdisciplinary approach. The development of the management science approach has led to the introduction of many scientists to actual military and business decision-making problems where they could apply their specialised knowledge and skill. Techniques such as linear programming, queuing theory, statistical decision theory and other analytical tools are now well-known.

In the last few decades there has been a dramatic growth in management science applications in government organisations particularly in defence and space activities.

In military and space programs, the term systems analysis is used to describe an integrative decision-making process using management science approaches. In other words, the management science approach seems to be effective where the problems are well structured and capable of mathematical and statistical interpretation.

In short, traditional management theory emphasised the structural and managerial sub-system and was concerned more with the development of principles. Behavioural scientists focused their attention on motivation, human relations and other psychological factors. The management scientists emphasised the quantitative techniques of decision making and control techniques.

Each has contributed substantially to the total quantum of knowledge but their attention in each case was almost exclusively directed to primary sub-systems. It is now becoming obvious that anyone who actually believes in the possibility of improving systems is faced with the problem of understanding the properties of the whole system.

F. THE SYSTEMS APPROACH

Organisations are complex human systems made up of psychological, sociological and economic elements which require intensive investigation. What is being sought is a model of, or a systematic way of thinking about, groups and group functioning that will lead to the improvement in the efficiency of group operations whatever the nature of those operations. At the same time the model must be such that membership in a group should provide an increasingly satisfying experience for each individual. The earlier organisation and management concepts have been improved by the contributions from the behavioural and the management sciences. However, the research and conceptual endeavours have, at times, led to divergent views. During the last few years there has emerged an approach which can serve as a basis of convergence of these various views—the systems approach, which facilitates unification in many fields of knowledge. It has been used as a broad framework of reference in the physical and social sciences. It can also be used as a framework for the integration of modern organisation and management theory.

1. The Nature of General Systems Theory

A system has been defined as an organised or complex whole; an assemblage or combination of things or parts forming a complex whole. The study of systems whether they are physical, biological or social, all tend to have a common conceptual basis; the principles of wholeness, organisation, and dynamic interaction. Awareness of the parallelism of ideas is now beginning to provide opportunities for the formulation and development of principles which hold for all systems in general. This has led to the development of a general systems theory—an attempt to develop scientific principles to understand and explain dynamic systems with highly interacting parts.

Clearly every system, whether physical, biological or human, is purposeful. The function of the digestive system in all creatures is to transform matter taken in as food into other forms of energy to support growth, movement and so on. Much more is learned about the digestive system, however, by looking beyond its function to its relation to other bodily systems and to the organism as a whole and to its adjustment to change to the external environment.

Systems can be considered in two ways: (1) closed or (2) open and in interaction with the environment. This distinction is important when considering organisations. Closed system concepts stem primarily from the physical sciences and are applicable to mechanistic systems. Many of the earlier concepts in the social sciences were closed system views because they considered the system under study as self-contained. Traditional management theories were primarily closed system views. The organisation was considered as sufficiently independent so that its problems could be analysed in terms of tasks to be performed, internal structure and formal relationships—without reference to the environment.

Organisations can be seen to possess certain characteristics common to all open systems. As social systems however, they differ in important ways from other types of open systems and therefore constitute a special category of open systems.

2. Common Characteristics

The characteristics of organisations common to all open systems are:

a. Open systems import some form of energy from their external environment. Organisations import energy in the form of supplies of material, labour and capital and must continue to do so if they are to survive.

b. Open systems convert energy they take in. Business organisations convert part of their raw materials, labour and capital into products and services.

c. Open systems export some product back to their environment. Business organisations export their finished products or carry out their services.
d. The process of energy in all open systems is cyclical in character. That is, there is a repetitive cyclical pattern of components. However, unlike biological and physical systems where the components are identifiable by their physical dimensions, the regenerative cycle of social systems consists of a chain of events. On the completion of one cycle there is the probability that the cycle will be repeated. However, the fact that social organisations are contrived by human beings, suggest that they can be established for an infinite variety of objectives and do not continue to follow the same pattern. Thus to study a social structure it is necessary to follow the chain of events from input of energy through its conversion to the completion of the cycle.
e. All forms of matter in the universe are subject to entropy—over a period of time they break down from an organised to a disorganised state. For example, in biological systems the organism lives, grows for a period of time but is subject to deterioration and death. A process similar to entropy occurs in open systems. The gradual erosion of rules and the breakdown of standard practices are common to all organisations. The only way in which the organisation can offset entropy is by continually importing material, energy and information in one form or another, transforming them and redistributing resources to the environment.
f. Open systems which survive are characterised by a steady state. A steady state (dynamic equilibrium) occurs while the system can still maintain its functions and perform effectively. Under this concept an organisation is able to adapt to changes in the environment and maintain a continual steady state. Open systems tend to import and store excess energy which serves to maintain the character of the system and to ensure survival. For example, a manufacturing firm may take over its supplier in order to secure its future.
g. The concept of feedback is important in understanding how a system maintains dynamic equilibrium. The process of feedback enables the system to receive information from its environment which helps it to adjust. Feedback is of vital importance in the complex organisation which must continually receive informational inputs from its environment.

It is important to note that all systems cannot absorb all forms of energy and some form of selectivity is employed. The mechanism for selecting inputs into the system is dictated by the nature of the functions performed. This selectivity is clearly illustrated in the digestive systems of living organisms. In the case of organisations it is the function of management to select, interpret and correct on the basis of feedback.

Open systems are characterised by the tendency to move in the direction of greater differentiation and a higher level of organisation. The growth and expansion needed to preserve an organisation in a steady state leads to the multiplication and elaboration of rôles as more specialised functions emerge. A business organisation may take over one of its major suppliers in order to ensure a continual source of raw materials. The trend towards mergers and conglomerates is an indication of this process. As organisations grow in size, there develops a move towards greater regulatory mechanisms such as rules, quotas, budgets and so on, and away from the simple personal interactions. The increased number of specialised departments and activities in complex business organisations is another example of differentiation and elaboration.

Open systems are also characterised by the principle of equi-finality. The concept of equi-finality says that the final results may be achieved with different initial conditions and in different ways. In other words, there is no direct cause and effect relationships between initial conditions and the final state as found in physical systems.

The principle of equi-finality can be observed in any business enterprise. There is usually no one best way of carrying out an operation or objective. However, the principle of equi-finality can be inhibited by regulatory mechanisms such as budgets, standards and other devices. This principle suggests that the management function is not necessarily one of seeking a rigid optimal solution but rather one of having available a variety of satisfactory solutions to decision problems.

3. Special Characteristics of Organisations

In addition to the general characteristics of open systems, Katz and Kahn (1966)* cite a number of properties distinctive to social organisations summarised as follows:

a. Organisations lack the fixed physical structure found in biological and other physical systems. As stated earlier the structure of a social system is the structure of relationships traced in events and therefore intimately tied to the functioning of the system. This distinctive aspect of organisations requires considerable energy if the system is to be maintained. Maintenance sub-systems are therefore created to perform this task. Examples of maintenance sub-systems in business organisations include departments such as Personnel and Public Relations.

b. Rôles performed by individuals are major elements of social systems and provide a basis for social structure. Within each social system there are a number of positions which individuals occupy. Associated with these positions there is a pattern of behaviour which is expected of the incumbent and this accepted pattern of behaviour constitutes the rôle which goes with that position. The rôle therefore establishes the structuring of activities and relationships and is necessary in all organisations if co-operative effort essential to organisational goals is to be achieved.

c. It was stated earlier that organisations are contrived systems. They are created to achieve a variety of purposes. Their structure is a structure of events and their effective co-ordination and control requires elaborate control mechanisms. The managerial function which provides these mechanisms (quotas, standards, budgets, etc.) is therefore of critical importance to organisations.

G. THE ORGANISATION VIEWED AS A TOTAL SYSTEM

Viewed as a total system, then, the organisation can be seen to comprise three major sub-systems, viz., the social sub-system with its emphasis on inter-personal relationships; the technical sub-system with its emphasis on machines, tasks and operating techniques and the managerial sub-system with its emphasis on the integration and co-ordination of the social and technical sub-systems and the relationship of the organisation to its environment.

1. The Inter-relationship of the Social and Technical Sub-Systems

Technology refers to knowledge about the performance of certain tasks and activities and includes machines, facilities and operating techniques. Technological constraints set the limit to work performed by organisations. Technology therefore determines the type of input and output from the system. The social system, however, determines the effectiveness and the efficiency of the utilisation of the technology. Both the technological and social sub-systems are inter-related. The technological system of an organisation is determined by the tasks to be performed, the specialised knowledge and skill required, the type of machinery and equipment involved, and the facilities available.

On the other hand, technology often determines the type of human input required and the relationships between jobs. Participants, however, possess aspirations,

* D. Katz and R. L. Kahn *The Social Psychology of Organizations*, John Wiley, New York, 1966.

expectations and other psychological values. Any change in the technical sub-system will have repercussions on the social system and conversely.

Task requirements and technology also have a bearing upon the organisational structure. Structure is concerned with the way in which tasks in the organisation are divided up and co-ordinated. The structure also concerns the pattern of authority, communications and work flow.

In this respect the organisation structure defines relationships between the technical and social sub-systems. Recognition of the inter-relationships existing between these primary sub-systems is an important step leading to the development of modern organisation theory.

2. The Integrating Rôle of the Management Sub-System

The function of the managerial sub-system is to direct the technology, to co-ordinate and integrate people and other resources, and to relate the organisation to its environment. In this respect management may be viewed as a system consisting of three levels—the technical or production level; the organisational (managerial) level and the institutional or community level (Parsons, 1960).*

The technical level: the technical system of an organisation involves the actual tasks performed. These tasks include not only physical work but also technical activities utilising knowledge. These technical activities include the production and distribution of products or services; specialist activities such as research and development, accounting, market research and operations research.

The second level, the organisational, is concerned with the co-ordination and integration of the task performance of the technical system.

The third level, the institutional, is involved in relating the activities of the organisation to its environment.

However, there are basic differences in the managerial approach at the various levels. The tasks at the technical level are of an economic-technical nature concerned with efficiency of output within the given level of technology. As such, managers at this level tend to adopt a closed-systems view. It is essentially a rôle of control. Their primary function is to see that the specific tasks are carried out effectively and efficiently. It is at this level that the contributions of management science are most valuable.

The primary function of the manager at the organisational level is co-ordination and integration of the technical task performance. It is their function to see that resources are obtained and used effectively and efficiently in the accomplishment of the organisation's objectives. Hence they must integrate the technical level with the institutional level. Management's concern at this level is with the psycho-social system of the organisation. The contributions of the behavioural sciences are most appropriate in this area.

The institutional management must have a broad conceptual framework of reference. This is the highest level of the management system and is involved in decisions concerning the objectives of the organisation, or changes in these objectives, on the resources used to attain these objectives and on the policies that are to govern the acquisition, use and disposition of these resources.

Management at this level should have an open-system view. Because of the uncertainties in the environment, decisions relate to unstructured, non-repetitive problems depending on external information, and subjective evaluations.

This three-level view of the management structure represents a major departure from the traditional view of a scalar chain of command. (That is, a line of authority extending from the top of the organisational level down to the technical level.)

* T. Parsons, *Structure and Process in Modern Societies*, The Free Press of Glencoe, New York, 1960.

However, this does not mean that the different managerial levels can operate independently, quite the contrary, they are interdependent.

It is the development of these new concepts of organisations and management that suggest a systems approach. The systems view suggests that management faces situations which are dynamic, and frequently uncertain. Management is not in full command of all the factors of production as suggested by the closed-systems approach of traditional theory. On the contrary, managerial activity is restrained by many environmental and internal forces. The technical, psycho-social and environmental systems all constrain the management system.

H. CONCLUSION AND AREAS FOR FURTHER ENQUIRY

The rôle of management has been stressed as the most critical factor in pushing advanced technology programs through to meet their goals. It has been suggested that the manager's primary rôle was one to integrate conflicting interest groups towards specific goal accomplishment. The point has been made that management should be charged with the responsibility for adapting the organisational resources to changing environments in order to maintain scientific and technological progress.

Perhaps a thought should be spared for the development of techniques to enable management to give consideration to the long term effects of "goals", "projects" and "missions" on the community, on the nation and on the general environment. A question might well be asked whether or not sufficient attention is paid to the long term effects of policies and decisions.

The growing complexities and diversities in today's advanced technological programs makes it necessary to conceptualise a system basis in order to integrate complex operations. Primarily a systems concept is a way of thinking about the job of management which provides a framework for visualising the internal and external factors affecting the organisation as an integrated whole. It provides for recognition of the proper rôle of functions in sub-systems. Systems are thought of as an organised or complex whole, assemblage or combination of things or parts forming a complex or unity whole. The systems concept which has become increasingly important in military management is an example of the application of this systems concept. Management is the primary force within the organisation which co-ordinates the activities of the sub-systems and relates them to the environment. Essentially management is the process of relating human and material resources into a total system for objective accomplishment.

It has been pointed out that the primary reasons for the emergence and application of the systems concept has been advancing technologies and increasing industrial complexities within society. These forces will continue, possibly accelerating in the future. The systems concept will allow more effective adaptation to scientific and technological environments. While this approach has been, and is currently being implemented in most of the more advanced technology industries, its use will spread to other industries in the future. One of the major changes within business organisations will be the breakdown of functional specialisations geared to optimising performance of particular departments. There will be growing use of organisational structures designed around projects and information decision systems. The systems concept calls for the integration into a separate organisational system of all those activities which are related to particular projects or programs. Business organisation must be considered as a sub-system of a larger environmental system. Even in industry or inter-industry, systems must be reorganised as sub-elements of an economic system and the economic system should be regarded as part of the society in general. The point that must be emphasised is that the system concept is primarily a way of thinking, a mental frame of reference which can be utilised by management in performing its traditional primary functions of planning, organising, and controlling operations. These activities have been and will continue to be fundamental functions in management process. The systems concept provides a new framework for carrying out the integration of these activities.

From what has already been said in this chapter it is obvious that the last word has not been written on organisations and their management. Current theories must be regarded only as approximations—mere beginnings in the quest for more viable theories of organisations. This chapter will conclude with some comments on the deficiencies in current understanding, the answers to which present a challenge to all who seek more efficient and effective contributions from human systems.

1. Some Areas for Further Research and Enquiry

a. Organisational structure. The traditional theorists emphasised the formal, rational structure determined very largely by the nature of tasks to be performed. The behaviour scientists stressed the importance of informal groupings and sought to embody the concept that it is human activities and human relations that should be structured not tasks alone. The problem is that both schools of thought seek to arrive at one pattern.

Some writers are asking whether it is not possible to develop more than one structure in an organisation, as organisations are often composed of a number of sub-organisations. Successful performance of these sub-organisations requires for their efficient performance a structure (i.e., relationships) tailor-made to fit the tasks confronting the unit. Thus sub-organisations carrying out routine repetitive tasks are best established as a group with one pattern of relationships while tasks of a non-routine or of a problem-solving nature are better tackled in a group with a different set of relationships.

b. Current theories of organisation deal inadequately with communications. Communications provide the means of planning the goals of the organisation, deciding upon courses of action and measuring the performance of action taken. Information handling and processing systems in organisations need to be developed which recognise that different types of information are needed at different points of the structure. In the process of developing such a system, changes will be made in the organisation, in information handling and even in the decision process.

c. The concept of a formal hierarchical authority is common in most current models of organisation. While it is recognised that this concept is basic to organised effort and achieves unity and compliance it has some deficiencies. Compliance to hierarchical authority inhibits innovation and creativity. In the dynamic environment of today it is imperative that these two forces should be encouraged.

d. Now, more than ever, there is a need for more flexible organisations. But from the point of view of the participants, flexibility means change and change affects stability of status and rôle, therefore it is resisted. A great deal of thought will have to be given to the dynamics of stability and change in organisations. In addition, the techniques for introducing change in order to achieve greater acceptance by members will need to be improved.

e. Perhaps most of all there is an educational challenge. Those who engage in higher educational pursuits should be entitled to a training which should be sufficiently basic, broad and flexible to enable them to change skills successfully several times in the course of their lives.

The need to produce broadly trained, adaptable people will become increasingly important at all levels, but special attention should be given to the education of future leaders in government, industry and commerce. What is needed is a new breed of administrators. A breed who will have an interest in the humanities—in philosophy, economics, history and the behavioural sciences and also an understanding of the physical sciences, their history and philosophy and the directions in which they are moving.

I. REVIEW QUESTIONS

1. Define organisation—do you agree that future scientific achievements depend on man's ability to develop more effective organisations? Why or why not?
2. What are the major factors contributing to the complexity of organisations? Discuss the impact of these factors on management practice.
3. What are the major distinctions between the contributions to organisational theory of the behavioural sciences and the management sciences? Illustrate your answer with references from current scientific, sociological and management periodicals.
4. Define systems. Why do modern views of organisation adopt a systems approach?
5. What is the difference between a closed and an open system? Which model is more appropriate for a business organisation?

J. EXERCISES

1. Select three organisations and evaluate the degree to which they adhere to traditional, behavioural or scientific management concepts.
2. Select a large-scale organisation and evaluate the degree to which it adheres to a systems concept of organisation and management.
3. Do you agree with the statement made by Sir Leon Bagrit quoted at the beginning of this chapter? Why or why not?

SUGGESTED READING MATERIAL

ARGYRIS, C. *Integrating the Individual and the Organization,* John Wiley and Sons, New York, 1964.

BAGRIT, L. *The Age of Automation,* Pelican Books, London, 1966.

BENNIS, W. G. *Changing Organizations*, McGraw-Hill, New York, 1966.

VON BERTALANFFY, L. *Problems of Life*, John Wiley and Sons, New York, 1962.

BOULDING, K. E. *The Impact of the Social Sciences,* Rutgers University Press, New Jersey, 1966.

BOULDING, K. E. "General Systems Theory: The Skeleton of Science", *Management Science,* April, 1956.

CHURCHMAN, G. W. *Challenge to Reason*, McGraw-Hill, New York, 1968.

CYERT, R. M. and MARCH, J. G. *A Behavioural Theory of the Firm*, Prentice-Hall, New Jersey, 1963.

GALBRAITH, J. K. *The New Industrial State*, Houghton Mifflin Company, Boston, 1967.

GROSS, B. M. *The Managing of Organizations*. The Free Press of Glencoe, New York, 1964, Vol. I.

KAST, F. E. and ROSENWEIG, J. E. *Organization and Management—A Systems Approach*, McGraw-Hill, New York, 1970.

KATZ, D. and KAHN, R. L. *The Social Psychology of Organizations*, John Wiley and Sons, New York, 1966.

KOONTZ, H. and O'DONNELL, C. *Principles of Management*, 4th ed., McGraw-Hill, New York, 1968.

MARSH, J. G. and SIMON, H. A. *Organizations*, John Wiley and Sons, New York, 1958.

MCGREGOR, D. *The Human Side of Enterprise,* McGraw-Hill, New York, 1960.

PARSONS, T. *Structure and Process in Modern Societies,* The Free Press of Glencoe, New York, 1960.

STORER, N. W. *The Social System of Science*, Holt Rinehard & Winston, New York, 1966.

CHAPTER 10

Learning, Adaptive and Goal-Seeking Systems

A. E. Karbowiak and
R. M. Huey

"This is an approximate idea only. Our maps always maintain an element of mystery. ."

From second stanza "Christopher Colombus"
W. Hart-Smith.

A. INTRODUCTION

Examples of systems of moderate complexity have been given in Chapter 1. In Part II of this book we consider in greater detail some key problems surrounding the design of sophisticated systems. But, be it a machine capable of playing chess or a radar defence system (this example is discussed in more detail in Chapter 23), the principal problems stem from our (or the system's) inability to predict with precision the future. By this we mean that if a radar defence system is to function properly it must be able to prepare its actions *ahead* of events to come. More particularly, to hunt a target successfully the system must be able to *predict* with precision the future trajectory of the target. Nevertheless, an interesting possibility exists that given sufficient storage facilities in the computer associated with the system we might be able to scan the data received and discover a definite pattern of behaviour of the target. This pattern would be associated with a definite path of the projectile and our strategy of hunting the enemy aircraft would be thereafter greatly simplified. In fact from there onwards a 100 per cent success could be assured. We note that to ensure a success the system must be programmed so that it is capable of "learning". This example serves to illustrate a system which, though variable, operates according to a definite pattern.

Frequently the target will not follow a definite pattern at all, and under such conditions the data representing the position of the target will not follow any definite pattern (such as a circular path) but is more likely to represent a stochastic process.* The problem is then more difficult: we have to decide, in view of the delay in transmission and the nature of the data received, on the best strategy for anticipating the future position of the target. The system must, therefore, be equipped with facilities suitable for processing statistical data and for extracting information from this kind of data. Hence follows the relevance of the statistical theory of communication to this kind of problem.

Characteristic of systems of this type is that, during the period of activity there is a flow of data (information) between the components, parts or functional units of a system. Information theory is concerned with this flow of information, its relation to the overall performance of the system and the goals envisaged. Even with systems of moderate size there is an immense variety of ways in which various components and functional units of the system can be inter-connected. The problems associated

* By a stochastic process we mean one which is characterised by a sequence of random events.

with optimum design of systems are, therefore, of extreme complexity, the ultimate aim being to design a system which can reach the goals in some optimum manner.

In view of the implied complexity of the problems associated with design of systems, a systems engineer needs models to help him to comprehend the problems involved. Models can be mathematical or analogue * but to have a useful model the problem needs to be formulated. However, a complex system cannot be adequately formulated until it is well understood, presumably by means of suitable models and at the same time it cannot be adequately understood, until most of the problems associated with the system have been solved. This, therefore, is the dilemma of the systems engineer.

Systems theory can be extended, through interaction with other disciplines, to include human operators as parts of a complete system. From the information theory point of view, however, such extension is trivial in so far that the human operator is usually replaced by an input/output device having a mathematical expression which adequately describes his behaviour.

This chapter is devoted to the examination of some of the problems with systems of this nature.

B. ADAPTIVE CONTROL SYSTEMS

Essential to the functioning of any automatic system is some form of a sensor, which senses the output of the system and feeds the information to some other parts of the system to make the system behave in accordance with a built-in criterion.

A feedback system as used in automatic control is a particularly simple illustration of such an arrangement (for details refer to Chapter 12). The aim here is to automatically sense and correct deviation from a chosen state. Figure 10.1 represents a block diagram of such a system.

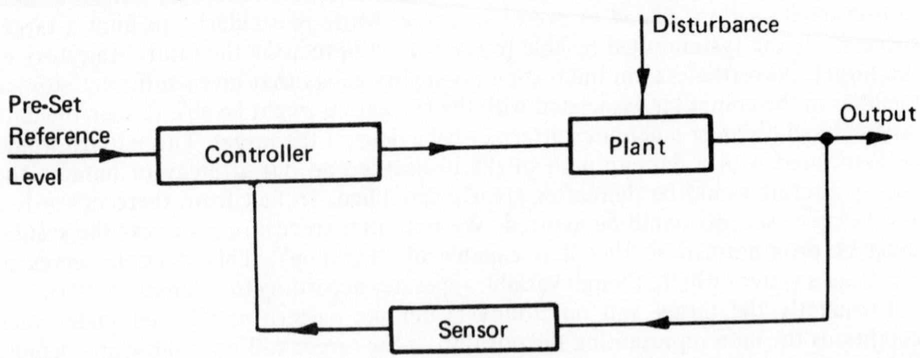

FIG. 10.1 Feedback Control System: the Feedback Attempts to Regulate the Output to conform to Preset Reference Level (A Fixed Criterion for Best Performance).

Such systems can be made very accurate in their steady state behaviour. However, to ensure that their transient response is well-behaved can be a major problem. The transient behaviour (i.e., the response to a sudden change in circumstances followed by the efforts of the control system to cope with this) will depend on factors such as:

(1) Properties of the Plant.
(2) Previous state of the Output.
(3) Nature of the Disturbance.
(4) Properties (i.e., parameters) of the Controller.

* By analogue we mean, for example, that a hydraulic machine or mechanical device can have an almost perfect model in a suitable electrical circuit. Such an electrical circuit would be called the analogue of the corresponding hydraulic or mechanical system. See also Chapters 7 and 26.

In view of the multiplicity of the factors involved, it is unlikely that the transient behaviour will always be optimum.

At the same time what is meant by "optimum" could be judged in a variety of ways, e.g.

(1) Shortest time (to settle down to the new steady state).
(2) Least expenditure of energy (to achieve the new steady state).
(3) Avoidance of undesirable stresses in a machine.
(4) Avoidance of overheating a component or device.

It is possible for a human operator to make judgments of the above sort and to decide to alter the characteristics (parameters) of the controller in order to do better in some new situation (i.e., to achieve something nearer to "optimal" control).

It would also be possible if we knew just how the human operator goes about making such judgments, to program this into a computer (either analogue or digital, or a hybrid combination of both) and to arrange for the computer to adjust the parameters of the controller to achieve "better" control of the plant.

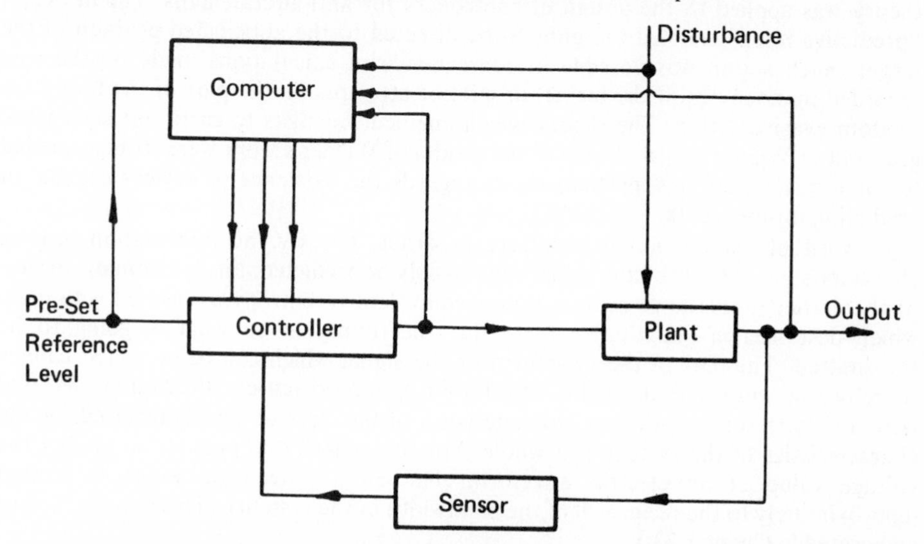

FIG 10.2 Adaptive Control System: Includes Feedback to Regulate Output to Pre-set Reference Level (Steady State) plus Computer Adjustment of Controller to Best Cope with the Kind of Disturbance being Experienced (Steady State and Transient; an Adaptable Criterion for Best Performance).

Figure 10.2 illustrates a more ambitious attempt at solving the control problem. When this system is compared with that shown in Fig. 10.1, we observe that one block, labelled computer, has been added. The connections have been arranged so that the computer inputs inform it of all the significant variables in the rest of the system. In the example shown these are:

(1) Pre-set Reference Level.
(2) Controller Output.
(3) State of the Plant Output.
(4) Disturbances to which the Plant is being Subjected.

The output of the computer is shown by several (three, in Fig. 10.2) connections to the controller. Although not shown on the diagram it is arranged that these inputs from the computer, will vary the properties (i.e., the parameters) of the controller.

Such a system would be called an adaptive control system because it adapts its parameters and thus attempts to improve its performance. However, it is necessary for the designer to anticipate the nature of the possible deficiencies (i.e., the way in which control could fall below "optimum") and to ensure that these can be reduced by the "program" he writes for the computer, which thus decides the manner in which the computer's answers will be used to vary the controller characteristics.

In some cases the computer might only collect the data from as few as two points. In other cases the computer might be fed with data collected from several dozen points.

Programmed computer control of complex systems such as the one described with reference to a radar defence system would appear to be the only practical solution to the many problems facing the designer.

C. PREDICTORS AND LEARNING MACHINES

A system of the type described in the previous section needs, to function properly, to be able to anticipate the immediate future of its inputs. The devices which enable the system to make anticipatory actions came to be known under the name of "predictors". The term came into use during the Second World War when system theory was applied to the design of controllers for anti-aircraft guns. The device, a "predictive filter" enabled the guns to be directed to the anticipated position of the target. Such action was possible in consequence of calculations made on the past recorded manoeuvres of the target, in spite of attempts on the part of the target for random evasive action. The theory which enabled scientists to carry out such work grew out of Wiener's work. Some of the results of Wiener's work were strange and led to some misconceptions, particularly as regards the existence of devices capable of predicting future events.

A word of explanation is therefore in order. Clearly, no information bearing characteristics or traits of any signal can possibly be predicted; this is axiomatic in view of the earlier discussions; information is equivalent to entropy associated with data, which describes in quantitative form the uncertainty as to what is going to be transmitted. The part of the waveform or the signal which can be predicted cannot therefore be information, in the information theoretic sense. "Prediction" as used here, refers to some stationary characteristics of the class of signals received, or the characteristics of the system as a whole. For this reason it is possible to predict the voltage value of an electric waveform, but on a time scale which is limited (approximately to the reciprocal of the bandwidth of the system). (This point is further elaborated in Chapter 23.)

In the same way, it is possible to predict the position of the moving target such as an enemy aircraft, but only on a time scale which is compatible with the manoeuvrability characteristics of the target. It is not possible to predict the noise voltage of a waveform at the next sampling point, nor the acceleration of the military target at the next instant of time. The rate at which a sudden acceleration can be translated into non-predictable changes in position is limited by the inertia and moments of inertia of the aircraft. While the acceleration cannot be predicted, the velocity can be for a short time, and the position for a longer period in advance.

Whereas information cannot be predicted, it is possible to say something about the future behaviour of various integrals which are related to the information bearing variables (see experiment at the end of Chapter 23): this is a different proposition to predicting the unknown, the information.

Thus, it becomes apparent that there exists a very intimate relation between system theory and the theory of random or stochastic processes. Normally, theindependent variable in time and the random variable can be quantities such as voltage, or position of the target, or the economic data pertaining to the functioning of a commercial enterprise, etc. In all such cases, one of the purposes of the computer is to record the past events and to perform various calculations on the recorded data so as to deduce various statistical characteristics of the input signals. From the data so obtained, the computer subsequently evaluates the various statistical averages such as the mean position of the target, the expected direction of the movement of the target, etc. Thus, we see, it is these various statistical averages which can be predicted and the control system is acting in accordance with such statistical data so as to optimise the performance of the system as a whole. The statistical data so obtained virtually characterises the environment in which the system is operating. It is this statistical data, in variously processed forms, which is used subsequently for decision making purposes.

Systems which can store past events and process the data so obtained as described above are also known under the name of learning systems. In effect, the machine learns certain statistical characteristics of its environment.

In addition to being able to learn, the machine is also equipped with decision-making criteria (or policies). This enables the machine to act in accordance with the built-in criteria in a new environment in spite of changes in the statistical properties of the environment.

At least four separate processes are (or may be) involved in learning machines.

(1) Recognition of situations or patterns of information.
(2) Memory or storage of previous recognitions and of previous actions directed by the learning machine in these situations.
(3) Association of present recognition with previous situations, followed by reinforcement which increases with each such successful experience. (Leads either to selection of a strategy or to production of value judgments.)
(4) Inference by which completely new associations or new strategies may be created.

In addition we will need some arrangements to allow communication between machine and man and between machines.*

D. VARIABLE SYSTEMS AND LOSS OF INFORMATION

There is an aspect of machine-environment interaction that needs a little elaboration concerning the so-called variable or time-varying systems (c.f., the example of a radar tracking system). These systems can be modelled in several different ways, but the capacity of a variable system can be expressed quite simply using a slightly extended concept of entropy (see Chapter 11). In particular, if the variation in the properties of the system can be described by a definite mathematical relation, however complicated, then in absence of noise the capacity of the system will not be affected at all, unless the time variable operator representing the system is a many-to-one operator (that is, when one symbol eventually stands for several different things). This is because under such conditions the entropy is unaffected. In engineering terms, the particular mathematical relation could have been stored in the machine and used in the encoding process without affecting the information carrying capacity of the system.† Yet, the form of the presentation of the signal would be altered and could, through changes in the semantic content, confuse the observer.

* In the present-day jargon of computer engineers, these arrangements are sometimes known as peripheral equipment or interfaces. These terms came into usage in that order and indicate a change in emphasis. The use of the word "peripheral" indicates a pre-occupation with the computer as the central feature of a system. The use of the word "interface" connotes more emphasis on the system as a whole.

† But, if such mathematical relation is not known or its implementation would exceed the storage capacity of the system then this route is not open to exploitation. Such mathematical "information" is not an asset. This is in fact the problem when trying to computerise chess playing.

Again, in the last example, distinction must be drawn between the different aspects of a signal: (i) its entropy, and (ii) its form or semantic content. Whereas the second is of no concern in communication theory, the first one is.

If the system were variable in a stochastic sense then, clearly, there would be a corresponding entropy associated with it, H_s. Under such conditions, the capacity of the system would be reduced to

$$H = H_o - H_s \tag{10.1}$$

where H_o denotes the capacity of the system in the absence of the stochastic variations within it.

A further point worth observing as far as variable systems are concerned, is that such a system can be regarded as a constant system provided with an additional input port. Clearly, in such equivalent presentation, if the additional input ports contain signals which can be described by definite mathematical relations, then the system capacity will be unaffected; but, if the inputs to the above mentioned ports are noise waveforms then the capacity will be reduced in accordance with Equation 10.1.

At this juncture it is helpful to note that, from the point of view of communication theory, a variable system, in the presence of a statistically constant source (stationary environment), is equivalent to a constant system with statistically variable input (non-stationary), provided that the balance of entropies in the two cases is the same. Thus, a change in the form of the input signal is unimportant, but a change in the *a priori* and *a posteriori* probabilities is.

A variable system, or a constant system operating in a variable environment, needs for its proper functioning an adaptive mechanism: a "predictor" is its essential constituent, an example being a radar tracking system. The system as mentioned previously cannot predict the information, it can only collect data and adapt itself to conform to an optimum criterion, based on statistical averages and the information for this must come from the entropy of the received signals. The capacity of the system is therefore reduced and in a sufficiently rapidly varying environment the whole of the system's capacity might be occupied by the entropy needed to up-date the statistical averages. This is the breaking point of the system; for example, with the radar tracking system, the breaking point is reached when the standard deviation* of the target exceeds the radius of influence of the projectile. This could be on account of delays in transmission (capacity too small) or due to increased manoeuvrability of the target (rate of generation of independent data too high).

Frequently, it happens with adaptive systems that the reason for reaching a breaking point is not the lack of information capacity *per se*, but rather that the form (or semantics) of the signal is a function of time and it is the adaptive system (dealing with data) that interprets the signal as highly time dependent. In such cases a self-organising system—which can change its characteristics as the need arises—has a clear advantage.

An effective implementation of such devices (excepting some very simple cases) hinges on the discovery of efficient means of dealing with semantic characteristics.

E. PATTERNS AND SEMANTICS

All learning and adaptive machines have facilities which enable the systems to perform pattern recognition; here there are a number of facets of varying degrees of complexity.

* The standard deviation or variance of a set of numbers $x_1, x_2 \ldots x_n$ from their mean value $\bar{x}$ is defined as the average of all the terms $(x_i - \bar{x})^2$ where i takes in all the values from 1 to n successively. Notice that the positive and negative errors do not cancel out because the square of the error is taken each time. A set of widely scattered readings will have a larger variance than a closely clustered set.

In the first place, if we know the constraints then we can calculate the total number of distinct patterns. However, we might encounter several difficulties: the constraints may not be completely known or be too complex for adequate description, or the total number of possible distinct patterns may be prohibitively large.

For example, the class of patterns may be specified as that of sinusoidal waveforms, that is:

$$\psi_n = A_n \sin(\omega_n t + \phi_n) \tag{10.2}$$

Where ψ_n is a particular pattern which can be completely described by three numbers A_n, ω_n and ϕ_n. Thus:

$$\psi_n \triangleq (A_n, \omega_n, \phi_n) \tag{10.3}*$$

The problem of pattern recognition reduces to the determination of three numbers, in *this* case.

One of the functions performed by a machine may be pattern identification, where we ascribe to every pattern of the set a symbol (the name of the pattern) on a one-to-one correspondence. This is a problem of encoding and clearly possesses a solution, but may be complicated in that it might be difficult to implement in practice, unless a systematic method is conceived.

To illustrate the point, in the example given the set of three numbers (A, ω, ϕ) identifies a pattern out of a class of sinusoidal functions, on a one-to-one basis.

Pattern classification is another function which machines are frequently called upon to perform. It is also another example of encoding, whereby we ascribe one symbol to a group of patterns. As such it is an example of a many-to-one translation. The information necessary to accomplish the correct pattern classification into the various groups, is either built into the machine (for operation in a fixed environment) or is "learned" by the machine with the help of a "teacher". In this way the machine can be used or can be made to adapt to different environments.

Returning to the example given above, the machine might be asked to classify all sinusoids of frequency greater than ω_o as X and send them to port A and all other waveforms as Y and send them to port **B**.

One could regard such a machine as exhibiting an "intelligent behaviour", particularly so if by manipulating suitable knobs a "teacher" could "teach" the machine to adjust the selection point, ω_0, to a desired value. On reflection however, one sees that such a machine is just an embodiment of a simple two output port (X and Y) filter with none of the fascinations of a learning machine, and man (teacher)-machine interaction. One needs, therefore, to be careful when using terms with emotional connotations, such as intelligent machines or thinking machines. For, what does it mean to be intelligent or what is thinking? Many of the functions which we regarded many years ago as peculiar to people are now performed with greater efficiency by a machine (for example, most mathematical calculations). But if we admit, as we should, that thinking and intelligent behaviour imply emotional involvement then, as of today, the machine cannot even crudely imitate a human being.

Signal detection (with communication systems) is another example of a function following in the class of pattern classification schemes in that the process concerns separation of mutilated patterns into different classes according to a pre-arranged plan.

The purpose of a machine performing pattern recognition or classification functions is to reach a goal according to a prearranged plan. This plan is essential for the successful operation of the system. As such, it is an attempt at translating patterns present in the minds of planners (e.g., engineers) into a machine language, and the process involves measurement as an essential step in the execution of the various

* The symbol $\triangleq$ indicates that the left hand side of the equation is *defined by the right hand side.*

functions by the machine. It follows, therefore, that the machine can classify only such items as are amenable to quantitative description. Frequently it is this step—i.e., conversion into a quantitative or digital form —that is a source of difficulty. Thus, emotions and feelings which are the essence of human communication cannot be communicated effectively by a machine directly, simply because we know of no suitable scale of measures.

Much of human communication is semantic in content, even though we think and act according to patterns. But whereas communication theory deals with entropy of patterns, humans are frequently more concerned with terms like simple, complex, beautiful, ugly, etc. What is simple for a human, is frequently complex or impossible to implement by a machine, and vice versa. On the evidence to date it appears that whereas humans are particularly adept at classification according to form (or semantics), machines are more suitable for dealing with quantitative data. In this sense machines complement humans, but a great deal needs yet to be done to solve the interface problem; that is, effective methods of translating semantics into a quantitative language.

Frequently, pattern classification can be accomplished by machines as well as by humans but the strategy employed by humans appears to differ fundamentally from that used by machines. Humans, in particular, have a remarkable ability for pattern recognition in that they can select at a glance out of a maze of billions of possible alternatives, those which are "worthwhile", thereby reducing the search procedure considerably. (But note that "worthwhile" implies emotional involvement; none of the machines invented to date have such capabilities.)

Pattern recognition is a province of information theory, irrespective of the effectiveness of the identification procedure, or strategy. This is the same class of problems as detection of signal (given class) in noise. However, the problem of establishing whether there is a signal (class not given) in a noisy background, or whether there is a pattern (class not specified) on a radar screen, is a much more difficult problem and often outside the province of communication theory.

Thus it appears, at least in principle, that given a sufficiently large and complex system we can classify patterns, *and* detect them in a noisy background, and we can design a system to adapt to a changing environment even though the machine might do it in a rather clumsy way. But the problem of devising a system which would search for a pattern, as yet unspecified and, having discovered it, would take appropriate action would appear to be beyond our present capabilities. The fundamental question to which, in the course of discovery, a scientist frequently seeks an answer is "does this series of measurements mean something?" or alternatively "is there a pattern or a law in this otherwise random set of data?" While history clearly shows that the bulk of great discoveries in science came in response to such questions, yet no machine could answer such questions (being semantic they are not amenable to quantitative scrutiny). The questions are not "scientific", they imply strong emotional involvement.

It is this constant search for elusive patterns which makes science and engineering so fascinating.

F. FUTURE USES OF LEARNING SYSTEMS

We are only just beginning to learn how to set about designing machines that can learn. This ability when it is achieved will create a new technological revolution.

Over the last couple of centuries we have seen the design, widespread construction and use of machines which have, in a sense, extended the capability of human and animal muscle power.

In the present century we will undoubtedly see the design and widespread use of machines which will extend the power of the human intellect.

It will be a very grave political problem as to whether the ownership and control of such machines should be in government or private hands, or in both.

It will be a very grave sociological problem to decide under what terms an individual or a corporate group (such as a large firm) should have access to or should monopolise such learning machines.

In order to analyse and understand such machines it will be necessary to be acquainted with a much wider range of mathematical processes than engineers have been accustomed to in the past. Some of these are included in the following list and are in addition to all those techniques which we have previously mentioned, in connection with other sorts of systems.

1. Group Theory.
2. Variational Calculus.
3. Symbolic Logic (including the use of Boolean algebra for processes other than simple binary logic).
4. Functions of Discrete Variables (i.e. compared to continuous variables).
5. Field theory of Discrete Sets (i.e. compared to field theory of a continuum).

These and other areas in mathematics will be needed in order that we can deal with both digital and analogue processing of data in machines which provide recognition, memory, association, re-inforcement and inference.

Systems theory in its broadest sense is the tool by which engineers, social and political scientists, military strategists and tacticians, commercial, industrial and agricultural managers, local, national and supra-national planners and executives may all play their rôle in the years and societies to come. It may be a tool which will influence humanity for good or ill as drastically as the industrial revolution or the discovery of atomic energy.

Looking back on the industrial revolution which provided a great increase of energy *per capita,* one may see that the actual utilisation of this energy remained however, by and large, under individual human control. One may ask, was it coincidental or was it necessary that mass production arose and prospered at the same time by subdivision of manual tasks into almost trivial repetitive processes.

The question must arise, will the present-day and future applications of systems engineering induce a similar subdivision of human perceptual and mental tasks into sets of strictly-controlled trivia? One hopes and believes that this will not be so. Indeed, the problem of ensuring that the new engineering, which is ushering in the twenty-first century, should truly enrich rather than impoverish human experience is indeed itself a problem in systems engineering of the greatest magnitude.

G. EPILOGUE

"The fault dear Brutus, is not in our stars, but in ourselves that we are underlings"
(Cassius in Shakespeare's *Julius Caesar*)

And what of the future? Is there no limit to the complexity and the sophistication of systems? The industrial revolution has made impacts in its time, both desirable and undesirable. Systems engineering is beginning to make its impact on the present century. Like the industrial revolution, systems engineering, while benefiting mankind, could have far-reaching and highly undesirable effects: we can already foresee some of the problems ahead, and should be concerned with them now. Let us examine a few.

From consideration of such factors many—even leaders of societies—have blamed disasters and misfortunes on developments in technology. Even such fantastic association as blaming bad weather or poor crops on the "wild developments" in radio links, or electric power systems, or the atomic bomb . . . It seems that this is to replace the scapegoats of the middle ages: the witches who were blamed for all sorts of evil. And, there is nothing new about it, Romans were afflicted by the same disease. Thus Tertullian (ca. 190AD) writes "They take the Christians to be the cause of every disaster to the State, of every misfortune to the people. If the Tiber reaches the wall, if the Nile does not reach the fields, if the sky does not move, or if the earth does, if there is a famine, if there is a plague, the cry is at once: 'The Christians to the Lions!' " It is prejudices of this kind that bring to light the educational level of our society.

Large systems are expensive. Any particular system may represent to a nation a substantial investment, and the whole economy of the country could be very profoundly affected by the malfunctioning or the failure of the system. Moreover, a significant downward trend in the economic situation of one country might have profound effects on others.

The environment, the world at large, is not an inertial frame of reference in which different systems could be said to operate substantially independently. But rather, like the building blocks of a system which form the whole (see later chapters), so the various systems are *not independent* but are *coupled* to each other through the medium of the environment. With large systems this coupling can profoundly modify the functioning of the system in a manner which may not have been foreseen by its designers. It would even be possible for the system to become unstable.

Furthermore, the coupling forces could be non-physical and be related to human aspects. As such they may not be measured and thus not taken into account in an unambiguous and calculable manner.

To make the point clearer, it needs to be observed that systems as we know them today are sophisticated goal-seeking machines which tend to optimise their performance in accordance with a criterion relating to a self-centred policy in a hostile environment. It is this kind of policy which must be criticised if a number of systems are compelled to work in a common environment.

It is not difficult to show that several systems of the type described and operating in a common resilient environment,* lead to an operation which is inherently unstable and therefore none of the component systems can optimise their performance. Moreover, the conclusion still stands, even if the coupling forces are small.

The corollary is inescapable: the *self-centred policy* leads to operation which is bad as far as all component systems are concerned, and *no one profits* by it. Perhaps a profound lesson can be learned from this. Are we thus doomed to disaster? The answer, fortunately is "No", provided that the self-centred folly is abandoned and that global optimisation policy is substituted. This works as follows: the various systems; together with the interaction forces, are considered on a global basis as a global system; so as to optimise the performance of the global system without it being unduly concerned about the local value of the pay-off of a particular sub-system: the benefits accruing from the global system would be re-distributed in an agreed, fair way: there would be no hostile environment, and the system would be inherently stable.

Undoubtedly global systems, whose operational criterion is that of genuine concern for all, would benefit mankind enormously: the alternative is global disaster. Which is it going to be?

The latter, the blunders of ignorance being the reward for blind self-centred policy, preoccupied with short-term gains, or the former, a global system based on a policy of genuine concern for each and every one of us, a triumph and a living monument to education?

"The earth is but one country and mankind its citizens."

(*Baha' Ullah*)

* By a resilient environment we mean one which substantially conserves (rather than dissipates) the energy associated with coupling forces. (c.f. the terms reactance and resistance used in electric circuit theory).

Part II

Elements of Systems

Foreword

You will have learned, in Part I, something of the ways in which systems engineers think about their occupation, and you have been given some indication of the very wide scope of systems theory and systems engineering.

Those who are prepared to delve somewhat deeper, will find a great deal of information in Part II. Not all of it will be easy to understand, as it is aimed at a level somewhat higher than Part I. The ten chapters in Part II deal with the technical foundations of information theory, operational methods for the analysis of dynamic systems, the meaning of systems statics, the physics and practice of semi-conductor devices and integrated circuits, the use of digital computers, the design and utilisation of logical information processing circuits and human management.

In brief, it is intended that these ten chapters give a sound basic introduction to the mathematical and physical techniques on which the semi-conductor revolution and systems theory were established during the 1950s and 1960s.

The reader who is tackling these concepts for the first time will have to work hard at compiling his own interpretation of the material presented, and at the task of attempting the exercises at the end of each chapter where, in addition, wide ranging suggestions for further reading are given. It is not to be expected that all readers will also study, understand and digest all the material in the many books listed. Some of these will be the subject of further detailed study in later years. However, the authors believe that there is considerable importance attached to even a cursory run over these works, at an early stage in a student's career.

It is most important too that many future practitioners in various professions who may not themselves be working directly in the field of systems engineering, should nevertheless have a sound grasp of the basic ideas and concepts of modern systems theory and technology.

It has been said that mathematics is the queen and servant of the sciences.* In rather the same way, systems engineering might be termed the genie and the workman of modern technologies. Some of the work mentioned in this part may not be treated in detail until later years of engineering or science courses. It is our hope that this brief introduction will make easier the more detailed, and more rigorous treatments presented to students in subsequent years. It is our hope also that for those students who are undertaking courses which do not lead on to more detailed studies of these matters, that they will absorb enough basic ideas from this text book to be able to understand and converse with the engineers and technologists who will have decided to specialise in systems.

If one is playing a game, it is not sufficient to do well by accident—one can hardly be pleased with oneself, unless one's achievements are planned on the basis of clear insight into the rules of the game. Part II gives an introduction to the "Rules of the game".

R.M.H.

* E. T. Bell, *Mathematics: Queen and Servant of the Sciences*. McGraw-Hill, N.Y. 1951.

CHAPTER 11

Signals, Information and Languages

A. E. Karbowiak

"When you have won, you will have everything."
From a speech attributed by the historian Tacitus to Suetonius, the Roman Military Governor of Britain. A.D. 61.

A. INTRODUCTION

In Chapter 3 we introduced background material against which we carried out a rather broad discussion, leading eventually to the definition of the scientific unit of information: one bit. One bit of information can be said to be the amount of information received in response to a binary question such as Yes/No or "Is it black or white?" We then went on to show that every branch of our knowledge, which is sufficiently advanced, can be subjected to the scientific scrutiny of Information Theory. In fact we can measure precisely, how much information is received under given conditions. We can also calculate the rate of flow of information (capacity of the channel) along a given channel of communication (capacity here would be measured in bits/sec.).

This chapter is devoted to the development of various concepts relating to the quantitative measure of information.

B. PATTERNS, CONSTRAINTS AND REDUNDANCY

Consider a binary set of 8×10 sites illustrated in Fig. 11.1(a) with the "grammar" of the language that any one site independently of the others can be either black or white. Clearly, therefore, to determine the complete pattern we need $8 \times 10 = 80$ bits of information (Yes/No decisions). Therefore, using the theory established in Chapter 3, the total number of patterns which the set can generate is

$$N_o = 2^{80} \approx 10^{24} \tag{11.1}$$

a very large number indeed. In general if the binary pattern can be resolved into $n \times k$ sites then the pattern generating capacity is given by

$$N_o = 2^{n \times k} \text{ patterns.} \tag{11.2}$$

The set just described is of particular significance in numerous applications (see examples given later). To this particular set we shall give the name the chequer-board pattern, and for the moment we shall regard it as the "mother set". We can derive from the chequer-board pattern, a large number of sub-sets by the simple expedient of introducing constraints. But before considering constraints in any greater detail, we can remark at this stage that the chequer-board pattern can be viewed as a pattern made up of a number of vertical bars with a given number of sites. Let there be n sites in each vertical bar and let the number of vertical bars be k. In the example just given we had 8 sites in a vertical bar and the complete pattern consists of 10 vertical bars. We can say that the information per vertical bar of the chequer-board pattern in the example given is 8 bits.

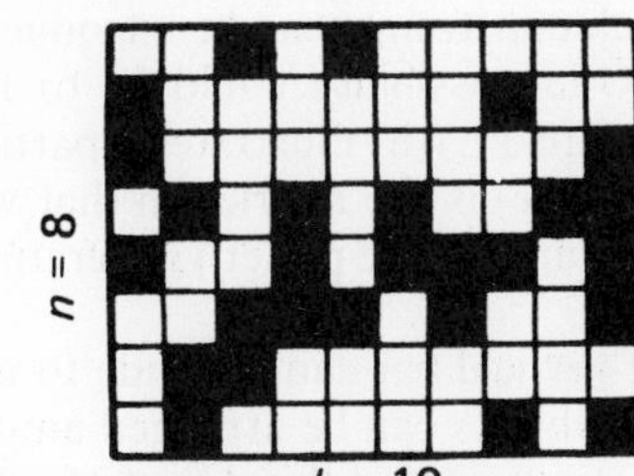

$N_b = 2^8 = 256$

$N = 2^{n.k} = 2^{80} \approx 10^{24}$

$C = n.k = 80$ bits

8 bits/column

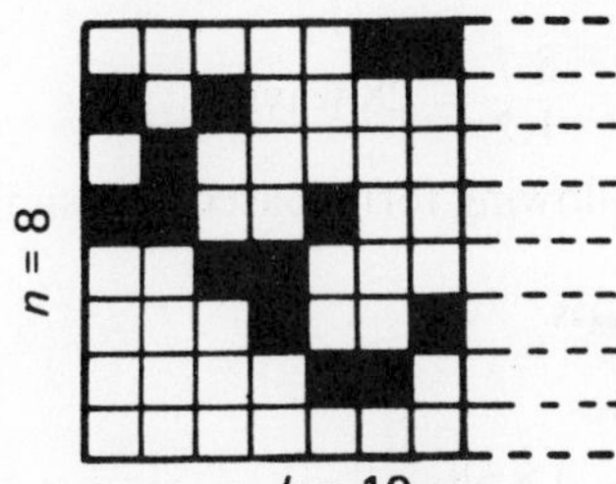

$$N_b = \frac{n\,(n-1)}{1.2} = C_2^n = 28$$

$N = 2^{k.\,\mathrm{lgg}\,28} \approx 2^{48} \approx 10^{15}$

$C = {}^{k.\,\mathrm{lgg}\,28} = 48$ bits

4.8 bits/column

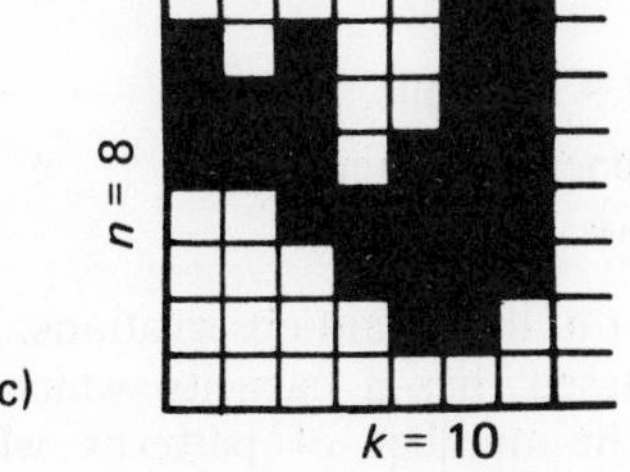

Same

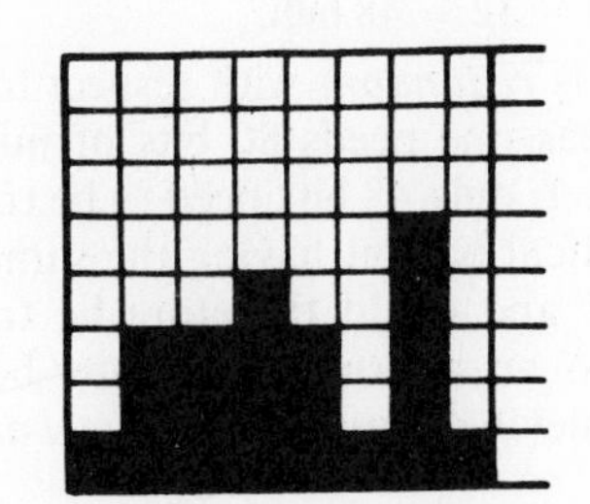

Vertical Bar Pattern

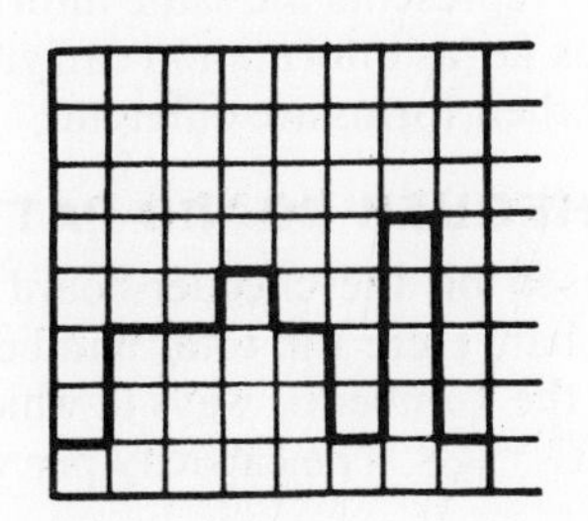

Quantised Wave Form

FIG. 11.1 Patterns, Constraints and Redundancy.

There exists a large variety of ways in which constraints can be introduced into the mother set to generate different sub-sets. One possibility could be to restrict the number of occupied sites per vertical bar. Figure 11.1(b) illustrates a particular case. Here the constraint (the grammar of the language) is the restriction that we agree to have all but two sites of every vertical bar unoccupied. The pattern generating capacity of this sub-set can be calculated as follows.

The number of distinct patterns which one vertical bar can be made to represent is the same as the number of ways in which two objects can be arranged among n sites. This is clearly given by the number of combinations of two objects out of n. Therefore, the general formula is

$$N_b = \frac{n(n-1)}{2} = C^n_2 = \frac{8.7}{1.2} = 28 \text{ ways} \tag{11.3}$$

The quantity N_b can be rewritten in the following form, based on a simple property of logarithms,

$$N_b = 2^{\lg 28}. \tag{11.4}$$

Therefore the capacity per vertical bar becomes

$$C_b = \lg 28 = 4.8 \text{ bits}. \tag{11.5}$$

However, since the capacity of the vertical bar sub-set is proportional to the number of vertical bars in the set, the capacity of the set is

$$C_1 = k.\lg 28 = 10 \times 4.8 = 48 \text{ bits}. \tag{11.6}$$

The total number of patterns which the set can generate is therefore:

$$N_1 = 2^{C_1} = 2^{k.\lg 28} \approx 2^{48} \approx 10^{15}. \tag{11.7}$$

With respect to this set, we can make a number of important observations. The first one is that the set with the constraints discussed, has a capacity which is very significantly smaller than the mother set. The number of patterns which the constrained set can generate is smaller than that of the mother set by a factor of 10^9 or in terms of bits, it amounts to

$$C_1 = C_0 - R_1 = 80 - 32 = 48 \text{ bits}. \tag{11.8}$$

We therefore say that, the constrained set is *redundant* with respect to the mother set by 32 bits. By this we mean, that whereas one needs 80 bits of information to determine a pattern belonging to the mother set, only 48 bits need to be transmitted to describe fully a pattern belonging to the vertical bar set having the same number of sites. The language is said to be redundant and could therefore be translated (or coded) on a one-to-one correspondence into an alternative chequer-board pattern having 48 sites instead of 80 sites, thereby achieving a *more economical description in terms of sites.*

We note, with respect to the pattern illustrated in Fig. 11.1(b), that in any one vertical bar, the squares located between the occupied sites carry no information and therefore the pattern illustrated in Fig. 11.1(c) represents the same information as the pattern shown in Fig. 11.1(b). The two sets, as far as information carrying capacity is concerned, are therefore equivalent, although their forms are different.

C. WAVEFORM: A SUB-SET OF A CHEQUER-BOARD PATTERN

Consider now a different constraint imposed on the chequer-board pattern. Let there be only one site occupied in a vertical bar. Here the total number of patterns which a vertical bar could generate is clearly the number of ways in which one object can be located in any one of the sites, that is $C_1^n = 8$. The capacity per vertical bar is therefore:

$$\lg n = \lg 8 = 3 \text{ bits}. \tag{11.9}$$

The capacity of the whole set is therefore:

$$C_2 = k.3 = 30 \text{ bits} \tag{11.10}$$

and the number of patterns which the set can generate is given by

$$N_2 = 2^{30} \approx 10^9. \tag{11.11}$$

We note that this set has higher redundancy than the set illustrated in Fig. 11.1(c), by a factor of 10^6. Or in other words, the pattern generating capacity of the present set, is smaller than that of the chequer-board pattern with the same number of sites by a factor of 10^{15}. The redundancy of the *bar* pattern with respect to the chequer-board pattern is clearly 50 bits. Therefore, the pattern could be translated without loss of information into a chequer-board pattern having only 30 sites.

Again since the sites below the occupied site carry no information, therefore all the sites below the occupied site can be filled in as shown in Fig. 11.1(d), without loss of information. The form of the pattern is different but the information carrying capacity is identical. The pattern illustrated in Fig. 11.1(d) we shall call the vertical bar pattern. A pattern such as the one illustrated in Fig. 11.1(e) has yet a different form but its capacity is identical with that of the vertical bar pattern.

We can now calculate the redundancy of the vertical bar pattern with respect to the chequer-board pattern by recalling that we can measure redundancy by the ratio R_s of the total number patterns which the two sets can generate, respectively that is the redundancy.

In general, the redundancy of a bar pattern with respect to a chequer-board pattern is given by.

$$R_s = \frac{2^{n.k}}{2^{k.\text{lgg}\,n}} = \left(\frac{2^n}{n}\right)^k > 1. \tag{11.12}$$

The set is redundant because a number of patterns contained in the chequer-board set is not used with the bar pattern.

The redundancy is more conveniently expressed in terms of bits, by taking the logarithm of R_s. Thus:

$$R = \text{lgg}\, R_s = k(n - \text{lgg}\, n). \tag{11.13}$$

This represents the information in bits (Yes/No decisions) which is lacking with the bar pattern as compared with the chequer-board pattern.

It will be apparent from the above discussion, that by imposing suitable constraints or inter-relations between the alternatives of a chequer-board array, one can realise a set with a much larger (or smaller) number of patterns, than that which can be generated by the chequer-board array. The particular constraints are important, because they define the "grammar" of the set in a unique way. On the other hand, unless the geometry of the set is of particular significance, no reference needs to be made to any particular form of an $n.k$ array, because the number of patterns which the set can generate is fully determined once the total number of sites together with the rules for distributing the attributes among the sites are given. What is of importance, is that every set possesses a finite number of distinct patterns which characterises the set. This number is called the capacity of the set, which can be calculated from the knowledge of the number of sites and the grammar (the mathematical constraints) of the set as discussed above.

When dealing with patterns arising in different engineering systems, one needs to study the properties of the physical system from which the class of patterns of relevance can be deduced, e.g., a television picture can be regarded as an array of $n.k$ points. Thus, if we neglect the various shades of grey*, the appropriate class is that

* See Experiment 2 at the end of the Chapter.

corresponding to a chequer-board pattern and the information capacity of the patterns can be calculated from the formula given above.

To take another example, if the $n.k$ array is formed by the signal voltage capable of assuming any one of n different values of voltage one in each of the k intervals of time, then clearly such a waveform is an embodiment of the bar pattern (Fig. 11.1(e)). The capacity of the system is therefore n^k distinct patterns in k intervals of time, or simply (k.lgg n) binary decisions. Any differences which might arise are purely configurational (therefore pertaining to form), and as such, need not be considered when calculating the capacity.

There are some simple physical reasons for thinking that a waveform is an example of a bar pattern (see Fig. 11.1(d) and (e)). This can be seen from the following reasoning. The physical attribute of voltage or current is that it must be a single valued variable, and consequently in any particular instant of time it can only assume one value (1 black point per vertical column). Hence, configurationally, a waveform can be identified with a bar pattern. As such it has a pattern capacity of n^k which, as we have seen previously, needs for its full description (c.f. result in Ch. 3).

$$C(k) = k.\text{lgg}\, n \text{ bits} \tag{11.14}$$

D. PATTERNS AND CONSTRAINTS

1. Coding

With reference to Figs 11.1(d) and 11.1(e), we noted that the bar pattern has a pattern generating capacity of only 30 bits, whereas the mother set, i.e., the chequer-board set, has a capacity of 80 bits. To identify which particular bar pattern has been transmitted, one would not take Yes/No decisions with respect to each site, as was done for the chequer-board pattern, since this would be inefficient, i.e., requiring as many as 80 bits every second. Instead one develops a different strategy in pattern identification. With the bar patterns, as shown, one needs to take only three decisions for every bar. The three decisions to be taken are as follows:

1. Is the height of the pattern more than the half of the total? (Yes/No)
2. Is it the upper quarter of the half determined in (1)? (Yes/No)
3. Is it in the upper eighth of the quarter determined in (2)? (Yes/No)

To give an example, consider the third vertical bar in Fig. 11.1(d). The answer to the above set of questions would give the following binary code

1. No — N
2. Yes — Y
3. No — N

The three answers taken together, can be put as N Y N which is the binary description of that particular bar.

2. Constraints

Constraint can be introduced into a language in a systematic manner to increase the reliability of communication. Let us introduce an additional constraint on the bar pattern consisting of the convention that we shall make use of only levels 2, 4, 6 and 8, and suppose that we transmit such patterns at a rate of 10 vertical bars per second, Fig. 11.2. Suppose also that the channel, due to imperfections (noise), can add or subtract from the transmitted height of a bar, one unit. Clearly, the set just described has an *error detecting property* built into it. Thus the receiver can decide whether the message received is correct or in error. This clearly follows from the property of the set in that if the received level is 2, 4, 6 or 8 units then the receiver would decide that the message has been correctly received. But, if an odd level were received, then the receiver would immediately decide that the received character is in error, and would request a retransmission of the message.

The essence of a language having the error-detecting property is to choose symbols in such a way that the effect of an error results in the conversion of the symbol from one belonging to the set to one that does not belong to the set, and it is on this basis that the receiver is able to make a decision whether the message received is correct or in error, and take action accordingly.

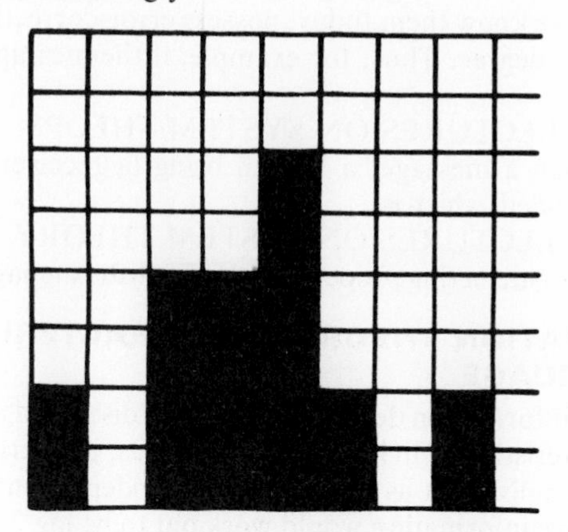

FIG. 11.2 Error Detecting Set.

The capacity of the set illustrated in Fig. 11.2, is clearly

$$C = 10 \operatorname{lgg} 4 = 20 \text{ bits/sec} \tag{11.15}$$

or in terms of total number of patterns this is

$$N = 2^{20} \approx 10^6. \tag{11.16}$$

It should be observed that the error-detecting property has been purchased at the expense of added redundancy. In the given example, it amounts to 10 bits per second. Therefore, the pattern generating capacity of the language has been reduced from 10^9 to 10^6 but we now can communicate in spite of imperfections in the communication channel.

Another possibility would be to agree to transmit only the levels 1, 4 and 7 (Fig. 11.3). Then, even if an error were picked up in transmission, the receiver would be able to extract the correct message, using its knowledge of the characteristics of the noise.

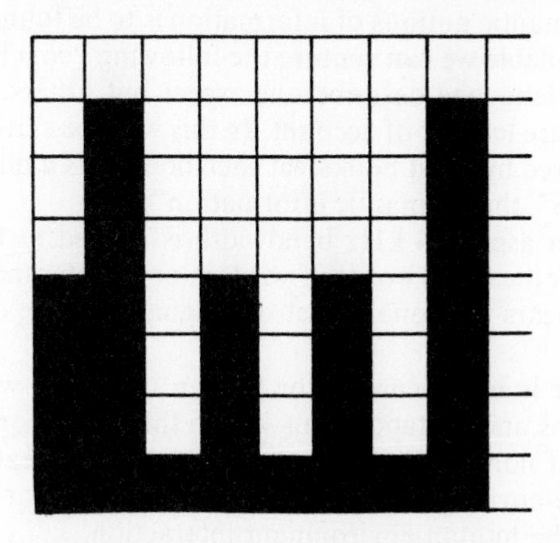

FIG. 11.3 Error Correcting Set.

The reason in this particular example could be that noise can at most add or subtract one unit to a vertical bar, while the levels used for communication are separated by three units. This knowledge (which would be stored in the system) enables the receiver to calculate which particular level was transmitted given a received message. This is an example of a crude *error-correcting* language.

Most languages as we know them today, possess error-correcting or error-detecting properties to a certain degree. Thus, for example, if the message received in English language, were

LECTURES ON SYSTFM THEOPY

Then on receiving such a message, a human being (a receiver) would immediately guess the message intended, which is

LECTURES ON SYSTEM THEORY

This illustrates the error-detecting property of the English language.

E. SOME INFORMATION THEORETIC CHARACTERISTICS OF ENGLISH LANGUAGE

Using measures of information devised on the lines discussed above, it is possible to assess various characteristics of living languages. Thus, as a crude approximation, if we were to consider English text as consisting of 25 independent symbols (letters and spaces) then the average information would work out to be lgg 25 = 4.64 bits/symbol. This, of course, is a greatly overrated figure in that it does not take into account the uneven distribution of the probability of occurrence of the symbols (c.f., Table 3.1). If this were considered, a figure of less than 4 bits/symbol would result, and much smaller figures would be obtained if conditional probabilities were included (effects such as that letter *h* is more likely to follow *t* than *q* or another *h*).

On this basis a machine can be made to produce "English words". It has been found that words so generated *resemble* English words but many are not to be found in the dictionary. Thus, some characteristics of spoken words have statistical traits but this is only part of the story.

If we encode English text by words, that is taking account of the relative frequencies of different words, about 2 bits per letter are required, and even this figure is reduced, if higher order approximations to the language are performed. Ultimately, the entropy of English text can be shown to be less than 1 bit per character. At this stage an "English text" produced by a machine resembles quite strongly written English, including good grammar in places, but lacking cohesion and sense. The text so produced wanders aimlessly and amounts to utterances with no story attached to it: perhaps the clue to semantic notions of information is to be found here.

On the evidence available we can venture the following "conclusion". The statistical properties of a living language describe one aspect but others, relating to the long-range inter-relations, are left out of account. In this way the statistical properties (1 bit per character) are shared by most books yet each book tells a different story betraying by its "phase sequence", the "semantic information".

To examine another aspect, 4 kHz bandwidth is needed to transmit human voice irrespective of whether it carries words of wisdom or an utterance of a senseless string of words. Thus it appears that one aspect of language can be described in statistical terms: its entropy.*

However, language is but a carrier for human thoughts, which are conveyed by long-range interactions and juxtapositions within the framework of the language and these properties are of non-stationary nature, the semantic features. From this point of view the machine-environment interaction would appear to be of a different character to that of the human-environment interaction.

* For an explanation of the meaning of the word entropy see Section G.

Deliberations of a similar nature apply with equal force to any scientific investigation, in particular to a measuring system and the flow of data therefrom.

We shall return to this particular aspect and shall examine it in a somewhat different way in Chapter 23.

F. DISCRETE AND CONTINUOUS SOURCES

All languages which we have discussed so far belong to the class of finite languages. Such languages are made up of a finite number of symbols (patterns), e.g., with the chequer-board set illustrated in Fig. 11.1(a) there are 256 (2^8) distinct patterns which can be accommodated in any one vertical bar. But there are only 8 distinct patterns which can be generated in one vertical bar of the set illustrated in Fig. 11.1(d). In fact the capacity of the language illustrated in Fig. 11.1(d) is only 30 bits per second. The pattern could therefore be *encoded* into an equivalent chequer-board pattern of three times ten sites in 1 second. This clearly must be the case because a chequer-board pattern of 3×10 sites has the capacity of 30 bits per sec. It is therefore possible to translate on a one-to-one correspondence any message represented by the vertical bar set of 8×10 into a chequer-board set of 3×10, without loss of information. This is a general property of all discrete languages whereby any message in one language can be translated into another language (also discrete in nature) on a one-to-one correspondence without loss of information provided the information carrying capacity of the two languages is the same. On the other hand it is not possible to translate on a one-to-one basis from a language of higher capacity into one of lower capacity. Any attempt to do so would result in loss of information because, of necessity, some many-to-one translations would have to be included.

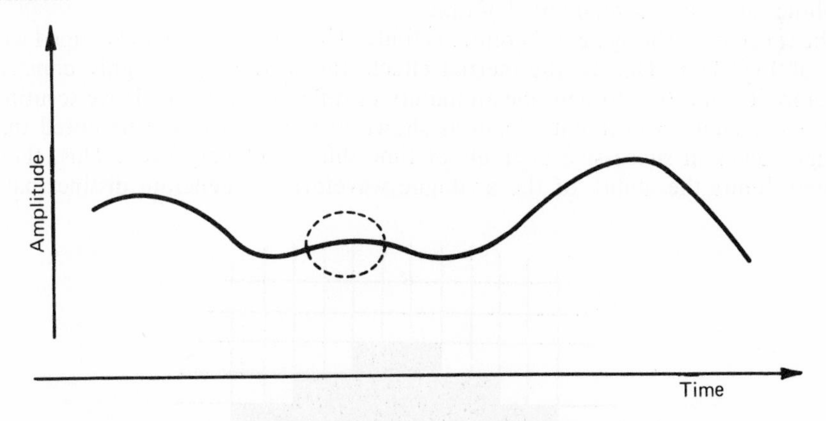

FIG. 11.4 An Analogue Waveform.

A waveform like that illustrated in Fig. 11.4 is a pattern belonging to a different class; the class of analogue signals. By an analogue signal, we mean a waveform which can assume in any particular interval of time, any of the values between specified limits. Thus, suppose the waveform represents an electric current as a function of time, then without further qualification, it is implied that the current can assume any of the values between certain limits, and therefore, at any particular time, the current variable can assume an *infinity* of distinct values. It therefore follows, that in a given interval of time and a given interval in amplitude an analogue waveform can, in principle, be made to represent an infinity of different patterns. It is an example of a continuous signal.

A source generating a continuous signal is known under the name of a continuous source. A continuous source therefore, in principle can generate an infinite number of distinct symbols, unlike the waveform shown in Fig. 11.1(e). This brings with it a mathematical difficulty in so far as we have defined a language as being made up of a *finite* set of symbols and finite number of rules for using them. Contrariwise, with a waveform, we have a source which can apparently generate an uncountable set of patterns and it therefore cannot be used for communication in the usual sense. So much for abstract notions concerning analogue waveforms.

When dealing with practical systems, however, we are normally concerned with waveforms as generated by physical apparatus. Physical apparatus, systems, as well as channels of communication, are characterised by various imperfections. There are in fact definite limits on what can be achieved. In particular, we have two factors in mind (c.f., Chapter 2):

1. A physical system, be it a machine or a communication system is not isolated from the rest of the world. It is therefore subjected to random fluctuations which have their origin in the atomistic nature of matter (noise).
2. Physical systems possess energy storing elements such as inductances and capacitances. The state of a physical system cannot therefore be altered significantly in an arbitrarily small interval of time.

These factors impose a limit on the amount of information which can be sent in a given time through a continuous channel. There is thus:

1. A minimum detectable amplitude, or energy change $\Delta E = \frac{E}{n}$ (noise limitations).
2. A minimum time $(\frac{1}{k}$ secs) which is required to make a detectable change in the state of the system (bandwidth limitations, or inertial effects).

For these reasons, a continuous channel can only *resolve* a finite number of amplitude steps in a given interval of time.

The capacity of the system therefore is finite. This can be further explained with the help of Fig. 11.5. Due to the inertial effects the waveform is highly constrained, therefore it is not free to describe an infinity of different patterns. If we scrutinise an analogue waveform in minute detail as shown in Fig. 11.5 it will be noted that the voltage values in successive intervals of time differ, but very little. This, therefore severely limits the ability of the analogue waveform to generate distinct patterns.

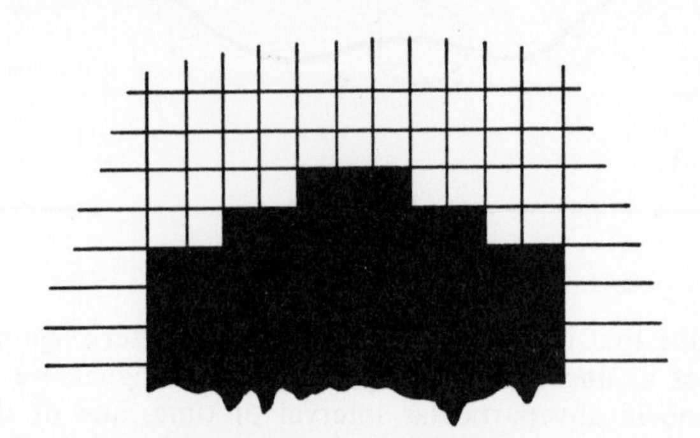

Fig. 11.5 Fragment of Waveform Shown in Fig. 11.4.

Moreover, due to the Brownian motion (factor kT, see Chapter 2) the equipment will not be able to resolve with accuracy voltage values which differ by less than a certain minimum value, and thus it can be shown that the waveform constrained in the above manner can be made to represent only a finite number of resolvable distinct shapes. Thus, we draw the important conclusion that an analogue waveform contains a finite amount of *independent* data. It is for this reason that every analogue waveform can be, by suitable processing, translated into an equivalent digital form without loss of information.

The technique for converting an analogue signal into a digital form is illustrated in Fig. 11.6. There are four steps in the process.

1. From consideration of the inertial factors (bandwidth limitation) the shortest correlation distance for an analogue waveform of the system is deduced. Let this be T. We then sample instantaneously the waveform values at time intervals which are shorter than or at most equal to the quantity T. In so doing, the analogue waveform becomes converted into a set of values erected at intervals of T.
2. From considerations of noise in the system we calculate the smallest amplitude step which the system can resolve.
3. We then measure the sampled values in terms of those steps. In this way, we convert the sampled values into a sequence of digits as shown in Fig. 11.6, Step 3.
4. The digital values so obtained are subsequently processed into a digital form convenient for use in the system, such as a binary language.

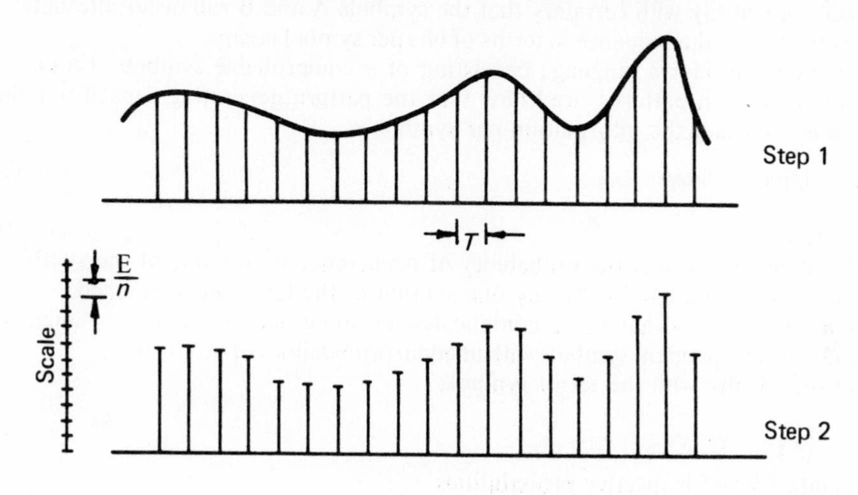

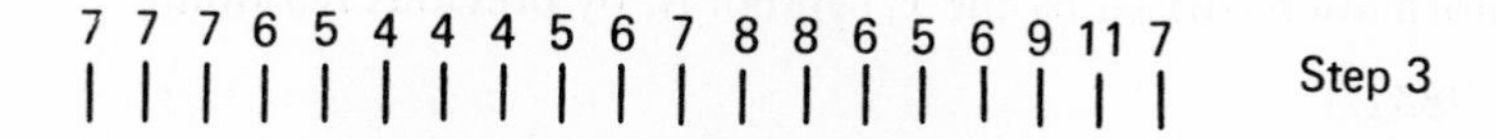

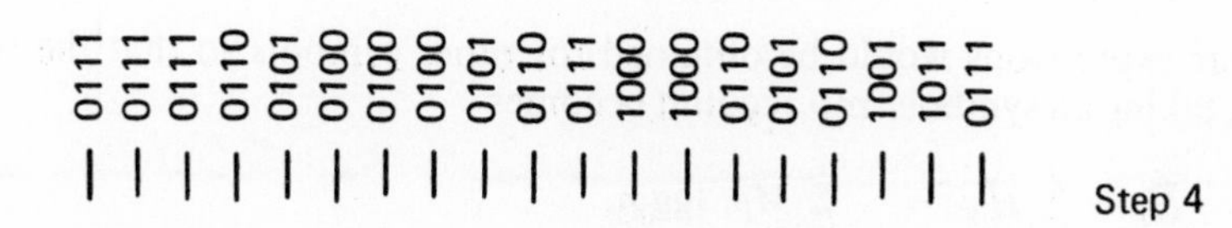

FIG. 11.6 Steps in Conversion from Analogue to Digital Form.

G. LANGUAGES, CONCEPT OF ENTROPY

As explained above, language is a set of symbols (letters, characters, etc.) making up a set, together with rules for using them (the grammar of the language). For example, a binary set consists of two symbols (A,B), and the grammar might be that the symbols occur with equal probability and that there are no inter-symbol constraints. Typically we would have a sequence like

$$\ldots ABBAAABABBABBBBABBAAA \ldots \tag{11.17}$$

Clearly, with such a sequence we cannot tell what the next member in the series is going to be. It could equally well be an A as a B. In fact the information needed to determine whether it is going to be A or B is just one bit. Therefore, the information per symbol of the sequence is one bit, being a characteristic of the language described.

On the other hand, the sequence

$$\ldots ABABABABABABABABABABABABABA \ldots \tag{11.18}$$

carries no information because from the knowledge of the pattern, given by the past data, we can say with certainty that the symbols A and B will occur alternately. The information of the sequence in terms of bits per symbol is zero.

Let us now consider a language consisting of n equiprobable symbols. From the earlier discussion on patterns, we know that the pattern generating capacity of the language is such that the information per symbol is

$$H_1 = \mathrm{lgg}\, n = -\mathrm{lgg}\, p \tag{11.19}$$

$$p = 1/n \tag{11.20}$$

In the above $p = 1/n$ is the probability of occurrence of any one of the symbols. H_1 is the information carried by any one symbol of the language so defined.

To consider a somewhat more complicated example, let us examine a language consisting of n independent symbols with unequal probabilities of occurrence.

More specifically with the set of symbols

$$\{x\} = x_1, x_2 \ldots x_i \ldots x_k \tag{11.21}$$

we associate a set of respective probabilities

$$\{p\} = p_1, p_2 \cdots p_i \cdots p_k \tag{11.22}$$

where

$$\sum_{i=1}^{k} p_i = 1 \tag{11.23}$$

the information carried by the x_i symbol is, by previous reasoning

$$-\mathrm{lgg}\, p_i . \tag{11.24}$$

Thus in a long sequence (M large) the total information conveyed by x_i will be

$$H_{iT} = -Mp_i \, \mathrm{lgg}\, p_i \tag{11.25}$$

since the symbol x_i will have occurred Mp_i times.*

Similar expressions would be obtained for other symbols so that the total information, taking all symbols into account becomes

$$C = H_T = \sum_{i=1}^{k} H_{iT} = -\sum_{i=1}^{k} Mp_i \, \mathrm{lgg}\, p_i . \tag{11.26}$$

* M being the total number of symbols in the sequence.

The average *information per symbol* of the language (assuming symbols to be independent) is

$$H = \lim_{M \to \infty} \frac{H_T}{M} = -\sum_{i=1}^{k} p_i \operatorname{lgg} p_i. \tag{11.27}$$

The average information per symbol is known as the *entropy* of the language and if logarithms to the base of 2 are used as in Equation 11.27, then the entropy is measured in bits per symbol.

The average number of patterns which a long sequence can generate is therefore given by

$$N = 2^C = 2^{M.H} \tag{11.28}$$

which shows that the number of patterns which a sequence can generate is an exponential function of the length of the sequence (c.f., Chapter 3).

Strictly speaking sequences of infinite length should be considered. This was the reason for formally using the limiting operation in Equation 11.27.

H. EXPERIMENTS

(1) Toss a coin (say 100 times) and record your results as a sequence of +1 (stands for heads) and −1 (stands for tails). View the result so obtained as a stochastic process and discuss its characteristics.

Plot your results in the form of a function $S_k(i) = +1 - 1 + 1 + 1 - 1 \ldots$ being the summation of the outcome of the original random process (the first integral). Then view $S_k(i)$ as a random function and compute the various statistical averages, also discuss fluctuation phenomena due to the finite size of the samples.

Then discuss problems relating to prediction, and ask questions of the nature: is it possible to predict from the past record whether the next event will be +1 or −1? Is it possible to predict from the knowledge of $S_k(j)$ the value of $S_k(j+1)$ or $S_k(j+2)$ etc?

Finally, questions of the form: if the coin were biased how could this *information* be obtained? If, instead of tossing the coin, one were to write down the "result of experiment" by reference to tables of random numbers (say +1 for even numbers and −1 for odd ones) could this be inferred from the examination of the "experimental" results? If, instead of tossing the coin, one were to write down the "result of experiment" by reference to a column of numbers picked at random out of a book in the library (the table of numbers could pertain, for example, to an exponential function) could this be inferred from an examination of the results, etc?

(2) Draw a chequer-board (10 × 10 sites) pattern of the type discussed in this Chapter. By scanning the pattern row by row, write down a sequence +1, −1, +1 . . . depending whether the particular square is black (+1) or white (−1). Discuss the results. Ask and try to answer questions of the type: could the sequences obtained in Experiment 2 resemble the sequences obtained in Experiment 1? Can a pattern be identified? Need the coding method (scanning convention) be stated to encode and decode the pattern? Can the presence of a pattern be detected? Are there any "simple" or "complex" patterns? Can the sequences be more efficiently encoded? How does one define classes of patterns, etc?

I. PROBLEM

Assume that the picture as transmitted in T.V. practice can be resolved into 500 × 500 sites and that 25 frames (complete pictures) are transmitted per second. Further, if it is known that under typical viewing conditions the human eye can resolve 64 shades of grey (between "white" and "black"), calculate the channel capacity to cope with the data generated by the system.

Solution:

Total number of sites $= 500 \times 500 = 2.5 \times 10^5$

64 shades of grey is equivalent to 64 attributes per site, therefore each site conveys lgg 64 = 6 bits of information.

A complete frame contains, therefore,

$$6 \times 2.5 \times 10^5 = 15 \times 10^5 \text{ bits of information.}$$

Consequently the rate of data transmission

$$= 25 \times 1.5 \times 10^5 = 3.75 \times 10^6 \text{ bits/sec.}$$

SUGGESTED READING MATERIAL

Morgan, D. and Bailey, B. *Thinking and Writing*. Rigby, 1966.

Cherry, C. *On Human Communication*: A Review, A Survey and a Criticism. John Wiley and Sons, New York, 1957.

Karbowiak, A. E. *Theory of Communication*. McGraw-Hill, 1963.

Miller, G. A. *Language & Communication*. McGraw-Hill, New York, 1963.

Pierce, J. R. *Symbols, Signals and Noise: The Nature and Process of Communication*. Hutchinson, 1962.

Emery, E., Ault, P. H. and Agee, W. K. *Introduction to Mass Communications*. Dodd, Mead & Co., 1965.

CHAPTER 12

Networks of Building Blocks: Dynamic Systems

R. M. Huey

"For the sake of persons of different types of mind, scientific truth should be presented in different forms, and should be regarded as equally scientific whether it appears in the robust form and vivid colouring of a physical illustration or in the tenuity and paleness of a symbolical expression".
James Clerk Maxwell, 1864.

A. INTRODUCTION

It is sometimes quite hard to tell, when using the word system, whether we are referring to the actual system itself or whether we are referring to the model of the system, which has been built up in our minds. In this chapter (and subsequent ones) the reader should consciously try to ask and to answer this question—are we referring to the actual system or to its model? Sometimes the distinction may be important, sometimes it may not matter.

This chapter and the one following deal in simple terms with some of the mathematical models which have been found useful for quantitative prediction of the behaviour of systems. The same mathematical models have been found useful for the other task of designing systems which will exhibit a desired behaviour.

B. METHODS AVAILABLE TO CALCULATE THE RESPONSE OF A DYNAMIC SYSTEM

The choice of the most suitable method of calculation will depend on two things:

(1) The kind of signal—the word "mode" is sometimes associated with this, and
(2) the way in which a mathematical model for the system has been built up.

We may say that the three things—signal, system and computational method must be compatible, i.e., they must all three fit together reasonably neatly.

Let us consider two examples.

EXAMPLE 1. A consideration of steady state response only.

First, when the input signal is a sinusoid[1] $E \cos \omega t$, then we may use the electrical engineer's ideas of impedance and *phasor algebra* in order to obtain the response of the system. This is described and taught as A.C.* circuit theory, and may be presented in undergraduate courses in an early Physics course and carried on in Electrical Engineering middle year courses. The basic signal of phasor algebra is the sinusoid of pre-specified frequency. It is an important point of the theory that the steady state response of a linear system to a sinusoidal input is another sinusoid of identical frequency.

Let us examine a block diagram which suggests the mathematical manipulation needed to obtain the response of a linear system to a sinusoidal input.

* Alternating current, by which is implied the particular kind of alternation described by a sinusoid of constant amplitude, frequency and phase angle.

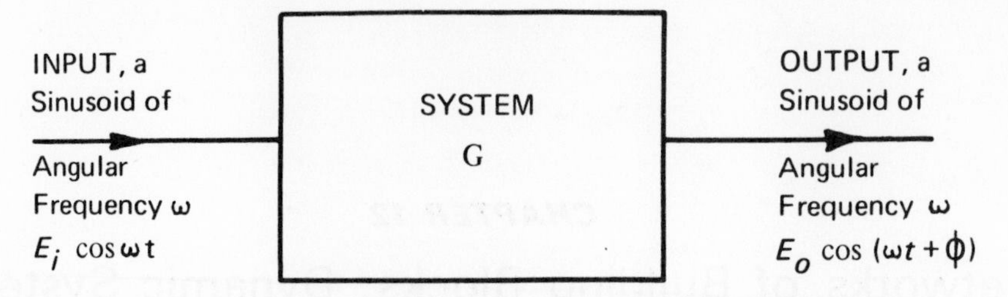

FIG. 12.1 Block Representation suitable for Phasor Algebra Technique.

Figure 12.1 is the same as the block diagram Fig. 5.2 or Fig. 5.5, except that we have specified the exact form or mode of the input and output signals. In this particular case the specification we have adopted for the signals has implied that we are concerned with (and only with) the steady-state portion of the response. By this limitation we have also limited the way in which the system function G can (and indeed must) be written down.

Let us suppose (and the reason for doing this appears later in this chapter) that we wish to be able to use a relation

$$(\text{OUTPUT}) = (\text{SYSTEM FUNCTION})\,(\text{INPUT}) \tag{12.1}$$

in order to calculate the response to a specified input. In operational terms this means

$$\begin{aligned}(\text{OUTPUT}) = {} & \text{result of operating on the } (\text{INPUT}) \\ & \text{with the operator } (\text{SYSTEM FUNCTION})\end{aligned} \tag{12.2}$$

and we also wish to be able to carry out the operation using the ordinary rules for algebraic multiplication. It will be shown that this is possible, provided that each of the three quantities (*Output*), (*Response*) and (*Input*) are written as complex numbers of a kind known as phasors[2].

Multiplication of these phasors is expressed in a phasor equation, i.e.,

$$\begin{pmatrix}\text{OUTPUT}\\ \text{PHASOR}\end{pmatrix} = \begin{pmatrix}\text{SYSTEM}\\ \text{FUNCTION PHASOR}\end{pmatrix} \times \begin{pmatrix}\text{INPUT}\\ \text{PHASOR}\end{pmatrix} \tag{12.3}$$

The two operations on the right hand side which are needed to obtain the answer for the left hand side are simply

(i) multiply the magnitudes,

and (ii) add the phase angles with due regard to sign.

Having established the idea of using Equation (12.3) in that form (i.e., for analysis, where output is the unknown quantity) we can turn it around to use for two other problems:

$$\begin{pmatrix}\text{SYSTEM}\\ \text{FUNCTION}\\ \text{PHASOR}\end{pmatrix} = \begin{pmatrix}\text{OUTPUT}\\ \text{PHASOR}\end{pmatrix} \div \begin{pmatrix}\text{INPUT}\\ \text{PHASOR}\end{pmatrix} \tag{12.4}$$

$$\begin{pmatrix}\text{INPUT}\\ \text{PHASOR}\end{pmatrix} = \begin{pmatrix}\text{OUTPUT}\\ \text{PHASOR}\end{pmatrix} \div \begin{pmatrix}\text{SYSTEM}\\ \text{FUNCTION}\\ \text{PHASOR}\end{pmatrix} \tag{12.5}$$

In each case we have written the unknown quantity on the left hand side. In Equations (12.4) and (12.5) these calculations are both possible and easy because simple rules can be made for phasor division.*

* To multiply one phasor by another we multiply their magnitudes and add their phase angles. To divide one phasor by another we divide their magnitudes and subtract their phase angles.

TABLE 12.1

Problem Usually Named As	Known Quantities	Unknown Quantities	Equation Needed
Analysis	Input and System Function	Output	(12.3)
Identification *or* Synthesis	Input and Output	System Function	(12.4)
Recovery *or* Inverse Analysis	Output and System Function	Input	(12.5)

The three equations concerned may be tabulated as shown in Table 12.1.

The identification problem usually refers to the process where output and input are measured on an actual system and the calculation is undertaken to identify the properties or coefficients of the system function. The synthesis problem is usually stated somewhat differently; given a specified input and a (specified) desired output, and also given a possible manner of connection of the components in the system, then let us calculate the magnitudes of the components involved. The synthesis problem is a design problem and in general does not have a unique answer since there may be many different ways of connecting up components which could achieve the desired result for a given system function. For one manner of connection (i.e., for a given circuit configuration) there is usually only one set of values for the components, if we have specified the problem correctly and provided that we have excluded negative values for the coefficients, and provided that the configuration does not result in indeterminate results in the arithmetic; the latter can occur for example in certain bridge or tee configurations.

A reminder may be useful before going on to the next section:

Remember all our discussions in this chapter have referred and will refer to linear systems.

EXAMPLE 2. A consideration of both transient and steady state responses.

Both the steady state and transient responses may be obtained by the use of the Laplace Transform method. This scheme of calculation is more general than that of phasor algebra. It is based on the theory of functions of a complex variable, which is well treated in the more advanced texts used for second and third year engineering courses.

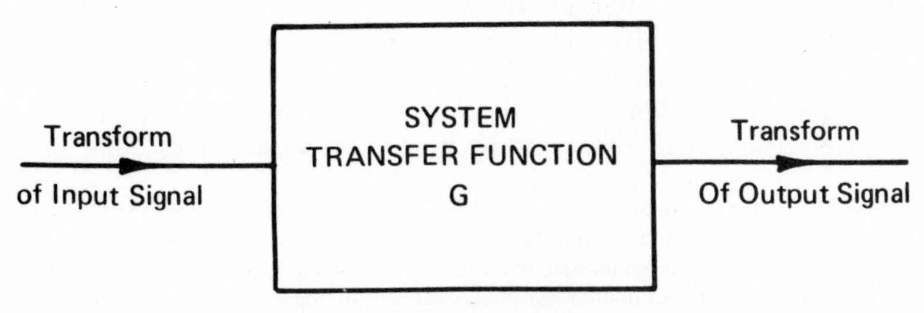

FIG. 12.2 Block Representation suitable for Laplace Transform Technique.

Let us draw a block diagram to represent what goes on in the Laplace Transform method for calculating system responses.

The similarity of Fig. 12.2 with Figs. 5.2, 5.5 and 12.1 should be noted. The meaning of Fig. 12.2 may be expressed by saying

$$\begin{pmatrix}\text{TRANSFORM}\\ \text{OF OUTPUT}\end{pmatrix} = \begin{pmatrix}\text{SYSTEM}\\ \text{TRANSFER}\\ \text{FUNCTION}\end{pmatrix} \begin{pmatrix}\text{TRANSFORM}\\ \text{OF INPUT}\end{pmatrix} \qquad (12.6)$$

or at somewhat greater length

$$\begin{pmatrix}\text{OUTPUT}\\ \text{TRANSFORM}\end{pmatrix} = \textit{result of operating on the} \begin{pmatrix}\text{INPUT}\\ \text{TRANSFORM}\end{pmatrix}$$

$$\textit{with the operator} \begin{pmatrix}\text{SYSTEM}\\ \text{TRANSFER}\\ \text{FUNCTION}\end{pmatrix} \qquad (12.7)$$

The basic signal or mode of time-variation which is associated with the Laplace transformation, is a damped sinusoid,[(3)] or, if we wish to build up more complicated cases, a summation of several different damped sinusoids.

Let us now try to gain some idea of the reason why the Laplace Transform method is so useful for system calculations. Consider the following analogy and work out also for yourself the way in which this analogy could be applied to illustrate also the behaviour of phasor algebra as a simple operational means of solving (i.e., computing) what would otherwise be a more difficult mathematical problem.

Addition (or subtraction) is much easier than multiplication (or division). Suppose that we wish to multiply p by q. Let us transform each quantity by taking logs. The problem is now to calculate

$$(\log p + \log q) \text{ in order to get } \log(pq).$$

The operation or process of multiplication has been transformed into the simpler operation or process of addition. The problem is completed by an inverse transformation being applied to $\log(pq)$ (i.e., taking antilogs), the final and desired result being the value of pq. This is the process employed in multiplying with the aid of a slide rule.

In calculating the response of a system to an input, we are faced with the problem of integrating the differential equation of the system so as to obtain both the Complementary Function and the Particular Integral, the two portions into which the total response may conveniently be divided.*

This integration can be carried out in several different ways. One such way is through the use of the Convolution Integral (sometimes also known as the Superposition Integral or Duhamel's Integral). Providing we can make a start by determining the response of the system to an impulse delivered at time $t = 0$, this so-called impulse response is reversed in time and then multiplied by the input signal and integrated between suitable limits. Readers who are competent at integration would no doubt manage to solve such problems and obtain correct answers.

However, it is usually much easier to take the Laplace transform of the input and to determine the transfer function of the system. The latter may be calculated from the rules of electric circuit theory or by more generalised rules quite simply and

* Reminder: the C.F. is the transient response; the P.I. is the steady state response. The C.F. results from solving the differential equation excluding the forcing function but including the initial conditions. The P.I. results from solving the differential equation including the forcing function but virtually excluding the initial conditions. Some degree of interchangeability between forcing function and initial conditions may be exercised arbitrarily in the way in which the problem is stated. See also Chapter 5.I.

directly from the differential equation itself. The former may be obtained from the voluminous tabulations of Laplace transforms which are readily available, provided we know the algebraic form of the input signal as a function of time. Having obtained these two quantities we can then apply the simple operation of multiplication instead of the more difficult operation of convolution integration. Using the initials L.T. to indicate Laplace transform and d.e. for differential equation we have

$$\begin{pmatrix}\text{L.T. OF}\\\text{RESPONSE}\end{pmatrix} = \begin{pmatrix}\text{TRANSFER FUNCTION}\\\text{corresponding to system d.e.}\end{pmatrix} \times \begin{pmatrix}\text{L.T.OF}\\\text{INPUT SIGNAL}\end{pmatrix} \qquad (12.8)$$

The above equation is written in terms of the complex variable $s = \sigma + j\omega$. The significance of σ is that it represents damping of a sinusoidal variation occurring at angular frequency ω (see Endnote (3)).

The transformation has taken the algebra over into "the complex frequency domain" or more simply "the frequency domain".

The final operation (analogous with taking antilogs) is then to get ourselves back into the "time-domain" (i.e., to recover the response or output signal as a function of time) and this may be done by applying an Inverse Laplace transformation[4] to the L.H.S. of Equation (12.8).

$$\text{Thus} \begin{pmatrix}\text{L.T. of response,}\\\text{say } F_o(s)\end{pmatrix} \text{ via } \begin{pmatrix}\text{Inverse}\\\text{L.T.}\end{pmatrix} \text{ yields } \begin{pmatrix}\text{Response as a time}\\\text{function, say } f_o(t)\end{pmatrix}$$

In addition to referring to the extensive tables of Laplace Transforms and inverse transforms, it is necessary for those learning the Laplace Transform method to understand and to utilise certain rules so that both transient (C.F.) and steady state (P.I.) portions of the solution may be identified and written down.

Reminder: The rules for using Laplace transform algebra are arranged to give both transient and steady state parts of the response. The rules for using phasor algebra are concerned with yielding only the steady state part of the response.

Another pair of names which correspond to these divisions are free and forced response

C.F. — transient — free response
P.I. — steady state — forced response.

The term steady state is based on the idea that after a suitable length of time the transient part will have died away to insignificance and only the steady state part will remain. During the early part of the response however both transient and steady state responses will exist and must be added together, with due regard for sign, to obtain the total response.

C. THE ASSEMBLY OF BLOCKS INTO BLOCK DIAGRAMS

We are now in a position to consider how we may build up systems by assembling single blocks in such a way that the output of one block forms the input of the next, see Fig. 12.3.

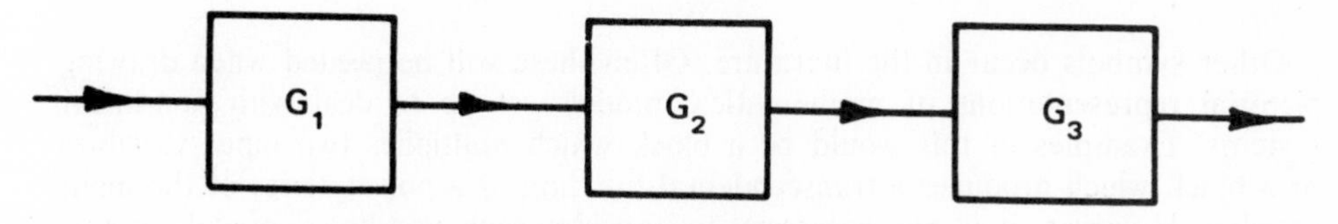

Fig. 12.3 Assembly of Three Blocks in Cascade.

An essential feature (or rule with which we must conform) is that each block is a closed sub-system whose input and output are its *only* channels for communication or exchange of energy with the remaining parts of the system.

Multiple inputs or outputs may be allowed if we can write the differential equations (of each subsystem) in matrix form so that vector arrays may be used for the group of independent (i.e., input) variables and for the group of dependent (i.e., output) variables. The condition then becomes that these two groups should form the only channels or connections from that block to the remaining parts of the system.

A simplified description of the system represented in Fig. 12.3 is possible and is shown in Fig. 12.4.

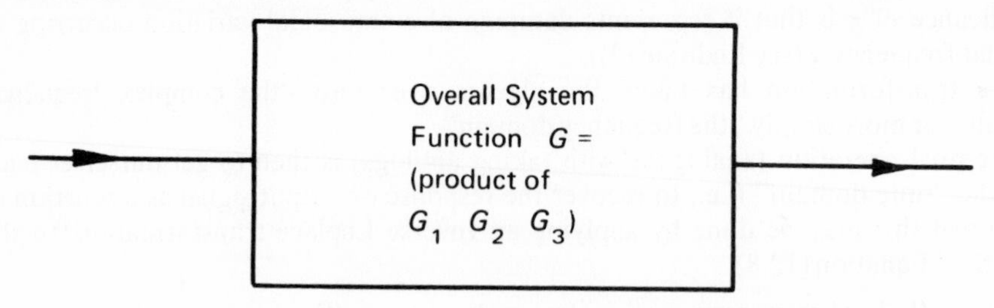

FIG. 12.4 Composition of System of Fig. 12.3 into a Single Block.

The simple composition (i.e., putting together) shown in Fig. 12.4 is possible because of the simple multiplicative property once the system function has been written as a function of the complex frequency s, that is as a transfer function. The inputs and outputs in order to be compatible with this simple multiplicative process must also be written as functions of the complex frequency s, that is as Laplace Transforms of the actual (i.e., time-domain) signals.

Two other formal symbols are needed to establish and complete our idea of a block diagram as a formal pictorial representation of a group of differential equations. The first is the adder or summer. This is a block with two inputs and one output (see Fig. 12.5). The output signal is the sum of the two input signals.

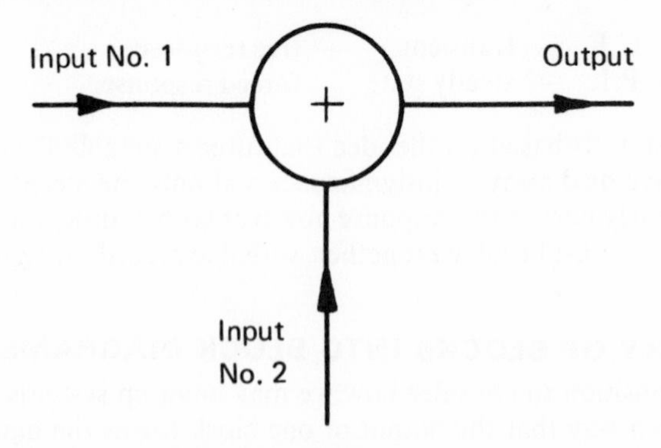

FIG. 12.5 Block Representation of an Adder (for summing two inputs).

Other symbols occur in the literature. Often these will be needed when drawing pictorial representations of mathematical models set up to deal with non-linear systems. Examples of this would be a block which multiplies two input variables or a block which produces a transcendental function or a power series of the input variable. However, it is not necessary to consider such non-linear models at this stage.

A square block may be used in place of a circle. By reversing the polarity (i.e., sign) of one input, the difference of two quantities may be represented.

Before describing the second symbol, consider how we sometimes go about solving a set of simultaneous equations. One process* widely used is to solve for a certain variable (let us call it x) in Equation No. 1 and then to substitute this solution (i.e., result or output) into Equation No. 2. The position before carrying out the substitution may be represented by Fig. 12.6.

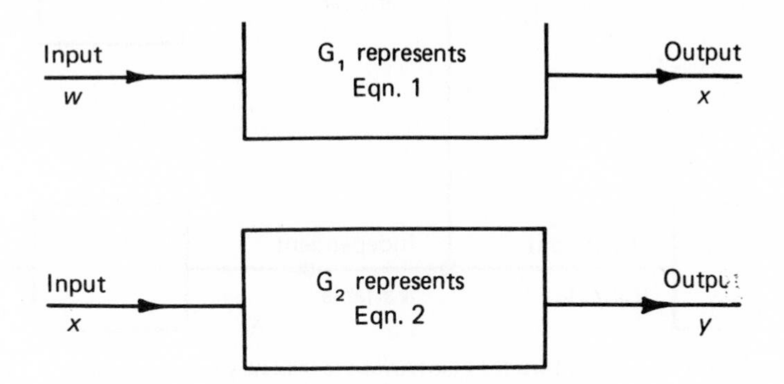

FIG. 12.6 Block Diagram representing Two Simultaneous Equations in Three Variables (w, x and y).

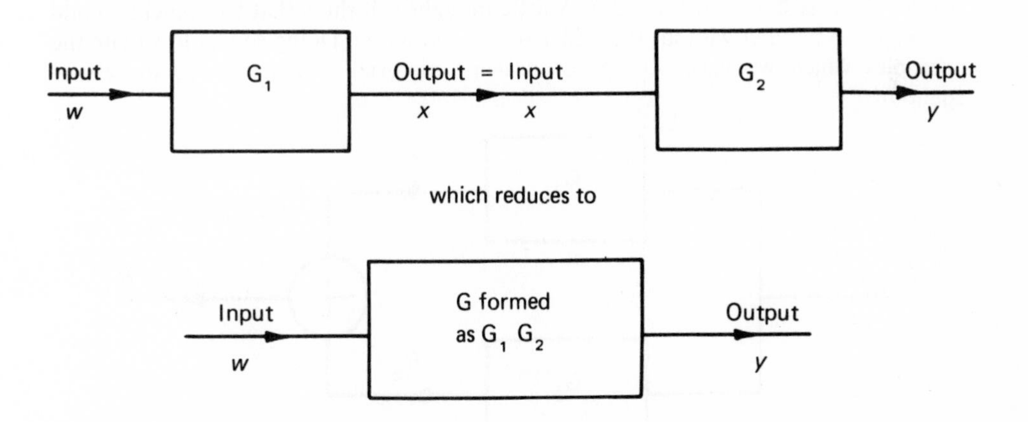

FIG. 12.7 Composition, in two steps, of Fig 12.6 into a Single Block, thus eliminating the Variable x.

The position after carrying out the substitution may be represented by Fig. 12.7, which shows first the process of and then the result of the substitution. The result is that the variable x has been eliminated.

The fact that block diagrams can be utilised in this way to solve quite complicated sets of simultaneous differential equations, depends of course on the formalism which demands that both signals and systems are written in a certain way (as described earlier in this chapter) so that blocks connected in series† may have their system functions multiplied together.

* The aim of the process is of course to eliminate one by one the variables in question, ending up with one equation so constituted that it can be solved without ambiguity for a single remaining dependent variable.

† Sometimes the phrase "inconcatenation" is used to describe this manner of connection.

Having thus set the stage, the second formal symbol, which is simply the directed line, can now be introduced. A line connecting two (or more) points simply indicates that the same variable quantity appears at these two (or more) points. The direction of the arrowhead indicates whether the variable in question is independent (arrowhead entering a block) or dependent (arrowhead leaving a block). See Fig. 12.8.

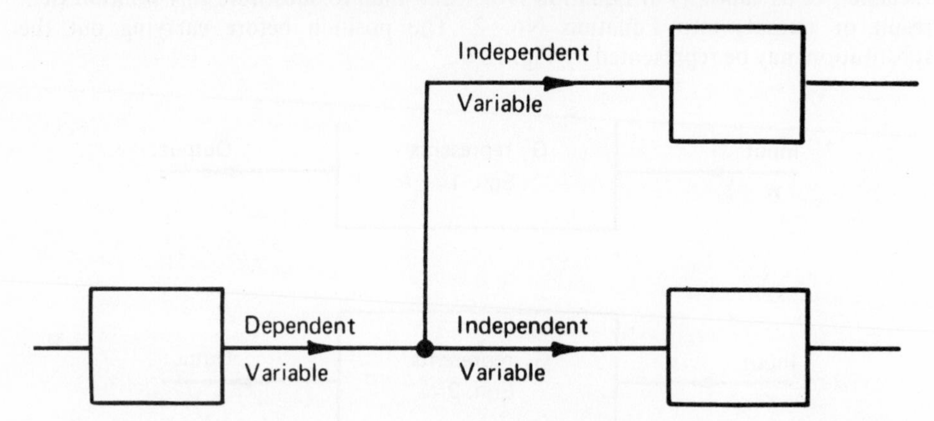

Fig. 12.8 Illustrating the use of a Directed Line (Arrow) to indicate whether a Variable is Dependent or Independent.

One other possible connection remains. In addition to connecting blocks (subsystems) in sequence or in series (concatenation) we may wish to connect blocks in parallel. This is shown in Fig. 12.9. A little thought will show that two blocks should not be put in parallel without the adder (or a subtractor). Doing so would violate the principles which we have set up for drawing pictorial representations of sets of equations.

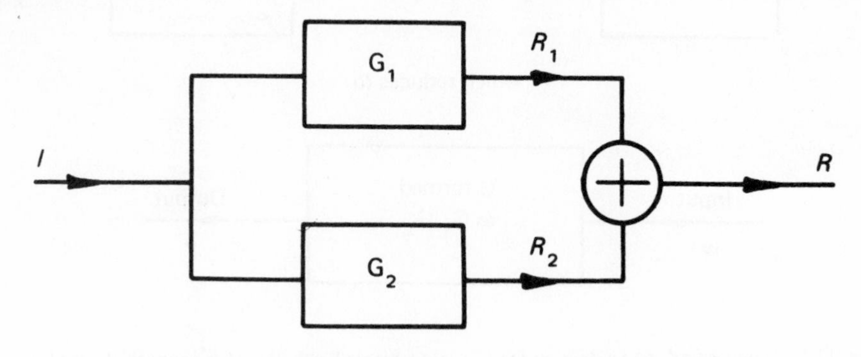

which reduces to

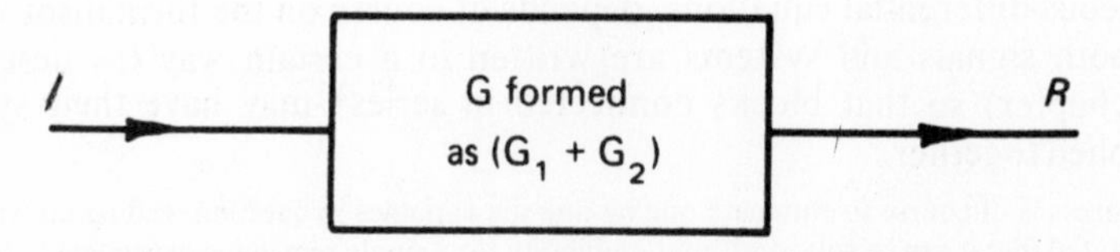

Fig. 12.9 Illustrating how Blocks may be combined in Parallel with the aid of an Adder.

Reminder: The representation as a block diagram of a number of sub-systems connected to form a larger system is subject to a few formal rules. These rules although very simple must *not* be disobeyed.

The block diagram representation is also very suitable for describing in pictorial form the mathematical model of a feedback control system. It should be noted that a block may be represented by either a rectangle or a circle. Whichever symbol is used, it is necessary to know what goes on within the block (i.e., within the sub-system or within the sub-set of equations) either in terms of a system function* or in terms of an equation (or set of equations). The significance of the circle or rectangle is simply that the sub-system is closed and is connected with the remainder of the system *only* by the designated variables (i.e., connecting lines as designated or shown on the diagram). The equations themselves will have been set up by applying suitable conservation laws.

There is one particular configuration of block diagram to which great importance must be attached. This is shown in Fig. 12.10 and it will be noticed that the chain of blocks has been turned back on itself to join up with the input. This configuration is known as a feedback system or feedback control system.

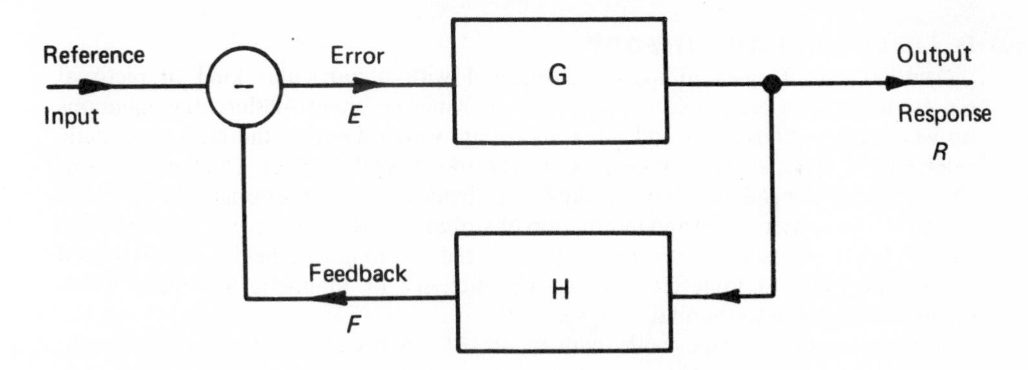

FIG. 12.10 Block Diagram of a Basic Feedback System.

Working with transformed signals (these can mean either Laplace-transformed signals or phasor signals) we can write down the equations

$$R = \mathrm{G}E \qquad \text{(from Block G)}$$

$$F = \mathrm{H}R \qquad \text{(from Block H)}$$

$$E = I\text{-}F \qquad \text{(from the subtractor)}$$

Eliminating E and F from these equations (readers should check this step)

$$R(\mathrm{H} + \frac{1}{\mathrm{G}}) = I$$

$$\text{or } \frac{R}{I} = \frac{1}{\mathrm{H}} \times \frac{1}{(1 + \frac{1}{\mathrm{GH}})} \qquad (12.9)$$

If the product G H (called the loop gain) is large then

$$\text{approximately } \frac{R}{I} \approx \frac{1}{\mathrm{H}} \qquad (12.10)$$

* This could be either a transfer function or a phasor response function. The latter is sometimes termed the frequency response. In either case the function for the block in question may be derived from the equations for the block.

This means that the ratio of output to input can be controlled very accurately to be $\frac{1}{H}$. This quantity can be made to be very stable using low power components of high stability. In other words, the response of a feedback control system can be made practically independent of the forward path (G) which may include a power amplifier of dubious stability, although of good energy-efficiency.

Sometimes the feedback path (H) is arranged so that H = 1. In that case $\frac{R}{I} \approx 1$ and can be made as close to unity as we please provided we can arrange a very high gain in the forward path (G). This arrangement is the basis of many automatic regulators.

The arrangement with both G and H represents a widely used system known as negative feedback which can stabilise the gain and decrease the distortion level in an amplifier or modulator. For example if H = 0.05 and if G were made large in value, then the ratio of R to $I \approx 20$ and this will be the effective gain of the amplifier with feedback. We have a choice which allows us to trade gain for improved stability, lower distortion and greater bandwidth.*

D. LINEAR CIRCUIT THEORY

Another set of rules intimately associated with a particular kind of pictorial representation has been built up as an aid to solving the integro-differential equations (in which both differentials and integrals occur) which describe the electromagnetic behaviour of various electrical devices or networks of such devices. The techniques of circuit theory depend (or may be shown to depend, if an appropriate viewpoint is taken) on the conservation and interaction of stored electrical energy (i.e., stored in an electric field), stored magnetic energy (i.e., stored in a magnetic field), and dissipated energy (i.e., due to imperfect electric conductivity or magnetic hysteresis which involves conversion to thermal energy).

Arrangements to cope with conversion of energy to and from chemical, mechanical, etc., forms of energy can be set up so that electric circuits may include such devices as batteries, electric motors, generators, etc. Because of its dependence on the law of conservation of energy, analogies based on electric circuit theory may be set up for mechanical systems, thermal systems, etc., subject of course to similar conditions which are imposed on electric circuits. These include for example not trying to operate the system at relativistic velocities nor at accelerations such that the rigidity of individual mechanical connections is impaired. The limitation to non-relativistic velocities also applies to electric circuits, of course if they are in motion, motion while over-large acceleration of electric charges will result in the radiation of electromagnetic energy, as from a radio transmitting aerial.

Indeed, wherever conservation of energy is the controlling conservation law in a system, we may usefully set up a circuit theory type of representation. Thus we may draw mechanical circuits, thermal circuits, acoustic circuits, etc., and use them to simplify the solution of many such problems.

Consider Fig. 12.11. This shows four electrical components with their terminals connected together by lengths of electrically conducting wire.

* Compare this idea with the description of "sensitivity" given in Chapter 26. In this case we are deliberately making the gain of the feedback system sensitive to H and insensitive to G. On the other hand the energy-efficiency can be made sensitive to the characteristics of G but relatively independent of H. The bandwidth and frequency response of a system may also be manipulated by choosing an appropriate frequency response for the block H. Those readers going on to an electrical engineering course will learn how to "trade-off" the overall gain of the feedback amplifier in order to increase bandwidth, or to decrease distortion.

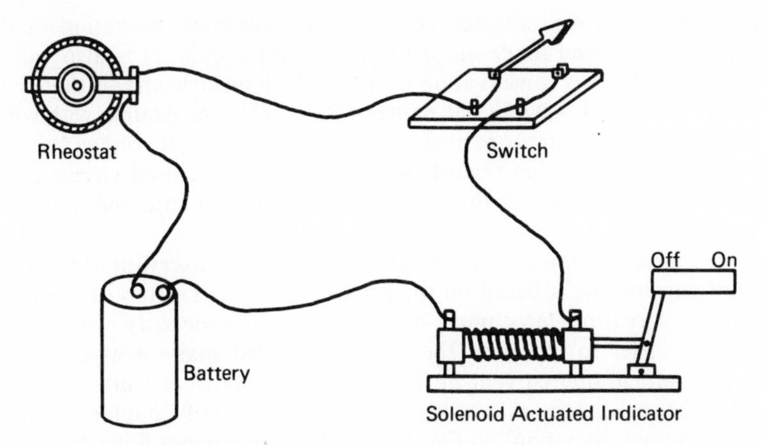

FIG. 12.11 Pictorial Representation: Electrical Components and Connections.

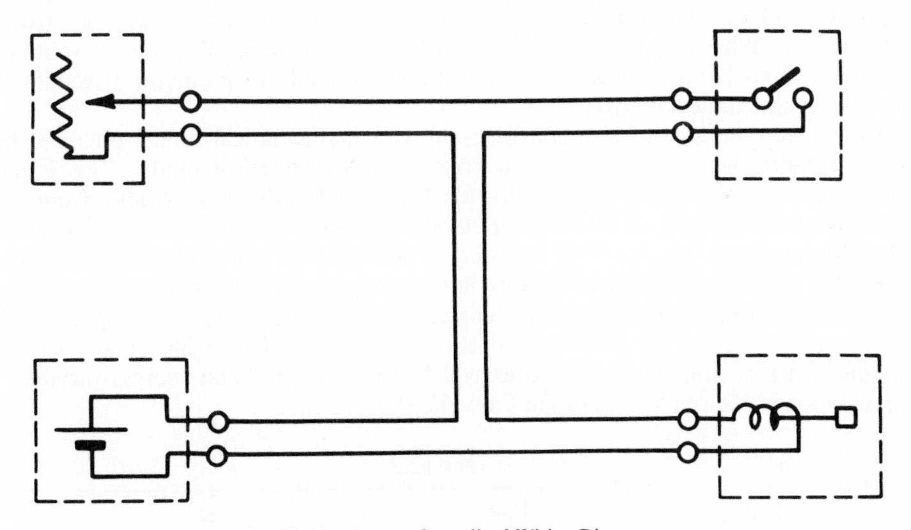

FIG. 12.12 A more formalised Wiring Diagram.

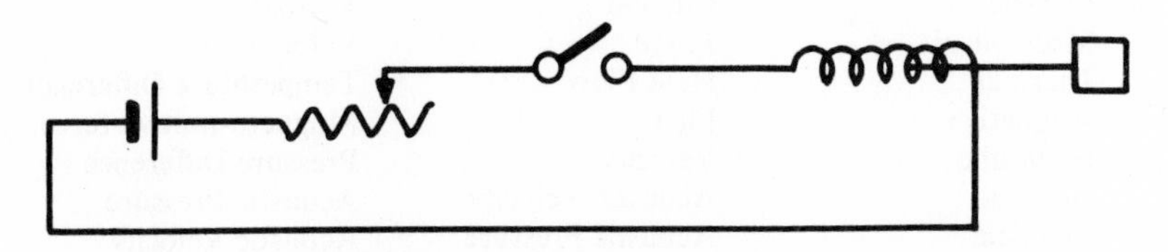

FIG. 12.13 Conventional Circuit Diagram ignores layout of Actual Wiring.

The description of this circuit has been redrawn in Fig. 12.12 as a wiring diagram which, incidentally, also suggests a neater layout for running the wires between components. Conventional symbols are used for each component.

In Fig. 12.13 the arrangement is redrawn as a conventional circuit diagram in which there is no attempt to display the relative positions of the components nor the layout (i.e., "run") of the wiring.

The last diagram is well adapted to explain the electrical operation of the small "system" which has been made up of the four components. It is also well adapted to display one of the two conservation laws on which electric circuit theory is based. These are the two laws known by the name of Kirchoff—one dealing with voltage, the other with current. As readers are no doubt aware, the Kirchoff voltage law states that the sum of voltages with due regard to sign around any closed circuit is zero: the Kirchoff current law states that the sum of currents flowing into and out of a current junction is zero.*

The Kirchoff voltage law is based on the principle of conservation of energy while the Kirchoff current law is based on the principle of conservation of electric charge. *Both* these conservation laws must be satisfied *simultaneously* within the closed system, and in order to represent the system we must make a wise choice of *two* variables, i.e. voltage *and* current in the case of an electric circuit. Readers should refer back to the earlier remarks on this near the beginning of Chapter 5.

There is no actual "junction" in Fig. 12.13, the same current flows through each of the four components; since the components are connected in series, the Kirchoff voltage law, which is evaluated by going around a closed loop, is applicable; the application of the Kirchoff current law in this particular case yields only trivial relationships. When there are parallel or shunt paths in a circuit diagram, it would be necessary to apply also the Kirchoff current law in building up our set of equations (i.e., our mathematical model).*

Historically the idea of a circuit diagram as a mathematical model preceded by many decades the idea of a block diagram as a mathematical model. This is not surprising since the circuit diagram for electric circuits follows so readily from the wiring diagram showing actual connections between electrical components of a system. The viewpoint that has just been put, may be summed up in a general way which is applicable not only to electric circuits but also to mechanical, thermal and other networks. Instead of speaking about current and voltage (which are applicable in electric circuits only) we make use of the terms "through-variable" and "across-variable". A tabulation of the quantities which are "through-" and "across-variables" for a few kinds of circuits is shown in Table 12.2.

TABLE 12.2

Kind of Circuit	The Through-variable is	The Across-variable is
Electric	Current	Voltage
Mechanical	Force	Velocity
Thermal	Heat Flow	Temperature Difference
Magnetic	Flux	Magneto-motive-force
Hydraulic	Velocity	Pressure Difference
Acoustic	Acoustic Velocity	Acoustic Pressure
Acoustic (inverse)	Acoustic Pressure	Acoustic Velocity

* Flow out of a junction is reckoned to be of opposite sign to flow into the function.

* Readers will learn later from more advanced textbooks that there are two alternative systematic ways of writing down a set of circuit equations, one set being known as the node equations (a node is a circuit diagram junction point) and the other as the loop equations (a loop is any closed path on the circuit diagram). A systematic scheme is desirable when we have to deal with complicated circuits in order to reduce labour and the possibility of making errors. Similar comments to these may be made for circuit theory applied to other dynamical systems, and all the well-developed theory used for electric circuits may be applied to other kinds of circuits. (See Table 12.2).

Please note that in certain textbooks inverse or so-called dual analogies to the ones listed above may be encountered. In such cases not only are the variables interchanged, but the circuit parameters must be re-specified, the circuit configuration looks totally different, and the actual terminals and connections have to be thought of in quite different ways. An example is the so-called "classical" mechanical circuit analogue where force corresponds to voltage. The mechanical circuit analogy shown in the table above is more useful and is sometimes called the Firestone analogue.

The essence of linear circuit theory lies in the specification of three passive circuit components. These would be capacitance, resistance, inductance (C, R and L) for an electric circuit; mass, damping coefficient, spring-constant (M, B and K) for a mechanical circuit. Each of these three is the coefficient (in some cases, one over the coefficient concerned) of a particular term in the integro-differential equation modelling the electric or mechanical system.

Taking these in turn,

C or M refer to the differential term in the model,

R or B refer to the undifferentiated term in the model, and

L or K refer to the integral term in the model.

This is illustrated in Table 12.3.

TABLE 12.3

ELECTRIC		MECHANICAL	
Parameter	Definition	Definition	Parameter
C	$i = C\frac{de}{dt}$	$f = M\frac{dv}{dt}$	M
R	$i = \frac{1}{R}e$	$f = Bv$	B
L	$i = \frac{1}{L}\int e\,dt$	$f = K\int v\,dt$	K

In looking at the above table we see that four of the parameters are directly the coefficients concerned. Two others form the coefficient by taking $\frac{1}{R}$ and $\frac{1}{L}$.

The action of four of these ideal components (C, M, and L, K) is to store energy. Thus energy may flow either into the component from the circuit or vice versa. The action of the other two ideal components (R,B) is to dissipate energy (usually into heat) whence it is no longer available to flow back into the circuit.

In addition to the *passive* components, a number of *active* components may be specified. An active component in a circuit corresponds to a forcing function (i.e., a driving source) in the mathematical model. Active devices such as transistors or vacuum tubes may be modelled in the circuit form* by making use of controlled sources; these are forcing functions which are themselves dependent on some other variable in the system.

The controlled source is also useful for modelling a transducer or transformer. A transformer connects two separate sections of an electric circuit by a shared magnetic coupling. A transducer will actually transform energy (usually in both directions) from one form into another.

* This procedure is sometimes described as using "an equivalent circuit" for the actual device. A particular device could thus be modelled by a variety of equivalent circuits, depending on what sort of problem we are interested in solving, or in what sort of signal we wish to apply to the system. See Lynch and Truxal (1961) at the end of this chapter.

Passive components and ideal sources will each have two terminals and one pair of variables (a through-variable and an across-variable). See Fig. 12.14(a).

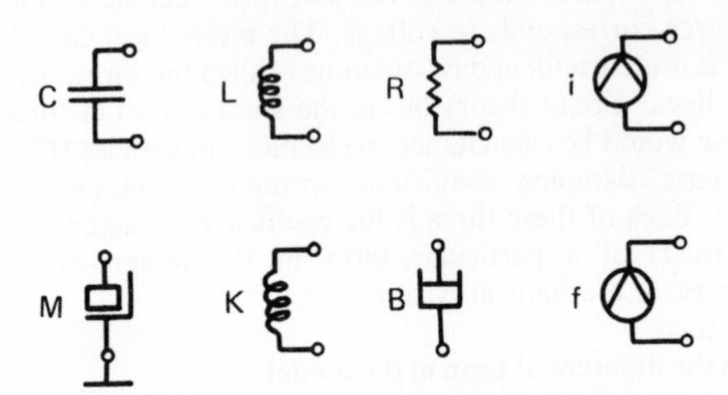

FIG. 12.14(a) Ideal Two-Terminal Components for Electrical (upper) and Mechanical (lower) Circuit Diagrams.

Transformers and transducers will usually have four terminals and two pairs of variables. See Fig. 12.14(b).

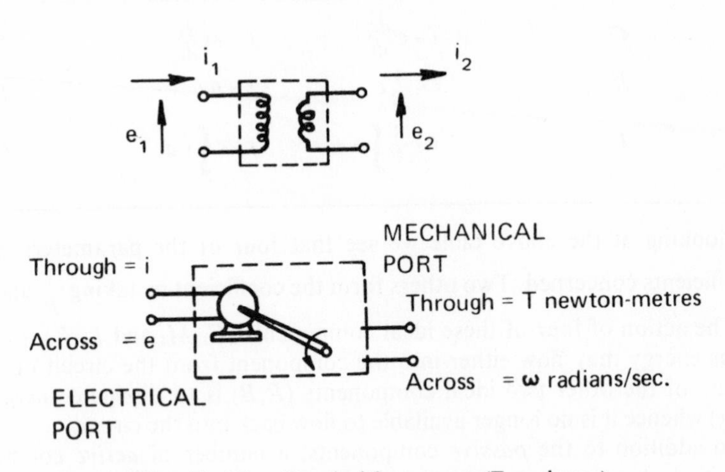

FIG. 12.14(b) Ideal Four-Terminal Components (Transducers).

To complete this brief section on linear circuit theory let us note that a differential equation of any order can always be manipulated into a set of simultaneous differential equations of lower order (e.g., first or second order), either by direct algebraic manipulation or by introducing subsidiary variables.

Correspondingly a circuit, by building up *L-C-R* meshes or loops (each of which is second order) or by using *L-R* or *C-R* meshes or loops (each of which is first order), can thus represent or model the dynamics of any lumped system no matter how complicated its interconnections.

We will go on in Chapter 14 to say something about distributed systems, i.e., those systems in which it is not permissible to "lump" the energy storage or energy dissipation effects into separable and discrete circuit parameters.

E. HIGHLIGHTS OF CHAPTER 12

(i) Methods of analysing linear systems have been developed using transformed equations and variables.

(ii) Two such methods involve the use of phasor algebra or the Laplace transform. Both these schemes make use of complex quantities.

(iii) The methods are useful because they replace more complicated operations by simple multiplication or division.

(iv) Associated with these methods is the idea of the block diagram as a graphical model of a dynamic system.

(v) Also associated with these methods is the idea of the circuit diagram as a graphical model of a dynamic system.

(vi) A circuit diagram may be constructed for any dynamic system possessing no more than two (coupled) mechanisms for energy storage and one mechanism for energy dissipation.

(vii) Circuit analogies are usually built up in terms of electric circuits, probably because electric circuit ideas were developed some decades prior to the ideas of mechanical, thermal, etc. circuits.

(viii) Either block diagram or circuit diagram models may be solved for transient or steady state responses.

(ix) The transient response is the part represented by the Complementary Function solution of the differential equation for the system.

(x) The steady state response is the part represented by the Particular Integral solution of the differential equation for the system.

(xi) The total response is obtained by adding together the transient and steady state responses.

F. EXERCISES

1. Two springs are available having spring constants K_1 and K_2 newtons per metre. Prove that when they are connected in parallel they may be replaced by a single spring with constant $(K_1 + K_2)$; or if in series, then by a single spring of constant
$$\frac{1}{1/K_1+1/K_2}$$

2. An input signal 30 sin 314t volts is applied to a black box whose transfer admittance expressed as a phasor quantity is magnitude 0.02 mhos, phase shift +0.5 radians. Determine the output current of the black box, in magnitude and phase. Write down the algebraic expression for the output signal.
HINT: Current (amps = voltage × admittance (mhos)

3. The input signal to a black box is $f(t) = 2e^{-4t}$ amperes. The transfer function of the black box is
$$\frac{10}{(s+2)} \text{ ohms}$$
Determine the Laplace Transform of the output signal. Write down the algebraic expression for output current of the black box expressed as a function of time.
(Assume that by reference to a table of Laplace Transforms you have discovered that
(a) the transform of e^{-at} is $\frac{1}{(s+a)}$
and (b) the inverse transform of $\frac{1}{(s+a)}$ is e^{-at}
HINT: If you encounter $\frac{1}{(s+a)(s+b)}$ expand it into partial fractions $\frac{A}{(s+a)} + \frac{B}{(s+b)}$.
The inverse transforms of each term in turn may then be written down.

4. Repeat Exercise (3) from Chapter 5. Assign symbols to the variables, write equations and try to manipulate these into the same form as those representing Fig. 12.10. What are the units in which the various transfer functions should be measured?

G. ENDNOTES

(1) Both $E \cos \omega t$ (where E and ω are constants) and $E \sin \omega t$ are described as sinusoids in the variable t, or sinusoidal in time. So also is the more general expression $E \cos (\omega t + \phi)$ called a sinusoid where ϕ is another constant known as the (initial, i.e., $t = 0$) phase angle. E is amplitude while ω is angular frequency in radians per second if t is measured in seconds. Also $\omega = 2\pi f$ where f is frequency in hertz or cycles per second.

(2) A phasor is a complex number usually expressed in polar form where the magnitude represents amplitude (base-line to peak) of a sinusoid and the angle represents the phase angle of the sinusoid. In the case of the phasor system function G of Fig. 12.1 the magnitude represents the ratio $\frac{E_o}{E_i}$, while the angle represents the shift in phase angle from (Input) to (Output). In Fig. 12.1 this shift in phase angle is $(+\phi)$ radians. The output is obtained from the input by use of the two rules

MAGNITUDE OF E_o = (MAGNITUDE OF G) times (MAGNITUDE OF E_i).

OUTPUT PHASE ANGLE = INPUT PHASE ANGLE + SYSTEM FUNCTION PHASE ANGLE.

(3) A damped sinusoid in the variable t (i.e., time) is a sinusoid whose amplitude decreases exponentially with time. It is described by $Ae^{\sigma t} \cos \omega t$. By a fairly simple rearrangement it may also be described by the expression Ae^{-st} where $s = (\sigma + j\omega)$ is called the complex frequency of the signal. Compare this with the artifice of representing $\cos \omega t$ by $ej^{\omega t}$. For a damped sinusoid σ must be negative. Positive values of σ will be interpreted as exponential growth of the signal. Here $j = \sqrt{-1}$.

(4) The functions of time, e.g., $f_o(t)$, are sometimes called originals. The corresponding transforms which will be functions of complex frequency, e.g., $F_o(s)$, are sometimes called images. The notation is commonly to employ lower case letters for time functions and the corresponding capital letter for transformed functions. Readers who are not aware of the lower case and capital letter symbols of the Greek alphabet should look up (and learn) such a list in the library.

SUGGESTED READING MATERIAL

LYNCH, W. A., and TRUXAL, J. G., *Introductory System Analysis*. McGraw-Hill, New York, 1961. (Preface; Chapters 1 and 2; highlights out of Chapter 3 and 4).

SEELY, S., *Electromechanical Energy Conversion*. First chapter only. McGraw-Hill, 1962.

CHAPTER 13

Networks of Building Blocks: Static Systems

R. M. Huey

"They were pursuing science for the sake of knowledge itself, and not for utilitarian applications. This is confirmed by the course of historical development itself. For nearly all the requisites both of comfort and social refinement had been secured before the quest for this form of enlightenment began."

Aristotle.

A. STATIC SYSTEMS

In the preceding chapter we outlined the theory of mathematical modelling and some of the techniques by which the behaviour of *dynamic* systems may be investigated. Corresponding problems for *static* systems are usually a good deal simpler and it is easy to see why. A static problem is simply one part of the more complicated dynamic problem, usually indeed the easier part. For an example of this, let us refer back to Chapter 5 where the concept of *signals* was introduced in Section I. It was stated there that the *complete* solution of a differential equation could be made up from adding together *two partial solutions,* one known as the complementary function and the other as the particular integral. We can regard the particular integral as representing the static portion of the solution, although there might be some argument about this simple cataloguing, in special cases. Of course it doesn't really matter a great deal what names we use to catalogue mathematical models and the results of operating on these models, *provided* that we ourselves understand what we are doing and provided, too, that the people with whom we are communicating understand what we are doing. We can choose for ourselves what conventions we will use in our descriptions, or even make up new conventions for ourselves. It is important however to let it be known what conventions we are using, and also it is important not to change horses in midstream unless we clearly say so.*

Let us therefore define what is meant by the phrase "static system" in this chapter. Briefly we mean one of three things:

(i) a system where the mathematical model is a set of equations where derivatives and integrals with respect to time, do not appear,

OR (ii) a system where we are interested only in evaluating the steady state response,

OR (iii) a system where actions proceed which are dependent on causes other than the regular passage of time—except for the fact that certain of these actions may need to occur sequentially.

Graphical representations such as block diagrams, signal flow graphs (see Chapter 5), computer or logical flow charts (see Chapter 18) or network diagrams of other

* Readers are invited to see whether they can detect any examples of this fault in this book and, if so, to write to the editors.

kinds, can be very useful as an aid to understanding both the static and dynamic behaviour systems.

B. BLOCK DIAGRAMS

When we are dealing with the dynamics of systems the *meaning* of block diagrams or signal flow graphs (i.e., the *conventions* we are using) are pretty well understood. These basic meanings have been outlined in Chapters 5 and 12 of this book with the warning that the rules ("conventions"), although very simple, must not be disobeyed (or changed in midstream). The same sort of block diagrams may be employed meaningfully to model static systems, if we so wish. However, when they are dealing with the statics of systems some authors and practitioners tend to use graphical representations with a great variety of underlying conventions not always stated, and sometimes even contradictory.

Basically there are two choices for the convention on which a satisfactory graphical representation of a system may be based

(i) that two operations in sequence on the diagram represent a multiplicative process while two operations in parallel on the diagram represent an additive process, or

(ii) vice versa.

In the example shown in Figs 5.6 and 5.7, three system functions in series are multiplied together to give the overall system function. In the next example in Fig. 5.9 two system functions shown in parallel are added together to give the overall system function. Clearly then the scheme of block diagrams, if it is used in accordance with the conventions set out in Chapter 5, represents a graphical system representation of the first kind.

In the next section we will give an example of a much used graphical system representation of the second kind.

C. TASK NETWORKS AND TIME MODELS

In a task network and time model such as the PERT* scheme uses, events or tasks such as "completion of layout drawings", "completion of building frame", "completion of computer calculations" are represented as closed boxes or circles on a diagram. For convenience each one is labelled with an appropriate alphanumeric code name which will be used later on as an identifier in a computer analysis. Each circle represents the time at which the designated task will be planned or estimated to be completed, although these exact times are not known before the graph is constructed—except in the case of the desired completion date of the *last* task in the whole project. Arrows or connecting lines are then drawn in reverse from the final task to each other task, which must immediately precede the final one, and so on until each task is completely identified with its predecessors.

Arrowheads indicate the sequence of events and each line is labelled† with the time which the next succeeding task should consume in that particular sequence of operations. All this needs is a good deal of care and judgment, but the very fact that the network is based on sequential tasks is a great help in making sure that no tasks are omitted from the planning.

In general there will be a multiplicity of directed paths on the diagram by which one could trace out a route from the first or initial event to the final or overall completion

* PERT is an acronym for Program Evaluation Review Technique developed in 1956 by a firm of consultants in the U.S.A. to control the Polaris submarine project for time and cost.

† In an actual PERT diagram, the lines would usually be labelled with optimistic and pessimistic estimates as well as the likely estimate, thus enabling planning against catastrophic performances as well as for likely completion dates.

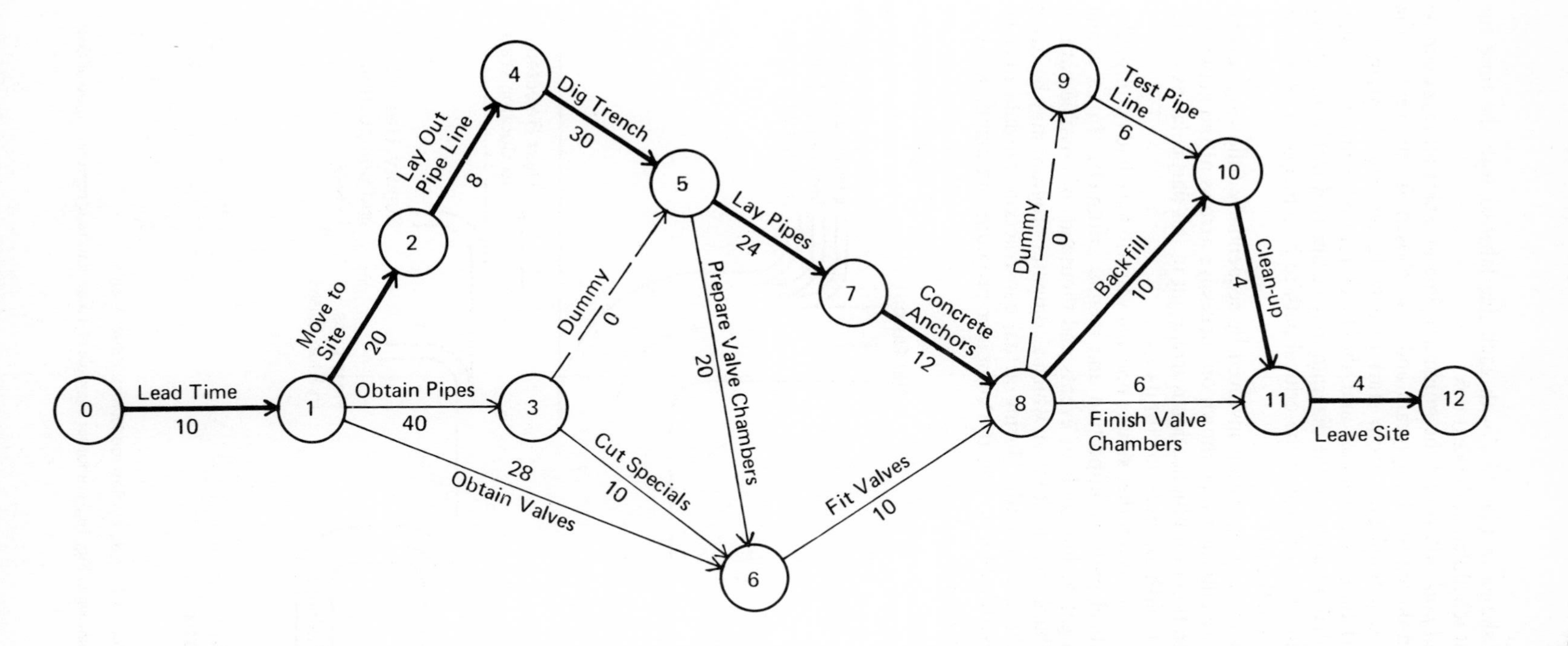

FIG. 13.1 Time Network Diagram for a Pipe-Line Construction Job. The diagram has been adapted from the example presented by J. M. Antill and R. W. Woodhead in their book *Critical Path Methods in Construction Practice*, John Wiley, New York, 1965 (see Chapter 4). In actual practice the same diagram would be utilised to help control the cost of the project, remembering that the cost of any one operation is related to the speed of execution. For example, the trench digging may be expedited by hiring an additional trench digging machine, at an increased cost. A dummy arrow indicates sequential dependency with zero time, e.g., "lay pipes" is dependent on "obtain-pipes" having occurred.

event. An example is shown in Fig. 13.1 with each line labelled with the time for completion of the event scheduled in the next box.

The so-called *critical path* will reveal the minimum time in which all tasks can be completed. It is shown dotted on the diagram and is so chosen as representing the most time-consuming sequential path from start to completion. The diagram pinpoints which estimated times are *critical*, and should therefore be more carefully scrutinised and controlled. In addition the diagram may be updated with actual times taken, and this will allow any change in the critical path to be anticipated as well as monitoring the progress of the job.

Obviously the mathematical operation involved by sequentially following several lines, is *addition* (i.e., we add the times shown on successive arrows of the network). The mathematical operation of multiplication is irrelevant in this particular example, so we do not need to establish a convention for its use.

The example we have used is a rather simple one and you will have had no difficulty in deciding on the critical path by inspection and mental arithmetic. In an actual project where there might be hundreds of events and thousands of possible paths it would be necessary to have an algorithm on which a computer program may be based in order to discover the critical path. During actual construction the data may be updated using actual performances, and the computer program run again, so as to monitor the progress of the job.

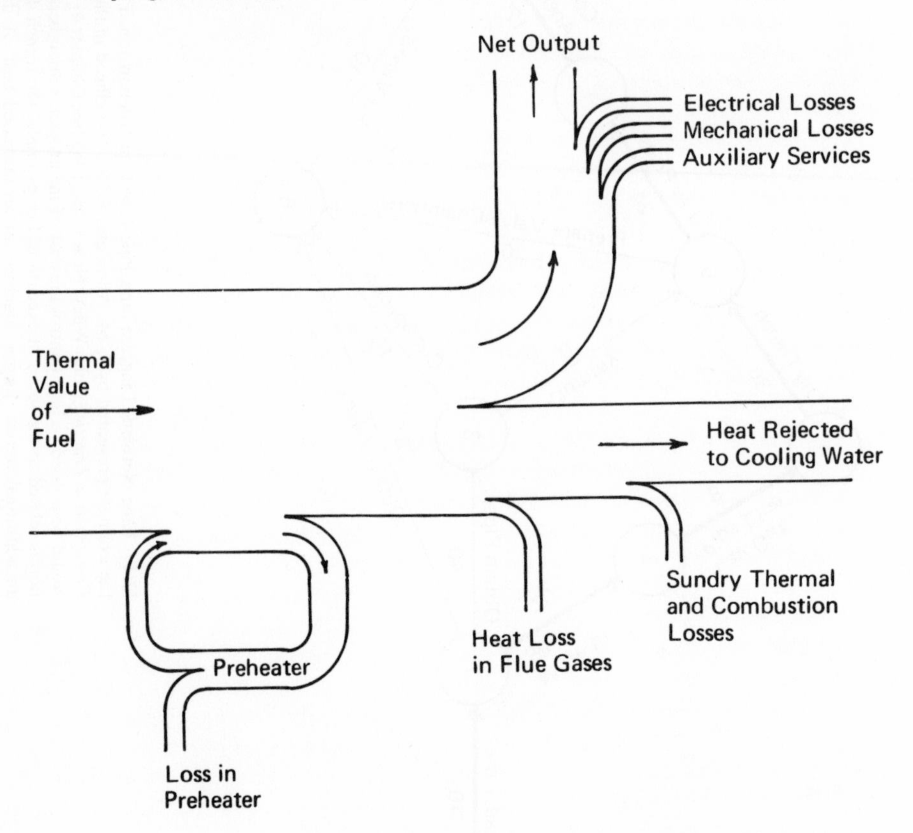

FIG. 13.2 Energy-flow model for a Power Station.

The current flow model shown in Fig. 16.5 is another example of a diagrammatic representation of this sort.

Thermal-energy flow models and cost-flow models are diagrammatic representations which are clearly based on the mathematical operation of addition (or subtraction) and, like the last example, the multiplication process is irrelevant to any quantity of interest to us. Figure 13.2 is an example of such an energy-flow diagram.

D. EVALUATION OF THE PERFORMANCE OF THE MATHEMATICAL MODEL

Let us assume that the investigator is well acquainted with the mathematics needed for solving the equations chosen to model both the static and dynamic systems in which he is interested. In the first place, this will mean an acquaintance with algebra, calculus both differential and integral, arithmetic and programming of a computer to carry out the arithmetic of the mathematical operations. If he is concerned with dynamic systems he may also need to be acquainted with complex variables and differential equations, and some or all of the short-cut methods of solving them (e.g., phasor algebra, Laplace Transforms, etc.) and the relevant programming ideas. If he is concerned with distributed dynamic systems he will need to understand partial differential equations, vector analysis, field theory and the related aspects of digital computation.

However it will not be much use to him even if he can work out answers to all his particular problems, unless he can interpret the meaning and value of his answers. For example, is *this* figure good? is *that* figure bad? Is that behaviour *optimum*? If not, do we really *wish* to *seek* for the optimum? What does optimum *really mean*? How much will it *cost*? How long will it *take*? Is it really *worth* it? Can we do better without over-running the *deadline*?

A complete hierarchy of systems engineering would include methods enabling us to formulate answers to these and similar questions. Methods adapted to this purpose do exist and indeed may be classified under headings which describe the mathematical model chosen to form the basis of the method. We will be content in this chapter in giving an example based on a linear algebraic equation, while hinting only at the existence of more sophisticated models which may include non-linear terms and mathematical operations other than addition or subtraction.

We must choose first of all a quantity which may be conserved, and for which the operations of addition and subtraction are meaningful. One such quantity is time, but we have already given an example of the task scheduling network, so let us choose another quantity. A possibility is money, expressed say in dollars. We will give two examples, the first being a decision as to which car one should purchase. We will assign values in dollars (spread over an operating life of two years) to the various performance and comfort capabilities of three alternative makes, *adding* together these quantities. We will then set out the initial and operating costs (for the same period of two years) and subtract these items from the first total.

We will then choose* the algorithm for decision as to whether A, B or C is the best buy, so that our decision will go to the one which shows the maximum *surplus* of benefit or value above the cost. Without going into details of how to estimate all the values and costs, suppose that we came up with the results set out in Table 13.1.

Clearly there is not much to choose between A and B: a small change in one of the estimates for either value or cost could change our decision. The algorithm (or system) is said to be *sensitive* with respect to the quantities we write into the problem.

* An alternative algorithm might be that we choose the one with the maximum *ratio* of value to cost. If we so elected, we would arrive at a different choice in the example shown above. The latter algorithm might make more sense if we did not have a large bank balance. Try to figure out why we make this statement.

Table 13.1

		Vehicle A	Vehicle B	Vehicle C
COSTS $	Initial	4000	3000	2000
	Two years operating	2000	1800	1500
	Total	6000	4800	3500
VALUES $ (BENEFITS)	Acceleration	1200	800	500
	Safety	1000	900	500
	Comfort, appearance	1500	1200	400
	Reliability	1500	1500	1200
	Ultimate Trade-in	2000	1500	800
	Total	7200	5900	3400
SURPLUS of VALUE over COST $		1200	1100	−100
RATIO of VALUE to COST		1.2	1.23	0.97

Dynamic systems are also said to be *sensitive* when their behaviour can be changed appreciably by a small change in one parameter. For either static or dynamic systems the sensitivity of a property S with respect to a parameter R is measured by the value of the differential $\frac{dS}{dR}$. If we can find or derive a mathematical expression linking S and R we can evaluate $\frac{dS}{dR}$. Even if the expression contains other variables we can still evaluate the partial differential $\frac{\partial S}{\partial R}$ on the assumption that all other independent parameters are held constant and this will give us a measure of the sensitivity of S with respect to changes in R.

In some cases we may wish to make the sensitivity small, in other cases it may be desirable to make the sensitivity large. In the last example we saw that the decision as to whether we should choose vehicle A or vehicle B was quite sensitive to small changes in our estimation of the values associated with various performance attributes. Since there is considerable uncertainty about how we might decide to place a dollar value on practically all the benefits attached to owning the car, we cannot be too sure that we have made the correct decision. On the other hand the rejection of vehicle C is quite clear cut—the demonstrable inferiority of vehicle C is clearly not sensitive to small changes in the estimated values.

A more specific engineering example might concern the design of a piece of electronic measuring equipment. Resistors with an accuracy of 0.1 or 0.2 per cent are very expensive. Resistors with an accuracy of 2 or 5 per cent are very cheap. High gain amplification using several transistors or an integrated circuit can be attained very cheaply. It will be less expensive overall to use a circuit with a much higher gain than is needed and to use a small number of components with a high accuracy placed judiciously in a feedback circuit (see Chapter 12, C for an instructive illustration of this) so that the *sensitivity* of the whole instrument is *very small* with respect to all component values *except* for the *chosen few* components. The alternative of using a single transistor without feedback plus expensive arrangements to stabilise temperature and operating voltages would, most likely, be less attractive. The principle applies in many ways that we can purposely make most of the sensitivities in a system quite small so that we don't have to worry too much about the values of the parameters involved —at least not to the degree where they are *expensively* accurate.

In other cases, as in a critical path scheduling task, rather than designing for small sensitivities we are concerned with discovering (or identifying) the *critical path* along which the variation in overall project time is dependent one-to-one on any variations in time taken for critical sub-tasks.

The reverse case can also crop up where we wish to maximise one sensitivity in a system while keeping others small. This is the case when designing a measuring instrument or sensor which must have an accurate calibrated response for the desired measurement, but without any appreciable response to other, undesired, quantities or signals.

When a number of sub-systems are combined together in a more elaborate network, the number of undesired variables and the number of parameters in the whole system (compared to the number of desired variables or signals) increases considerably. The number of terms of the nature $\frac{\partial S}{\partial R}$ where S is any variable and R is any parameter increases dramatically, as also does the number of transfer functions relating S_j to S_m where these two symbols stand for any two variables. It is important therefore in coupling together separate blocks or sub-systems to make sure (if this is possible) that *only desired signals* can flow from one block to another. This is a rarely attainable ideal, and in practice one is concerned to minimise rather than to eliminate completely the flow of undesired signals from one sub-system to another. A different way of stating the same philosophy is to say that we usually try to maximise the flow of the desired signals between various sub-systems, relative to all other (i.e., undesired signals). Clearly, in order to carry out such design procedures we must have a good understanding of (and mathematical models to describe) *both* the *desired* and the *undesired* actions within a system.

E. EXERCISES

1. Read the quotation from Aristotle at the beginning of this chapter. Discuss with your fellow students the two questions:
   a. Did Aristotle believe that technology should have reached a steady state in his day and that utilitarian applications were unnecessary?
   b. Do you think this may have been the beginning of the "false" division into pure and applied sciences, with the connotation that one was "better" than the other?
2. In the example illustrated in Fig. 13.1 assign "crash times" for each operation, i.e., the time in which that operation could be carried out if it were permissible to spend extra money. Insert these crash times on the diagram and see whether the critical path will alter. Try other times, slower as well as faster, and the effect on the critical path. In each case calculate the time to completion of the whole project.

SUGGESTED READING MATERIAL

ANTILL, J. M. and WOODHEAD, R. W. *Critical Path Methods in Construction Practice,* John Wiley and Sons, New York, 1965.

CHESTNUT, H. *Systems Engineering Tools*, John Wiley and Sons, New York, 1965.

CHESTNUT, H. *Systems Engineering Methods*. John Wiley and Sons, New York, 1967.

CHAPTER 14

Distributed Systems and Fields

R. M. Huey

". . . whose virtues are the definitions of the analytic mind . . ."

"The People", W. B. Yeats, 1919.

A. INTRODUCTION

In this chapter it will be pointed out that the notions of systems theory may be extended to what are generally known as distributed systems. In particular, we wish to emphasise the point that circuit theory can be used to tackle a large number of problems in distributed systems. However, when circuit theory fails (and even in some other special cases) we may have to call on the more powerful (and often more difficult) methods of field theory. A few of these methods will be named.

By a distributed system is meant a system in which the properties of the system are distributed in such a way that they cannot be lumped into separate packages with *each* package representing the effect of only one particular kind of energy storage (or energy dissipation).

If, instead of considering the energy within a dynamic system (i.e., in order to classify the difference between a distributed and a lumped system) we had focused our attention on the way in which a suddenly applied forcing function (e.g., in an electrical system, the closing of a switch) affects the various parts of the system, we would have come up with a different answer. This answer would have run something like this—in a lumped system there is some sort of instantaneous response (although perhaps the rate of rise is limited) even in the most remote part of the network; in a distributed system we must allow in addition a certain time (the so-called propagation time) for the initial disturbance to spread through the system; at a remote point nothing at all happens until after the propagation time has elapsed. In a lumped system the propagation time is reckoned as zero.

If, instead of considering energy interchange or ideas such as propagation time, we had focused our attention on the mathematical models for the two kinds of systems we would have seen a very clear and sharp dividing line.

Lumped systems are described by ordinary differential equations (or by sets of simultaneous such equations in more complicated cases).

Distributed systems are described by partial differential equations.[1]

Such problems are naturally somewhat more difficult to solve. Since many of the readers of this book have not yet dealt with the mathematical ideas and manipulations, we will proceed at first by thinking about some electric circuit theory problems which are "distributed" in one dimension only. Then it will be pointed out that there exist many distributed systems which are non-electrical in character but which may be modelled by the same group of partial differential equations.

A brief reference will be made at the end of the chapter to field theory problems where variations may occur in either two or three spatial dimensions.

B. THE UNIFORM TRANSMISSION LINE

One of the most common examples of the uniform electric transmission line is two long parallel circular metallic wires which have therefore a uniform spacing and diameter.

Such a system may store energy in either an electric field or in a magnetic field* and may dissipate energy due to the imperfect electrical conductivity and due also to the imperfect insulation afforded by the medium in which the wires are immersed. This specification leads us to the two partial differential equations

$$\frac{\partial v}{\partial x} = -Ri - L\frac{\partial i}{\partial t} \tag{14.1}$$

$$\frac{\partial i}{\partial x} = -Gv - C\frac{\partial v}{\partial t} \tag{14.2}$$

where the coefficients represent the properties per unit length suggested in Fig. 14.1, which is a circuit model of an infinitesimal length Δx of the line. Note that the coefficient G represents a conductance per unit length (mhos per metre) rather than a resistance.

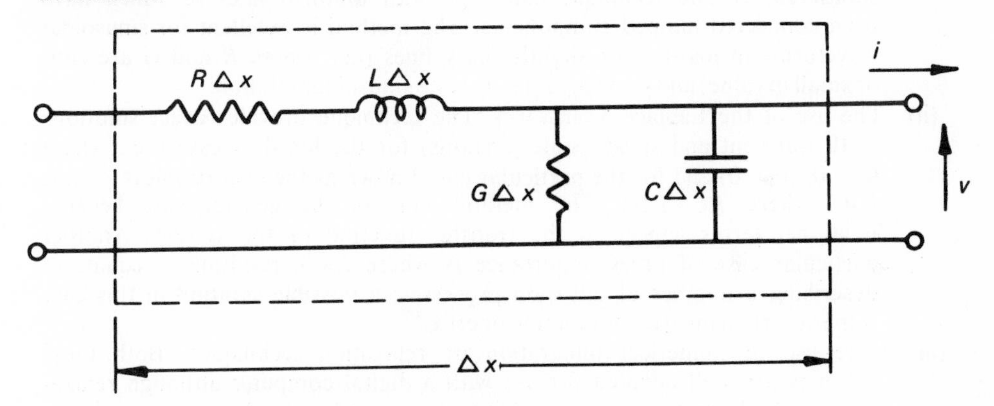

FIG. 14.1 Circuit Model of an Infinitesimal Length of Transmission Line.

C. THE TELEGRAPHER'S EQUATION

The above pair of equations may be manipulated (by choosing either to eliminate i or to eliminate v in the pair (14.1), (14.2)) into

$$\frac{\partial^2 v}{\partial x^2} = RGv + (RC + LG)\frac{\partial v}{\partial t} + LC\frac{\partial^2 v}{\partial t^2} \tag{14.3}$$

or into an identical equation in which v is replaced by i.

In the form (14.3) the telegraphers equation is a fairly general form of a class of equation known as wave equations. Space variations are written in one group, time variations in another group of terms (the L.H.S. and R.H.S. respectively in Equation (14.3). Also since the examples being considered are linear, all the coefficients in the equation will be constant quantities.

Interestingly the equation was named the telegrapher's equation since it was first formulated and solved by Lord Kelvin about a century ago in order to explain the limitations of signalling speed encountered on the early trans-Atlantic telegraph cables at that time.

* After reading more advanced texts it will be realised that the separation of energy storage into these two components is possible only in certain cases where we have uniform simple structures and/or particularly simple driving sources (forcing functions).

The equation is however of considerable importance elsewhere—it may be used as the mathematical model for many situations including—
(i) transmission of electric signals along a uniform line,
(ii) mechanical waves on stretched strings,
(iii) plane acoustic waves (gases, liquids, solids),
(iv) seismic waves (of other types besides the acoustic compression-rarefaction wave in a solid medium),
(v) heat conduction in solids,
(vi) mass diffusion (this applies to gases, liquids and solids),
(vii) charge diffusion in semi-conductors,
(viii) certain cases of the Schrodinger equation in quantum physics.

D. METHODS AVAILABLE FOR SOLUTION

Electrical engineers make use of a number of methods for solving the partial differential equations which they encounter and which may be classed as derivative from the telegrapher's equation. These methods include:

(i) The use of phasor algebra, with the addition of a graphical aid known as the Smith chart. The technique can cope with uniform lines to which have been connected lumped components. The method is excellent for sinusoidal waveforms in loss-less or slightly lossy lines (i.e., where R and G are zero or small in value) and provides a steady state sinusoidal solution.

(ii) The use of the Laplace Transform. The technique provides exact solutions (both transient and steady state portions) for the loss-less case (i.e., where $R = 0$, $G = 0$) and for the particular case known as the distortionless [2] line (i.e., where $RC = LG$). The method fails in the general case because irrational terms appear in the transfer function of the system. Another particular case of great importance is where $L = 0$, resulting in equations describing a number of diffusion processes; a possible solution in this case is the queerly named error function or $\operatorname{erf} x$. [3]

(iii) The use of numerical integration or relaxation techniques. Both these schemes are well adapted for use with a digital computer although relaxation methods may require excessive storage space in the computer if we concern ourselves with more than a few dozen points in the problem region.

(iv) The use of certain other mathematical procedures such as perturbation, the stationary phase principle, contour integration schemes other than Laplace Transforms, etc. Procedures such as these are not always specifically included in engineering undergraduate courses.

As the complexity and size of engineering systems increase, the value of other mathematical procedures will be recognized. Only occasionally will these procedures supplant previously used techniques for mathematical modelling. The scope of engineering will thus seem continuously to increase. One should not be discouraged by this. Most professional people find that it is not too hard to learn to live with this situation, i.e., continuous and rapid growth of concepts and ideas as well as hardware. Indeed it can be a strong source of intellectual stimulation and gratification.

E. AN HISTORICAL VIEWPOINT

Sometimes ideas and concepts permitting analysis and insight precede the construction and use of the hardware. However, it is a commonly held opinion that the invention often comes first and will be followed by appropriate analysis and mathematical models. For example may we quote some remarks made on the occasion of opening the fifteenth Brooklyn Polytechnic International Symposium in New York in 1967; a symposium which was devoted to Systems Theory.

Dr. Emmanuel Piave, Vice-President, IBM:

> "Industry will sort of roll its sleeves up and move ahead. It always does, whether it understands something or not. The market place will determine whether a thing works, whether it is viable or not."

Dr. Ernst Weber, President, Brooklyn Polytechnic Institute:

> "System Theory, particularly as it might be extrapolated to population problems, to economic and biological systems, and others, indicates the penetration of quantitative thought into areas to which engineering has already contributed such basic concepts as feedback, information content, realizability, modelling, etc. . . . The broadly ranging concepts and tools of System Theory may serve to accelerate the trends discernible now but still shrouded in early morning mist."

Dr. W. G. Shepherd, Vice President, Institute of Electrical and Electronic Engineers, N.Y.:

> "World War II alerted the technological world to the need for systems capability. The initial system efforts of that time were largely *ad hoc* and fragmented."

These extracts seem to be saying that analysis and insight follow construction and use. Let us give historical examples where the reverse has sometimes been true. For this, let us trace some developments in electrical transmission lines.

First of all, some dates:

About 1830 early attempts at telegraph transmission by people such as Gauss and Weber in Europe.

1844 the Morse telegraph in the U.S.

1851 undersea telegraph cable, England-France.

1866 undersea telegraph cable, trans-Atlantic.

1876 telephony, needing wider bandwidth capabilities in the transmission lines.

1886 invention of the polyphase induction motor by Tesla and the need (or the possibility?) for transmission of electric power over polyphase transmission lines leading to efficient central generating stations supplanting local generation.

1901 Marconi successfully transmits a radio signal across the Atlantic. Transmission lines needed to join the equipment to the aerials.

About 1940 Widespread use of hollow pipe waveguides when microwave sources of power became "commercially" available.

If one looks at these one may see successful telegraph lines antedating Kelvin's mathematical analysis of the telegrapher's equation (about 1866). However, one may also see the much greater success of later telegraph lines and telephone lines which had been "loaded" (i.e., with increased inductance) in order to increase the speed and frequency of operation.*

On the other hand Marconi's successful construction of radio stations came many years after the prediction (by Clerk Maxwell) and the demonstration (by Heinrich Hertz) of the existence of radio waves. Also we may note the theory of guided electromagnetic waves as available many years (early this century) before the application of waveguides in the early microwave radar systems of the Second World War.

* Readers should see what they can discover in the library about this. Among others look for mention of Kelvin (England), Bell (U.S.A.), Pupin (U.S.A.) concerning information transmission; look for mention of Tesla (U.S.A.), Siemens (Germany) and Ferranti (England) concerning power transmission.

This seems confusing unless we can understand that the application and development of modern technology proceeds rather in a leap-frog fashion, with advances in theory, discovery by trial and experiment, improvement by acquaintance and operating experience all contributing their share from time to time (and not necessarily in any fixed order nor in any fixed proportion) to the next developments.* The various phases mentioned here proceed, not necessarily in any predictable order but nevertheless each will contribute understandably to the later advances.

In field theory we are concerned with more complicated partial differential equations, which may be written in two or three spatial coordinates (or in one dimension, although we have discussed some special cases of this for Cartesian coordinates under the heading of the Telegrapher's Equation). Table 14.1 suggests that one way of looking at field theory problems is to classify the equations by the way in which variations with time appear in the equation—i.e., whether terms in $\frac{\partial^2}{\partial t^2}$ or $\frac{\partial}{\partial t}$ appear or not (in more complicated cases combinations of these terms may crop up).

TABLE 14.1

Class of Problem	Written in Variables	Name of Equation	Typical Application
STATIC problems	1, 2 or 3 space variables	Laplace and Poisson equations	High Voltage insulation.
DIFFUSION problems	1, 2 or 3 space variables and $\left(\frac{\partial}{\partial t}\right)$	The Diffusion equation	Diffusion of water through an earth dam.
WAVE problems Propagation of	1, 2 or 3 space variables and $\frac{\partial^2}{\partial t^2}$	The Wave equation (loss-less medium)	Radiation from aerials. Propagation of Sound underwater.

The actual equations tend to be rather complicated. For example the simplest one is the Laplace equation. Written in three spatial variables using spherical coordinates (r, θ, ψ), it is

$$r\frac{\partial^2\phi}{\partial r^2}+2\frac{\partial\phi}{\partial r}+\frac{1}{r}\frac{\partial^2\phi}{\partial\theta^2}+\frac{1}{r\tan\theta}\frac{\partial\phi}{\partial\theta}+\frac{1}{r\sin^2\theta}\frac{\partial^2\phi}{\partial\psi^2}=0$$

where ϕ is used to indicate a potential. A shorthand approach which is more generally

* Readers should think again of the analogy presented earlier by one of the other authors of this book, where the process of technical innovation was likened to the process of a child playing with its toys. One will have observed the actions of different children at play. Some occupy themselves with a busy series of random trials; others are more contemplative and make pictures in their minds before acting out the situations they have created in their imaginations; others again will seize on a theme, in which creation by imagination proceeds step by step, alternately with movements of the toys and differing involvements of the child itself with the patterns formed by the toys. Readers should use their imagination too to discern patterns of social behaviour created and carried on by adolescents and adults of their acquaintance.

applicable (i.e., to a variety of coordinate systems) is provided by a branch of applied mathematics known as vector analysis. Using the symbolism of vector analysis the last rather nasty looking equation reduces to $\nabla^2\phi = 0$ where the ∇^2 is known as the Laplacian operator [(4)].

If we are content with sinusoidal signals then the diffusion and wave equations may be reduced to a form known as the Helmholtz equation. Solutions to both the Laplace and the Helmholtz equations are tabulated in handbooks or in publications devoted to more individual problems.*

Certain special functions are also tabulated in handbooks for use in carrying through the solutions. These are associated with particular coordinate systems and you will certainly be familiar with the first one listed in Table 14.2.

TABLE 14.2

Coordinate Systems	Type of Special Function
Rectangular	Trigonometric (i.e., sin, cos)
Spherical	Legendre functions
Cylindrical (circular)	Bessel functions
Cylindrical (elliptic)	Mathieu functions
Cylindrical (parabolic)	Weber functions
Conical (elliptic)	Lamé functions

With these aids we may analyse quite a number of field problems, provided that the boundaries of the problem are not too complicated in shape. When the problem won't fit in with the above analytical methods (e.g., because the hardware doesn't fit onto one of the available and usable coordinate systems) it may be possible to go to numerical methods or to a method using the complex variable known as conformal mapping (often suitable for two-dimensional problems) or else to experimental methods (suitable for two- and three-dimensional problems). Experimental methods include the use of electrically conducting paper or electrolytically conducting liquids in a "tank".

G. THE MAXWELL EQUATIONS

Using the formalism of vector analysis (i.e., the notation, plus the ideas conveyed by the notation which will be encountered a little later in undergraduate courses) the fundamental actions of practically the whole of electromagnetism may be written down as a set of four simultaneous partial differential equations.

These are the simultaneous equations for electric and magnetic fields, $\bar{E}$ and $\bar{H}$.

Note ρ = charge density, ϵ = permittivity, μ = permeability

$$\text{Curl}\,\bar{E} = -\mu\frac{\partial\bar{H}}{\partial t} \qquad \text{div}\,\bar{E} = \rho/\epsilon$$

$$\text{Curl}\,\bar{H} = \sigma\bar{E} + \frac{\partial\bar{E}}{\partial t} \qquad \text{div}\,\bar{H} = 0$$

It was an outstanding feat to have written down in such a compact form such an enormously generalised set of laws.

Maxwell's equations form the basis of nearly every analytical attack on electromagnetic problems. Most textbooks on electromagnetic theory consider only

* Note: One-dimensional field problems may also be solved by a direct integration of the Laplace or Helmholtz equation.

rigid conductors, either stationary or moving. However, texts dealing with flexible and fluid conductors have also been written, an example being Shercliff (1965) referred to in the suggested further reading.

H. HIGHLIGHTS OF CHAPTER 14

(i) In distributed systems, the propagation time or travelling time of a signal is significant.
(ii) Distributed systems are described by partial differential equations.
(iii) The telegrapher's equation for a uniform transmission line is also a model for many other diffusion and wave processes.

I. EXERCISES

1. Derive Equation (14.3) from Equations (14.1), (14.2).
2. On page 337 of reference Ramo, Whinnery and van Duzer (1965), a distinction is made between a plane wave travelling in a "good" dielectric and in a good conductor. Try to build up the analogy between these cases and a uniform transmission line. To what ranges of value for R, L, G and C would these two cases correspond?
3. In this chapter is given a list of dates signifying steps in the successful technical development of electrical transmission lines. Make a similar list with approximate dates to show alternate steps of theory and practice contributing to the development of one or two other modern technologies.

J. ENDNOTES

1. Some of the readers of this book may not yet have encountered partial differentials. These operators are used to describe the behaviour of functions which depend on more than one independent variable. In the case of distributed systems we have to consider not only variation with time, but also variation with spatial position. Partial differentials are written using one form of the Greek small delta, i.e., $\frac{\partial}{\partial t}\,\frac{\partial^2}{\partial t^2}$ etc., whereas ordinary differentials are written with our normal letter "dee" such as $\frac{d}{dt}$ etc.

   Suppose we were concerned with variations in time and in one space coordinate, say x. We would be concerned with partial differentials with respect to t and partial differentials with respect to x. The notation and the meaning would be

   $\frac{\partial}{\partial t}$ meaning $\left.\frac{\partial}{\partial t}\right|_{x\,=\,\text{const.}}$ i.e., rate of time-variation when x is held constant,

   and $\frac{\partial}{\partial x}$ meaning $\left.\frac{\partial}{\partial x}\right|_{t\,=\,\text{const.}}$ i.e., rate of variation in x direction when t is held constant.

2. In the general case where $RC \neq LG$ all waveforms (except a few special cases such as sinusoids) will become distorted during transmission through a distributed system. Typically, rapidly changing portions of a waveform become spread out or dispersed in time (i.e., slowed down).
3. A brief historical account of the development of the method nowadays known as the Laplace Transform method is given by Carslaw and Jaeger, pp. 297-8. They point out for example (pp. 50-2) that there are at least seven types of solution for the simple diffusion equation $\frac{\partial^2 v}{\partial x^2} - \frac{1}{K}\frac{\partial v}{\partial t} = 0.$

   See reference at the end of this chapter.
4. Field theory problems are often expressed by using the operators and symbolism of vector analysis.

SUGGESTED READING MATERIAL

MOORE, R. K. *Wave and Diffusion Analogies*. McGraw-Hill, 1964. Chapters 1, 2, 8, 10.

CARSLAW, H. S. and JAEGER, J. C. *Conduction of Heat in Solids*. Oxford University Press, 1959. Pages 1-3, 50-2, 297-8.

MAGNUSSON, P. *Transmission Lines and Wave Propagation*. Allyn and Bacon, 1964. Preface, Introduction and some of Chapter 7, particularly the overview on p. 160.

RAMO, S. WHINNERY, J. R. and VAN DUZER, T. *Fields and Waves in Communication Electronics*. Wiley International, New York, 1965. Pages 330-7.

SHERCLIFF, J. A. *A Textbook of Magnethohydrodynamics*. Pergamon, 1965. Chapter 1 and a small portion of Chapter 2.

CHAPTER 15

Materials

L. W. Davies

"... We are such stuff
As dreams are made on, and our little life
Is rounded with a sleep."

Shakespeare, The Tempest.

A. INTRODUCTION

Shown in Fig. 15.1 is the conventional division of matter into the three states: solid, liquid and gas. Because our principal interest here is in electrical properties, we shall focus our attention on the solids. Nevertheless, it should be noted that electrical engineers do indeed make use of materials in the form of both liquids and gases; for example, liquid insulators may be used in the manufacture of transformers to increase the rate of heat removal (by convection) from copper conductors, while gases may be employed as insulators of improved voltage breakdown rating.

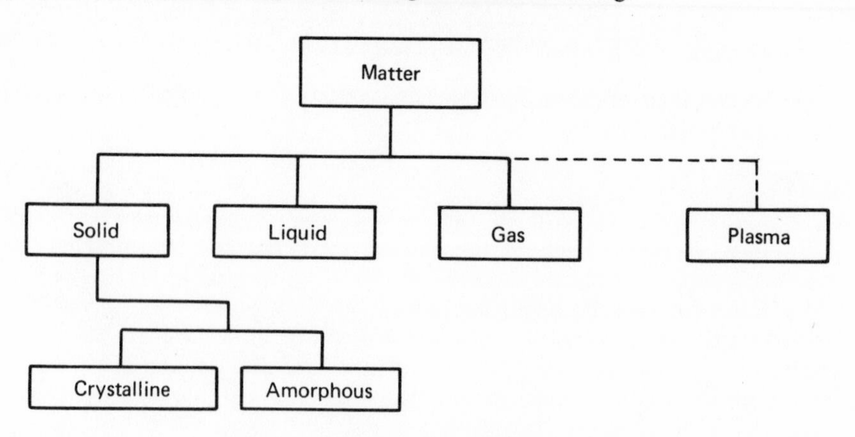

FIG. 15.1 The States of Matter.

A fourth "state" of matter—plasma—is also indicated in Fig. 15.1. The term was first used to describe a highly ionised gas, such as is found in fluorescent gas discharge lamps, whose properties differ markedly from those of a non-ionised gas because of large equal densities of positive and negative charge-carriers. The many interesting properties of gas plasmas find wide application in electrical engineering, as also to some extent do the properties of plasmas of electrons and holes found in semiconductors. Unfortunately, space precludes a detailed discussion of them.

If we make a microscopic examination of common salt (sodium chloride, NaCl) we find that the crystalline particles of the material tend to have similar cubic shapes, even though they may differ in size. If salt is dissolved in water and allowed to recrystallise, we again find small cubes, with flat faces. Even when larger crystals are broken up into fragments, the fragments also tend to have the same cubic shape. From all these observations it can be concluded that in sodium chloride the atoms of sodium and chlorine have a regular, periodic arrangement in space. This conclusion has been confirmed by observations of the diffraction of X-radiation by the regularly arranged planes of atoms in crystals of NaCl, and of many other materials. The X-ray observations give a value of 2.8 Å* for the interatomic spacing in NaCl.

Not all solids are crystalline, however. Common glass, for example, which is a mixture of oxides of silicon, sodium and other elements, solidifies in a form in which there is no long-range regularity in its structure. We call such materials amorphous. It is possible for some materials to exist in either form: for example crystals of quartz ($\alpha - SiO_2$) when melted and resolidified are found to have been transformed to the amorphous form known as (fused) silica. The chemical composition SiO_2 is retained, but the long-range order of the crystal has been destroyed in the amorphous glass.

B. CLASSIFICATION OF CRYSTALS

It is convenient for many purposes to classify crystals according to properties of particular interest. Crystallographers, for example, are primarily interested in the symmetry relations of the arrangement in space of the atoms or molecules making up the crystal, and classify accordingly. From the point of view of the electrical engineer, interested as he is primarily in the conductivity and other electrical properties of crystal, it is convenient to classify crystals according to the forces which bind the constituent atoms together. This brings us rather naturally to a consideration of the distribution throughout the crystal of electrons, for electrostatic forces are the ones principally concerned in forming bonds in a crystal, and in order that a current may flow in a crystal there must be some re-arrangement in the crystal of charge-carriers, i.e., of the electrons.

Let us proceed then to enumerate the various classes of bonding mechanism that have been observed in crystals.

1. Ionic Crystals

This is perhaps the simplest of the crystal binding mechanisms to visualise. Consider NaCl as a typical ionic crystal. The sodium atoms each have a single valence electron in the outermost electron "shell" of the atom, while chlorine atoms have an outer shell which is almost complete but lacking one electron. In solution in water, for example, there is a tendency for the sodium atoms to lose their valence electrons and become positively charged ions (NA^+) with a closed-shell structure, and for the chlorine atoms to take up an electron and become negatively charged ions (Cl^-), also with a closed-shell structure, i.e., with a charge distribution which is spherically symmetrical.

On crystallisation, there are forces of attraction between oppositely charged ions, and of repulsion between ions carrying like charges. The solid crystal NaCl forms with the structure shown in perspective in Fig. 15.2. There is a nett binding force holding the crystal together, because the Cl^- ion at the centre of the structure shown is subject to attractive forces between it and its 6 nearest-neighbour Na^+ ions, which outweigh the forces of repulsion between the central Cl^- ion and its 12 next-nearest-neighbour ions of like charge. We recall that the electrostatic forces of attraction, or repulsion, are inversely proportional to the square of the separation between the ions concerned.

*1 Å = 10^{-10} metres, is named the Ångstrom unit. Another widely used small unit of length is the micron (micro-metre). Note that 1 micron = 10^{-6} m. = 10^4Å.

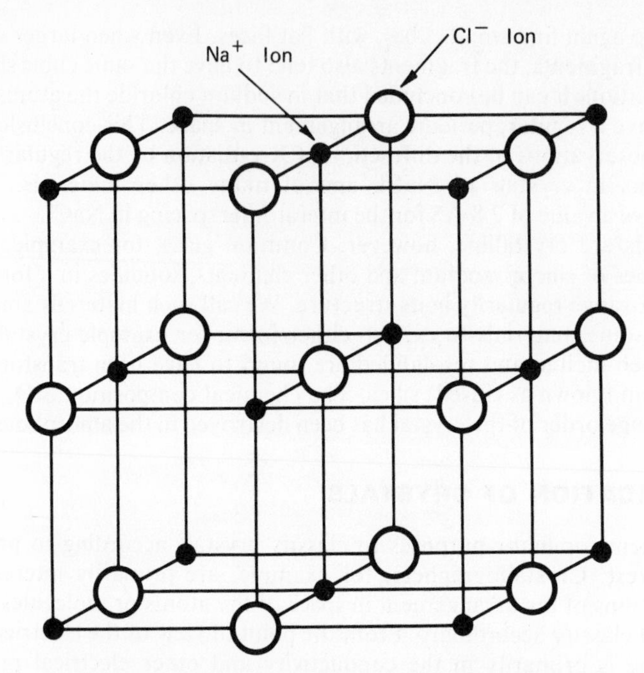

FIG. 15.2 Structure of a Crystal of Sodium Chloride (NaCl).

A feature of ionic crystals such as NaCl is that there is a very low probability of finding electrons in the space between the ions. Consequently ionic crystals are insulators, for there are no electrons available to transport charge through the crystal.

2. Metallic Crystals

Atoms of elements which form metals, of which sodium is a simple example, have one or more valence electrons which are readily detached from each atom as it moves into a growing metallic crystal on solidification. The resultant picture is one of positive ions, regularly arranged in space and "floating in a sea of electrons", as it has been picturesquely described. Once again the forces of repulsion between ions of like sign are outweighed by the attractive forces between ions and the near-uniform distribution of electrons between the ions, and there is a nett force of cohesion as a result. There is a number of possible arrangements in space of the ions, as exemplified by metals such as copper, zinc and chromium, to name three; a schematic diagram of a metal structure is shown in Fig. 15.3, with the electrons shown as a relatively uniformly distributed background of mobile charge.

It is clear from Fig. 15.3 that the electrons are no longer associated with particular atoms in a metal. There is a high probability that an electron will be found in the space between the ions of the crystal, in strong contrast to the situation in an ionic crystal. This contrast is directly related to the observed contrast in the electrical conductivities of metals and ionic crystals. The distribution of electrons throughout a metal, ready in position to transport electric current in the presence of an electric field, gives rise to conductivities of approximately 10^5 ohm^{-1} cm^{-1}. These are orders of magnitude larger than the conductivities of order 10^{-10} ohm^{-1} cm^{-1} encountered in insulating ionic crystals.

Because the detachment of the electrons from fixed ions in a metal, as shown in Fig. 15.3, is not due to the thermal vibrations of the ions about their mean position, the electrons remain free, down to very low temperatures. It follows that metals are good conductors at all temperatures.

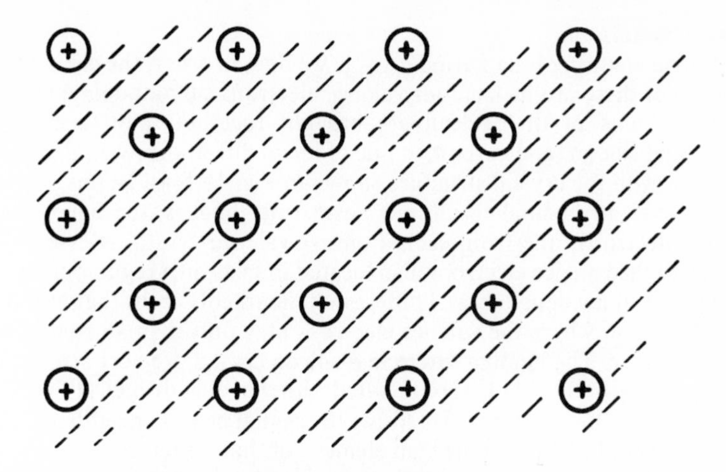

FIG. 15.3 Schematic Structure of a Metal.

3. Molecular Crystals

It is known that hydrogen gas consists of molecules H_2, each molecule being made up of two protons (the hydrogen nuclei) covalently bound together by two electrons. As the temperature of hydrogen is lowered at atmospheric pressure it first liquefies at 20° K,* and then solidifies as a crystal at 14° K. In the crystal structure, shown in Fig. 15.4 with the electron valence bonds represented by line pairs, it can be seen that the molecular structure has been retained in the solid. The binding forces are weak electrostatic forces between neighbouring molecules. Their weakness is indicated by the relatively small amounts of thermal energy required to melt the crystal at only 14° K.

From the schematic diagram of Fig. 15.4 it is clear that the electrons in crystalline hydrogen, a molecular crystal, are essentially located within each molecule and are therefore unable to move through the crystal under the influence of an externally applied electric field. Molecular crystals thus tend to be electrical insulators.

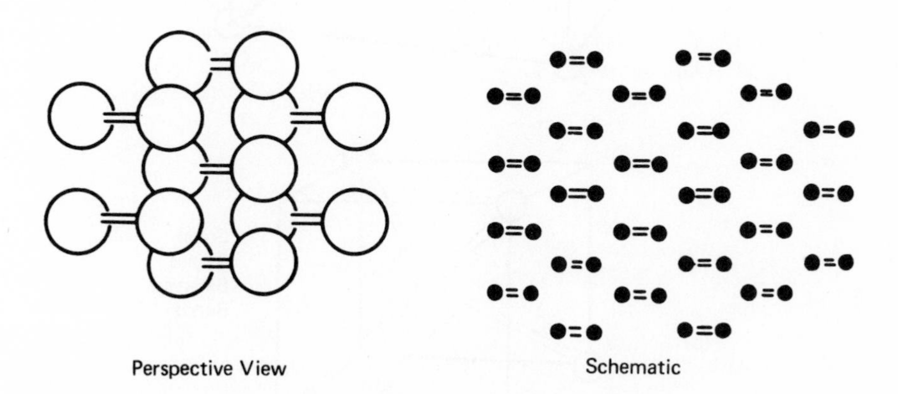

FIG. 15.4 The Structure of Solid Hydrogen.

* ° K signifies temperature measured on the Kelvin scale, i.e. temperature above the so-called absolute zero, which is approximately −273° Centigrade.

4. Valence Crystals

In the fourth column of the Periodic Table we find, in order, the elements C, Si, Ge, Sn, Pb. Each of these atoms has four valence electrons surrounding a closed electron shell. When atoms of tin or lead are brought together they form a metal, but germanium and silicon (and carbon, in one of its modifications, viz., diamond) solidify from the melt with the crystal structure shown in Fig. 15.5(a). In this valence crystal, each ion is covalently bound to its four nearest neighbours, the bonds consisting of two valence electrons, one from each atom. A valence crystal is essentially a giant molecule, with the valence electrons also located in space in a regular array.

The structure of a valence crystal differs quite markedly from that of the molecular crystal of Fig. 15.4, where the valence electrons play only a small role in holding the molecules together. The binding forces in a silicon crystal are very strong, as testified by the high melting point (1414° C) compared with that of hydrogen (14° K).

In order to assist the reader in visualising the spatial nearest-neighbour relationships in the silicon crystal of Fig. 15.5(a), an element of that structure is shown in detail in Fig. 15.5(b). There is one ion located at the centre of a cube, and its four nearest-neighbours are located equidistant at the cube corners shown. It may readily be shown that the four valence bonds to the central ion make equal angles with one another.

The semiconducting electrical conductivity of valence crystals, and of silicon, is discussed in Section C.2.

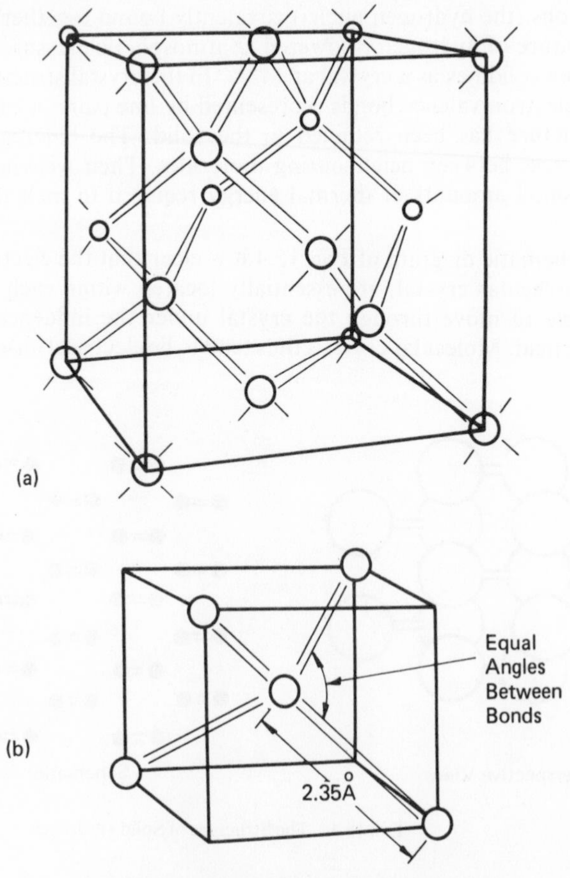

Fig. 15.5 (a) Covalent Crystal Structure.
(b) Nearest-Neighbour Relationships in Silicon.

C. MATERIAL PROPERTIES

In the design of electrical and electronic equipment, engineers naturally make use of virtually all known properties of materials. In this introductory description of materials it is not possible to elaborate in great detail. We focus our attention primarily on electrical conductivity, for that leads naturally to the important properties of semiconductors, and of semiconductor devices, which have already been shown to have a strong influence on the directions in which electronic engineering is currently making progress. Other properties will subsequently be touched on briefly in Section C.3.

1. Electrical Conductivity

Consider a metal, in which the electrons are free to move throughout the periodic crystal lattice. The electron density will be of the same order as the number density of atoms in the crystal, say 10^{23} cm^{-3}.

If we apply a potential difference across a portion of the metal crystal, a current will flow. The electrons are accelerated by the electric field E associated with the potential difference, and in a direction of course opposite to that in which the current is said to flow, according to a convention of long usage. The velocity of each electron does not increase indefinitely when the field E is applied: the electrons are randomly scattered by departures of the crystal lattice from perfect periodicity. Such imperfections may take the form of impurities, or of defects in the crystal structure, but at room temperature the departures from perfection are principally the thermal vibrations of the ions of the crystal about their mean positions in the lattice.

The force on each electron may be written:

$$(-q)\ E = m_n\ a, \qquad (15.1)$$

where $-q$ is the charge on an electron of mass m_n, and a is the acceleration of the electron due to the electric field E. Let us make the simplifying assumption that each electron has a time of flight τ between collisions with the lattice, and that after each collision the electron starts with zero velocity. Then the distance travelled by each electron between collisions is $\frac{1}{2}a\tau^2$, and the average velocity $\bar{v}_n$ due to the presence of the electric field is therefore:

$$\bar{v}_n = \text{distance/time} = \tfrac{1}{2}a\tau^2/\tau = \tfrac{1}{2}a\tau. \qquad (15.2)$$

Substituting the value of a given above,

$$\bar{v}_n = -(q\tau/2m_n)\ E. \qquad (15.3)$$

We have thus shown that the average "drift" velocity of the electrons, $\bar{v}_n$, is proportional to the electric field strength E; we write

$$\bar{v}_n = -\mu_n E, \qquad (15.4)$$

where the constant μ is known as the electron mobility. The electrons drift in a direction anti-parallel to the electric field E, so that there is a negative sign in Equations (15.3) and (15.4).

The density of current carried by the electrons is given by the flux of electrons (number per unit area per unit time) multiplied by the charge on each electron, that is, the current density is

$$i = n\bar{v}_n\,(-q). \qquad (15.5)$$

Using Equation (15.4) we therefore have

$$i = nq\mu_n E. \qquad (15.6)$$

This may be re-written

$$i = \sigma E, \tag{15.7}$$

where the electrical conductivity

$$\sigma = nq\mu_n. \tag{15.8}$$

On the basis of the simplifying assumptions made above we have thus derived Ohm's Law (Equation 15.7), and have obtained a theoretical expression for the electrical conductivity of a metal. The true situation is complicated further by the random thermal motion of the electrons, but we shall not discuss such complexities. The conductivity is seen to be dependent both on the density of charge-carriers n and on their mobility μ_n.

2. Conduction in Valence Crystals

In Section B. it was concluded that ionic and molecular crystals are insulators, while metals are good conductors at all temperatures. What of the conductivity of a valence crystal?

According to the model of a silicon crystal shown in Fig. 15.5, the application of an electric field would not lead to conduction. This is because the electrons, located in valence bonds throughout the crystal, are not free to move. It is important to note, however, that the model of Fig. 15.5 depicts the situation at the absolute zero of temperature, for the atoms are not vibrating about their mean positions. As the temperature is increased the thermal vibrations of the atoms will in fact disrupt a small proportion of the valence bonds; electrons are then set free to move through the crystal and to contribute to electrical conduction. The density of free electrons, and hence the conductivity, increases strongly with increasing temperature.

For convenience of discussion the valence crystal structure of Fig. 15.5 may be shown schematically in two dimensions (Fig. 15.6). Each atom has four nearest-neighbours to which it is covalently bound. The disruption of a bond, creating an electron free to move through the crystal, is also illustrated in Fig. 15.6.

We note immediately from Fig. 15.6 that a further mechanism exists for the transport of current through a valence crystal, over and above the conduction due to the electrons freed from the valence bonds. The electric field also acts on the electrons still remaining in the valence bonds, and it is possible for an electron to move from an adjacent bond into the vacant bond position. By a sequence of such movements, electrons in the valence bonds can also contribute to conduction, when there is a vacancy present. Alternatively, attention may be fixed on the vacancy itself; the electric field causes the vacancy to move. This movement is in a direction opposite to that of the electron freed from the valence bond, that is, in the direction a positive charge would move.

Thus there are two essentially independent modes of conduction in a valence crystal: conduction by the electrons freed from the valence bonds (density n), and by the vacancies in the valence bond structure which we call "holes" (density p). The concept of a hole is simply a convenient method of describing the motion in an electric field of all the valence electrons in the crystal when there is one such electron missing; the density of holes is denoted by p to remind us that the holes move in an electric field as though they were *positively* charged.

The above discussion is a simplification of a much more elegant and informative treatment using wave mechanics. It does, however, give us some insight into the processes of conduction in valence crystals such as silicon and germanium, in spite of the imperfections of such a simplified treatment.

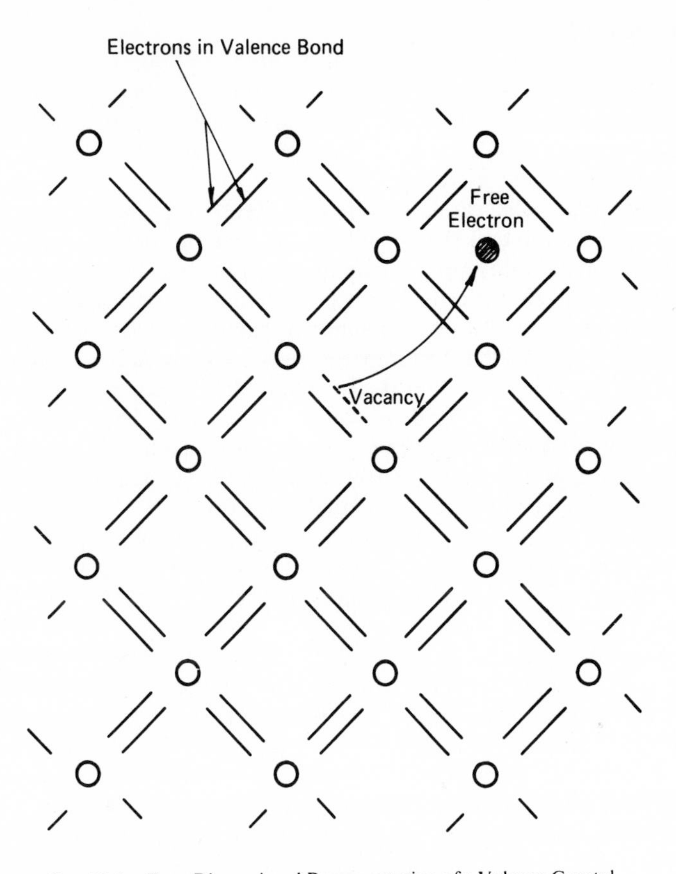

FIG. 15.6 Two-Dimensional Representation of a Valence Crystal.

If the electron and hole mobilities are respectively μ_n, μ_p the conductivity of a valence crystal is given by

$$\sigma = nq\mu_n + pq\mu_p. \tag{15.9}$$

In the case of the pure or intrinsic crystal of Figs 15.5 and 15.6 the electron and hole densities are equal, since they are created in pairs. In the next chapter we see that it is possible to vary the electron and hole densities by incorporating impurities in a valence crystal.

The electron and hole density

$$n = p \tag{15.10}$$

in an intrinsic valence crystal is determined by two competing processes at any particular temperature. On the one hand we have the continuous creation of electron-hole pairs by the thermal vibrations of the lattice, while on the other it must be recognised that electrons and holes may spontaneously recombine, the energy involved being transformed into heat or radiation. It turns out that the equilibrium densities n and p increase exponentially with temperature. The mobilities are relatively weakly dependent on temperature. The densities n, p being very much less than the number density of atoms in the crystal, the conductivity σ, as well as being an exponential function of temperature, is therefore intermediate between that of metals and that of insulators. This is the observed pattern of behaviour in the materials known as semiconductors.

3. Engineering Properties

Regrettably it is not possible in this book to do more than follow up in some detail the electrical properties of materials, and to show how these properties have been applied to the invention of useful devices, particularly in the semiconductor field. Nevertheless our story would be quite incomplete if we did not refer, in passing, to other aspects of material properties.

We have seen that solid crystals are either electrically insulating, conducting or semiconducting, to an extent dependent on the binding mechanism. In the case of some metals and metal alloys there is an abrupt transition to infinite conductivity as the temperature falls below a critical transition temperature, usually less than 20° K. A current set up in a closed loop of a superconductor, as such materials are known, will flow indefinitely, provided the temperature is not allowed to rise above the critical temperature and provided also that the magnetic field strength, where the current is circulating, is kept below a critical value. There have been recent engineering applications of superconductors to the construction of high-field magnets, which are of some importance, and further engineering advances can be expected as materials of higher transition temperature are formulated in the future.

In all materials other than superconductors, dissipation of energy is associated with the flow of current. In order to dispose of heat generated in this way, provision must be made for its removal by thermal conduction. Therefore the thermal conductivity of materials is of importance also. Since thermal conductivities are finite, increases in temperature can be expected in any devices (other than superconductors) in which current flows. Thus engineers have concern also for the temperature dependence of properties; e.g., the dependence with temperature of mechanical strength, ferromagnetism, cathodoluminescence, or whatever is of importance to the application concerned. Ferromagnetic materials, to take one example, lose their spontaneous magnetic moment and become paramagnetic above a temperature characteristic of each material (known as the Curie temperature), and care must be taken therefore to ensure adequate dissipation of heat to prevent the Curie temperature being reached if the ferromagnetic property is to be retained under operating conditions.

In materials in which non-uniformities of composition have been introduced by design, such as the cathode of a vacuum tube, or a $p - n$ junction in a semiconductor, another fundamental consequence of increases in temperature must be taken into account. At temperatures above a few hundred degrees Centigrade, the thermal vibrations of atoms become so intense that some of them leave their positions in the crystal lattice, creating vacancies, and diffuse through the crystal under the influence of any gradients in concentration which may be present. The desired non-uniformities in composition thus tend to become less pronounced, and this is of course the mechanism of failure in cathodes of vacuum tubes which was referred to in Chapter 4.

Solid-state diffusion processes of this kind may also be employed to advantage however. Heating a semiconductor crystal in the presence of impurity elements leads to solid-state diffusion of the impurity into the crystal, and one is in this way able to produce a controlled distribution of impurity in the crystal which is of use in device fabrication, as in the fabrication of silicon integrated circuits (see Chapter 17). The rate of diffusion increases exponentially with temperature, so that with control of time and temperature one is able to achieve an extraordinarily close control of the distribution of impurity in the crystal.

D. PROBLEMS

1. Given that the atom spacing in a silicon crystal is 2.35Å, with the aid of Fig. 15.5, show that the number density of silicon atoms is 5.10^{22} cm $^{-3}$.

2. By taking plane sections of the cube of Fig. 15.5 (b), prove that the angle between any two valence bonds in a silicon crystal is 109.5°.
3. Construct portion of a model of a silicon crystal, using lozenges to represent the atomic ions and a toothpick to represent each valence bond. Make use of Fig. 15.5(b) to achieve the correct bond angles.
4. Calculate the electrical conductivity of an intrinsic silicon crystal at 300° K, when it is known that the electron and hole densities are 1.5×10^{10} cm^{-3}, and the mobilities are respectively 1350 and 480 cm^2 $V^{-1} s^{-1}$.
   (The magnitude of the charge on an electron is 1.6×10^{-19}C).

SUGGESTED READING MATERIAL

Materials. A Scientific American Book. W. H. Freeman and Co., San Francisco, 1968. (Reprint of *The Scientific American*, **217**, No. 3, September, 1967).

CHAPTER 16

Semiconductors

L. W. Davies

"Behold, the half was not told me."

1 Kings, 10:7.

In Chapter 4 the semiconductor revolution was described in somewhat general terms. Here we propose to discuss semiconductor devices so that a deeper understanding may be achieved of the semiconducting elements germanium and silicon and their overwhelming influence in bringing about this revolution. We first investigate the electrical properties of semiconductors, placing particular emphasis on the properties of slightly impure semiconductors, and on their non-equilibrium properties.

A. INTRINSIC SEMICONDUCTORS

The conduction mechanisms in pure, or intrinsic, semiconducting valence crystals were described in Chapter 15, B.4. Briefly recapitulating that discussion, both germanium and silicon crystallise with the structure shown in Fig. 15.5(a): a regular array of electrons in valence bonds extends throughout the space occupied by the crystal. In an intrinsic semiconductor at non-zero temperatures some of the valence bonds are disrupted, giving rise to equal densities of conduction electrons and holes. Electrons and holes are charge-carriers of negative and positive sign respectively, moving in opposite directions in the presence of an electric field to make quite independent contributions to conduction in the crystal. The electrical conductivity is thus given by the sum of two terms:

$$\sigma = nq\mu_n + pq\mu_p. \tag{16.1}$$

In an intrinsic semiconductor the electron and hole densities n and p are equal, and increase strongly with increasing temperature. We write, for the case of an intrinsic semiconductor,

$$n = p = n_i(T), \tag{16.2}$$

where the subscript i denotes intrinsic. In the case of silicon, for example,

$$n_i(300^\circ\,\mathrm{K}) = 1.5 \times 10^{10}\,\mathrm{cm}^{-3}. \tag{16.3}$$

B. EXTRINSIC SEMICONDUCTORS

The electrical conductivity of any semiconductor may be profoundly modified by the presence in the crystal of quite small amounts of impurity. Consider for example, a silicon crystal which has grown from a melt in which there was present a small amount of arsenic. Each arsenic atom has five valence electrons, which surround a closed electron shell. In the growth of the crystal, arsenic atoms are incorporated substitutionally, taking the place of occasional silicon atoms in the lattice, as shown schematically in Fig. 16.1. Four of the five valence electrons are shared with neighbouring silicon atoms to form covalent bonds, and the fifth electron remains bound to the arsenic ion by electrostatic forces of attraction, at very low temperatures.

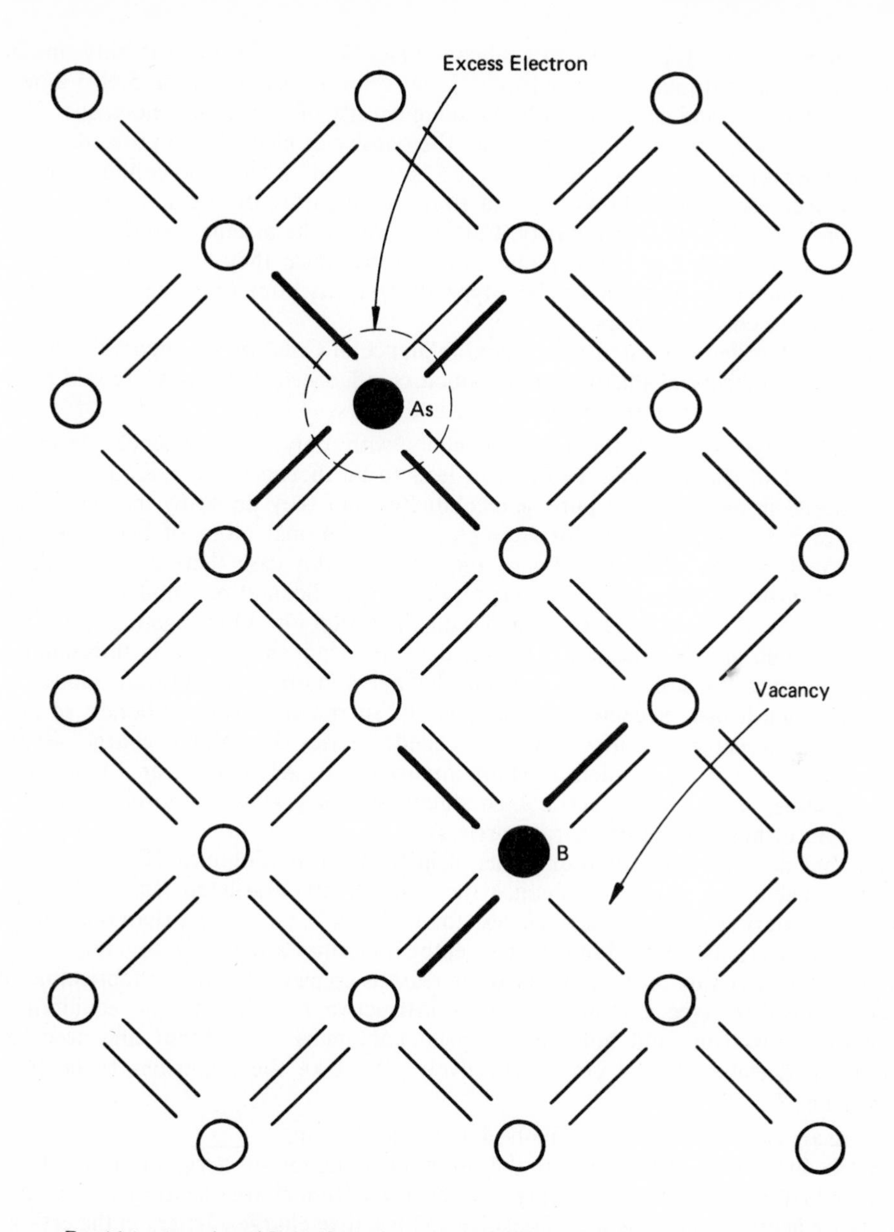

FIG. 16.1 Arsenic and Boron Impurity Atoms in a Silicon Crystal at Low Temperature.

The specific dielectric constant of silicon ($\varepsilon_r = 11.7$) is quite high, and as a consequence the energy required to detach the fifth valence electron from the arsenic ion is small, being only 0.04 eV. At temperatures above 50° K virtually all such electrons are detached from their parent arsenic ions, being then free to move through the crystal as conduction electrons. The positively charged arsenic ion remains behind, fixed in position in the lattice. In this way arsenic impurity atoms (or atoms of any of the elements P, Sb, Bi of the Vth column of the Periodic Table) can increase the equilibrium density n of conduction electrons in the crystal. Such elements are referred to as donor impurities, each atom "donating" a conduction electron to the crystal.

The number density of atoms in a silicon crystal is 5×10^{22} cm^{-3}. If only one in every 10^8 of these atoms is replaced by a donor impurity there will be 5.10^{14} donor atoms in each cm^3, and approximately the same density of conduction electrons. This electron density is very much greater than the density of electrons in pure silicon at room temperature, namely n_i (300° K) $= 1.5 \times 10^{10}$ cm^{-3}. Thus the conductivity of silicon is increased some 33,000 times at room temperature by the addition of only 10^{-8} atom fraction of donor impurity. Furthermore, in the arsenic-doped crystal the flow of current is almost entirely due to electrons, since they are so much more numerous. Such a crystal is said to be *n*-type, drawing attention to the negative charge on the effective charge-carriers.

We have thus demonstrated a profound influence of Column V* impurities on the electrical conductivity of silicon. Semiconductors influenced in this way are said to be extrinsic, in contrast to intrinsic (pure) semiconductors.

As might be expected, impurity elements with only three valence electrons (Column III of the Periodic Table) also lead to an increase in conductivity when incorporated in an otherwise pure semiconductor, but they do so by increasing the density of holes, p. Consider for example a substitutional atom of boron, shown schematically in the silicon crystal of Fig. 16.1. In this case there are only three valence electrons to share with the four neighbouring silicon atoms, and a vacancy in the valence bond structure is associated with the boron ion. Once again the binding energy is small, and the vacancy is readily detached from the boron ion to become a mobile vacancy, or hole. There remains behind a boron ion plus an additional electron which it has "accepted" to complete its surrounding valence bonds, so that the ionised acceptor impurity (as it is called) carries a negative charge. Semiconductors which contain such acceptor impurities (B, Al, Ga, In or Tl) are said to be *p*-type, since their electrical conductivity arises almost entirely from the movement of holes (positive charge carriers).

We thus see that very small amounts of impurity from Columns III or V of the Periodic Table can strongly influence the conductivity of silicon. In view of the difficulty in purifying silicon to levels less than 10^{10}cm^{-3}, i.e., to less than one part in 5.10^{12}, it is likely that both donor and acceptor impurities are present together in any given crystal. For this reason, and also for reasons connected with the fabrication of silicon transistors (see Chapter 17), it is instructive to calculate the equilibrium densities of electrons and holes in a crystal containing both donor and acceptor impurities, densities N_D, N_A cm^{-3} respectively. We take the impurities to be fully ionised, i.e., $T > 50°$ K.

There are two basic equations involved in the calculation.

In the first place, we note that the addition of donor or acceptor atoms to a semiconductor crystal does not imply any departure from charge neutrality, since the atoms are neutral. The densities of positive and negative charge-carriers in the crystal may, therefore, be equated, yielding

$$p + N_D = n + N_A, \tag{16.4}$$

when the donors and acceptors are fully ionised, as assumed.

Secondly, it turns out from considerations of thermodynamic detailed balance, in a semiconductor in equilibrium, that the product of electron and hole densities is constant, independently of the amount of impurity present. Thus we may write

$$np = \text{constant}. \tag{16.5}$$

* i.e. Column V of the Periodic Table of chemical elements. (See G. H. Aylward and T. J. V. Findlay, *S. I. Chemical Data.* John Wiley and Sons, Sydney, 1971.)

This constant is evaluated by noting that in an intrinsic crystal both n and p are equal to $n_i(T)$. Thus in general for any density of impurities we have

$$np = n_i^2(T). \tag{16.6}$$

With the aid of the above equations we may calculate the equilibrium electron and hole densities for any given N_D, N_A and temperature. For example, suppose $N_D > N_A$ in a given silicon crystal; substituting for p in equation (16.4) we find

$$n^2 - (N_D - N_A)n - n_i^2 = 0, \tag{16.7}$$

a quadratic equation in n with only one positive solution:

$$n = \tfrac{1}{2}(N_D - N_A)[1 + \{1 + 4n_i^2/(N_D - N_A)^2\}^{\frac{1}{2}}]. \tag{16.8}$$

From Equation (16.8) it can be seen that when $(N_D - N_A) \gg n_i$ the electron density is closely equal to the effective donor density, $(N_D - N_A)$; part of the actual donor density N_D goes toward "compensating" the acceptor atoms present. The compensation of electrically active impurities is made use of in transistor fabrication.

C. *p-n* JUNCTIONS

An interesting situation arises—one of great importance to semiconductor device physics—when we consider the properties of a monocrystalline semiconductor containing adjacent p-type and n-type regions, formed as a result of a non-uniform distribution of donor or acceptor impurities. A p-n junction arising from abrupt discontinuities in donor and acceptor distributions is illustrated in Fig. 16.2.

Measurements of the hole density p in the crystal of Fig. 16.2 would give values $p \cong N_A$ in the p-region, and very small values indeed (approximately n_i^2/N_D) in the n-region. It must be concluded that some influence associated with the p-n junction is preventing the diffusion of holes from p- to n-region under the influence of the gradient* in hole density which exists. There is in fact a contact potential difference V_o between p-and n-regions, the n-region being positive with respect to the p-region and thereby constraining almost all the holes to remain in the p-region. The potential barrier, of height V_o, is at the same time of the required polarity to constrain the electrons to the n-region (see Fig. 16.2). In equilibrium, there is an exact balance of the hole currents (and also of the electron currents) in the two directions through the junction. The current that arises from the small number of holes able to surmount the potential barrier and diffuse into the n-region is exactly balanced by a current of holes, thermally generated in the n-region, which diffuse the potential barrier and are thence accelerated to the left and into the p-region. A similar argument applies to the electrons.

The varying electrostatic potential in the vicinity of the junction has associated with it a space charge layer, also shown in Fig. 16.2. The space charge to left and right of the junction, is made up of ionised acceptors and donors respectively; the densities of electrons and holes in the space charge layer are negligibly small, and for this reason it is also referred to as a depletion layer. As the depletion layer is a region of low conductivity which supports an electric field, as also is the dielectric in a parallel-plate capacitor of conventional construction, it follows that all p-n junctions in semiconductors have capacitance associated with them.

When metallic contacts are made to p- and n-regions of the junction (as illustrated in Fig. 16.2), and a voltage V applied externally, the balance in the two components of hole and electron current through the junction which exists in equilibrium is destroyed. Suppose a positive voltage V to be applied so as to make the p-region more

* Particles of matter tend to diffuse from regions of higher density into regions of lower density, and will do so unless other forces prevent the action of diffusion.

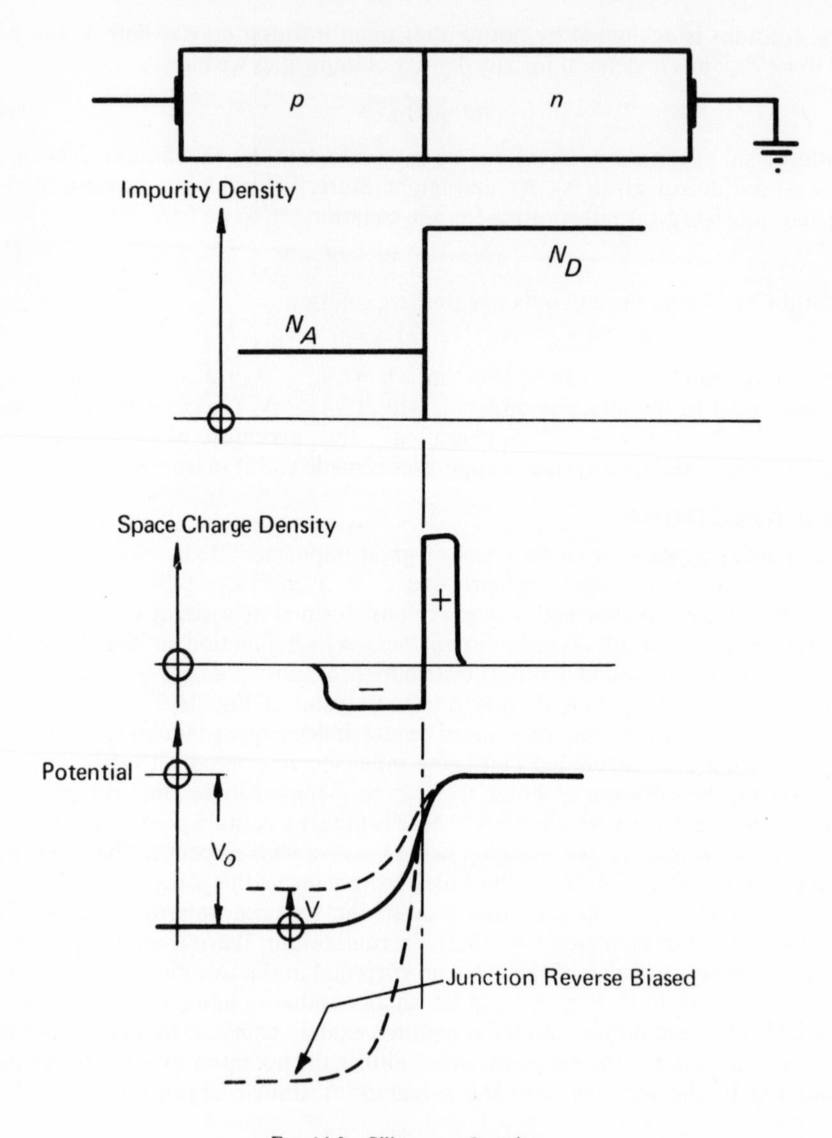

FIG. 16.2 Silicon *p-n* Junction.

positive with respect to the *n*-region, thereby lowering the height of the potential barrier from V_o to $(V_o - V)$. Many more electrons and holes are then able to surmount the barrier and diffuse respectively into the *p*- and *n*-regions, where they eventually recombine. The current through the junction therefore increases rapidly (indeed, exponentially) with increasing V of this polarity, and we say that the junction is biased in the forward or easy direction of current flow.

On the other hand, if the junction is biased in the reverse direction by a voltage of reverse polarity (i.e., $V < 0$), the height of the potential barrier is increased. Holes from the *p*-region, and electrons from the *n*-region, can no longer surmount the barrier: we are left with the two components of current, due to the thermal generation of holes in the *n*-region and of electrons in the *p*-region. In principle, these two components are independent of the applied voltage.

The current-voltage characteristic of a *p-n* junction is shown in Fig. 16.3, and is given by the equation:

$$I = I_s \{ exp(qV/kT) - 1 \}, \qquad (16.9)$$

where k is Boltzmann's constant and $-I_s$ is the saturation value of I, i.e., the current which flows for negative values of V such that

$$-V \gg kT/q. \qquad (16.10)$$

(At room temperature, $kT/q = 25\text{mV}$). It is seen that a *p-n* junction is very nearly an ideal rectifier. We note that the saturation current I_s is a strong (exponential) function of temperature, since it arises from the thermal generation of electron-hole pairs by the vibrating atoms of the lattice.

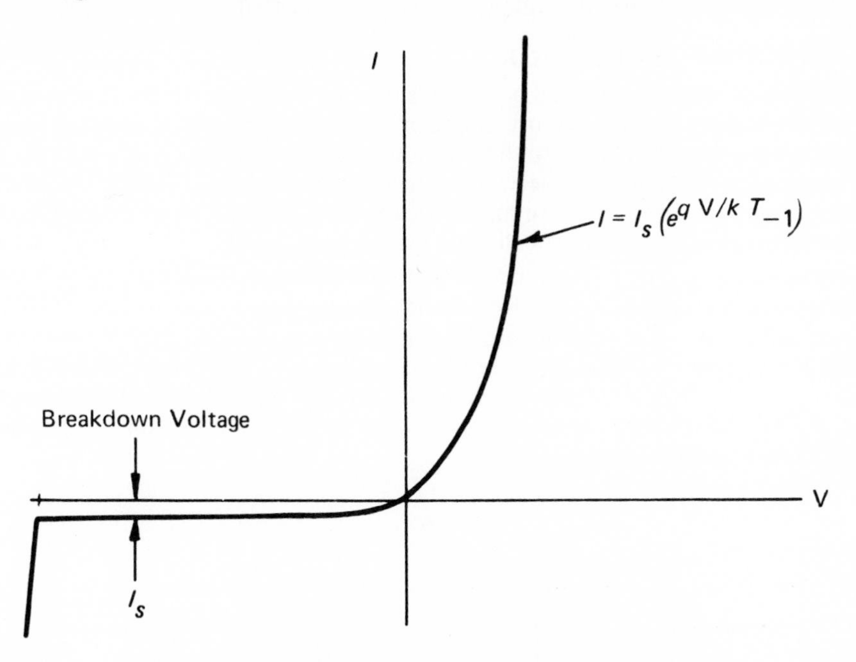

FIG. 16.3 Current-Voltage Characteristic of a *p-n* Junction Diode.

Light which falls on a semiconductor may also create electron-hole pairs by the disruption of valence bonds, provided the photons have sufficient energy. In silicon the energy required is 1.1 eV or more, corresponding to photons of wavelength less than 1.1μ (11,000 A).* If the electrons and holes are created near a reverse biased *p-n* junction, some of them will be accelerated across the junction by the electric field in the depletion layer and the current will rise above I_s. A *p-n* junction operated in this mode, so that it "collects" charge-carriers in excess of equilibrium in its vicinity, makes a useful detector of radiation (photodiode).

Even if no external bias is applied to a *p-n* junction, the built-in potential barrier is of sufficient height to separate electrons and holes which are created nearby in excess of equilibrium. Holes and electrons pass into the *p*- and *n*-region respectively. In this way sunlight incident on the *p-n* junction of a solar cell creates charge carriers which, when separated by the potential barrier, set up an electromotive force. If a resistive

* The wavelength of a photon containing a certain quantum of energy is proportional to the reciprocal of that amount of energy. Photons of visible light (from about 6500Å for red to about 4000Å for violet) also possess energies greater than 1.1eV. The symbol μ is used to indicate length in microns.

load is connected across an illuminated solar cell a current will flow, and there is a direct transfer of energy from sunlight to electric power dissipated in the load. If the *p-n* junction is arranged to lie about 1μ below the surface of the silicon crystal, where most of the electron-hole pairs are created by sunlight incident normally on the surface, conversion efficiencies as high as 10 per cent are readily attained. The importance of solar cells to satellite communication systems has already been referred to in Chapter 4.

When a negative voltage V is applied to a *p-n* junction there is a widening of the depletion layer. As the depletion layer widens, the capacitance associated with the junction decreases: the equivalent "parallel plates" are being drawn further apart. Thus a further interesting property of a *p-n* junction is that it has a capacitance C(V) which is a function of the voltage applied to the junction.

D. JUNCTION TRANSISTOR

As we have seen, the potential barrier in a *p-n* junction is lowered by the application of a bias voltage in the forward direction. The electrons which are then able to surmount the barrier diffuse into the *p*-region adjacent to the junction, where they are in excess of equilibrium. The electrons diffuse further into the *p*-region, recombining with holes at the same time; we have a steady-state distribution of excess electrons which have been injected electrically into the *p*-type material.

We have also seen that a *p-n* junction biased, in the reverse direction, will collect any excess carriers which arrive in its vicinity. It follows that if a reverse biased junction is placed sufficiently close to a forward biased junction in the same crystal, as shown in the *n-p-n* structure of Fig. 16.4, the current flowing to the reverse biased junction can be controlled by the current which we allow to flow in the forward biased junction. As will be shown, the *n-p-n* structure, which we call a transistor, is capable of amplifying a signal.

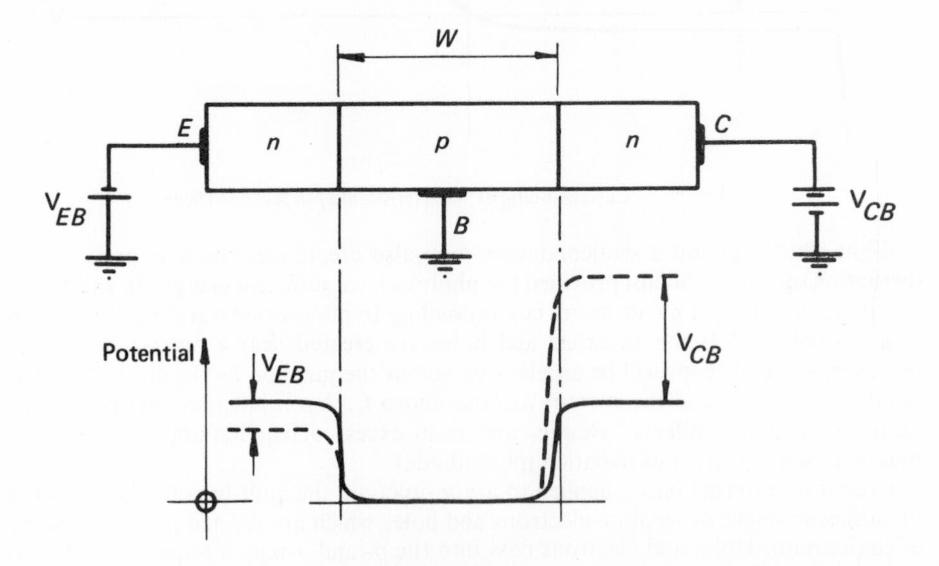

Fig. 16.4 *n-p-n* Junction Transistor.

The distribution of electrostatic potential in equilibrium in the *n-p-n* transistor of Fig. 16.4 is shown by the full line. We put the central base region (*p*-type) at zero potential by grounding it. When a positive voltage V_{CB} is applied between the right hand electrode (the collector) and the base, the collector junction is biased in the reverse direction and only the saturation current flows; the increase in collector barrier height is shown by the right hand dashed line. If now a negative voltage V_{EB} is applied between the left hand electrode (the emitter) and the base, the reduced height of the emitter potential barrier allows electrons to flow into the base region, where they are in excess of equilibrium. If the width W of the base region is sufficiently small, i.e., is less than some tens of microns in the case of a silicon transistor, almost all the electrons diffuse to the collector junction before they are able to recombine. We are thus able to enhance the current flowing in the collector-base circuit, by means of a current flowing in the emitter-base circuit. In a well designed *n-p-n* transistor almost all the current flowing through the emitter junction is carried by electrons, almost all of which reach the collector before recombining with holes in the base. The current I_B flowing in the base lead is consequently only one or two percent of the current I_E flowing in the emitter lead.

Junction transistors are often operated in the "common emitter" configuration of Fig. 16.5; a positive voltage V_{BE} between the base and the grounded emitter determines the base current I_B, which in turn controls the much larger collector current. The bias voltage V_{CE} applied between collector and emitter is positive. The distribution of current between the three electrodes is also shown in Fig. 16.5. An input signal current in the base lead causes an amplified current to flow in the collector circuit.

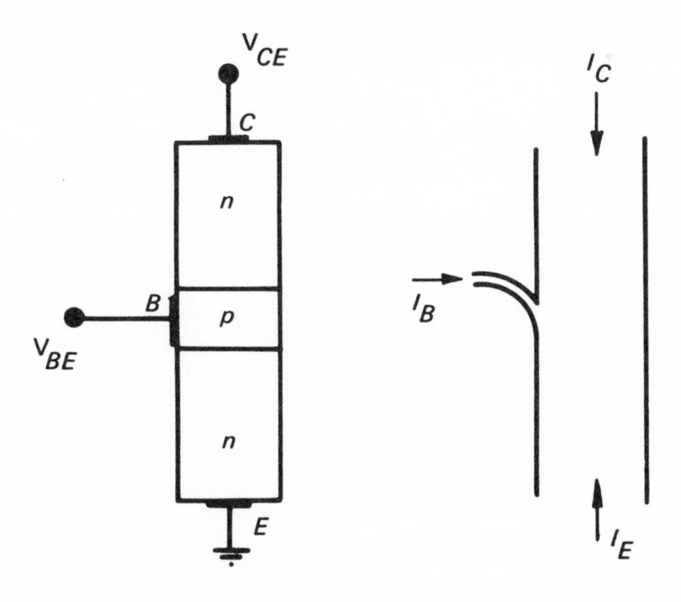

FIG. 16.5 *n-p-n* Transistor in Common-Emitter Connection, and Distribution of Current between Electrodes.

It is clear that there are two possible types of transistor: the *n-p-n* configuration discussed above, and a *p-n-p* configuration in which holes are injected into the *n*- type base by the emitter and diffuse to the collector. In this respect transistors are more versatile than vacuum triodes, in which the current that is controlled can only be carried by electrons.

E. PROBLEMS

1. Calculate the electron and hole densities in a crystal of silicon at 300° K which contains 10^{15} cm^{-3} substitutional gallium atoms.
   (For silicon, n_i(300° K) = 1.5×10^{10} cm^{-3}).
2. Calculate the ratio of electron and hole densities in a germanium crystal at 300° K containing 5×10^{13} cm^{-3} substitutional indium atoms; take n_i (300° K) to be 2.3×10^{13} cm^{-3} in germanium. Is the crystal n-type or p-type?

SUGGESTED READING MATERIAL

ROSE, R. M. SHEPARD, L. A. and WULFF, J. *The Structure and Properties of Materials* Volume IV: "Electronic Properties". (Chapter 5 and 6). John Wiley and Sons, New York, 1966.

DAVIES, L. W. and WILTSHIRE, H. R. *Semiconductors and Transistors, and Further Experiments in Electronics*. Amalgamated Wireless Valve Co. Pty. Ltd., Sydney, 1965.

CHAPTER 17

Integrated Circuits

L. W. Davies

"All are but parts of one stupendous whole
Whose body nature is, and God the soul."

Alexander Pope.

A. INTRODUCTION

In the previous chapter the operation of an *n-p-n* transistor was discussed in the form of an idealised monocrystalline structure with contiguous *n*-, *p*- and *n*-regions. Nothing has been said to this point on the question of how one might fabricate such a device. Severe technological problems had to be overcome before the first junction transistors of germanium were fabricated in 1951, particularly in view of the narrow width of the central base region, and the close tolerances to be met in impurity content of the crystal.

No detailed historical development will be given here of the successive improvements in manufactured forms of the junction transistor, culminating in the so-called *Planar* silicon transistor. Nevertheless, it is important to keep in mind that the advanced form of transistor, manufactured today, has depended for its development on a long series of technological advances in the processing of semiconductor materials, and that these advances, in turn, have depended on a deepening knowledge of the basic physical properties of materials. Basic physics, process technology and semiconductor electronics have advanced together, and will no doubt continue to do so in the future.

In a sense, integrated circuits were a natural and consequential development of the Planar process for the fabrication of silicon transistors. The advances achieved in switching speed of the new devices, and in reliability, made it essential to consider a more compact and reliable method of interconnection if full advantage were to be taken of the situation. We, therefore, preface our discussion of integrated circuits with a brief description of the fundamentals of the Planar process for fabricating silicon transistors.

B. SOLID-STATE DIFFUSION IN SILICON

Reference was made in Chapter 15, C.3 to the diffusion processes which occur in solids at high temperatures. If a crystal of silicon containing a non-uniform distribution of impurity atoms is heated to temperatures around 900°C or above, the impurity atoms begin to migrate in the crystal in such a way as to tend to remove any gradient in their concentration. The changes in concentration arise from the random thermal motion of the atoms of silicon and impurity at these high temperatures.

The same processes occur when a silicon crystal is exposed to an atmosphere of impurity atoms at high temperature, for in this case there is a steep gradient in impurity density, which is located at the surface of the crystal. The impurities condense on the surface, and subsequently migrate into the silicon by solid-state diffusion. The ultimate depth of penetration may be very closely controlled by precise control of the temperature of the process, and of the time for which it is allowed to proceed.

Consider a uniformly doped *n*-type silicon crystal heated, say for one hour at 1150°C in the presence of an atmosphere of boron atoms. (We neglect some of the details of the technological process, such as the exact composition of the atmosphere in which the process takes place). At the end of this time, examination of the surface layers of the crystal would show a distribution of boron atoms (acceptors) given by the curve labelled N_B in Fig. 17.1, which gives the impurity content in the crystal (plotted on a logarithmic scale) as a function of distance (plotted on a linear scale). It is seen that there is a layer at the surface, 2.5μ deep, in which the density of boron atoms N_B is everywhere greater than the initial donor density N_D, so that we now have a *p*-type layer on an *n*-type crystal.

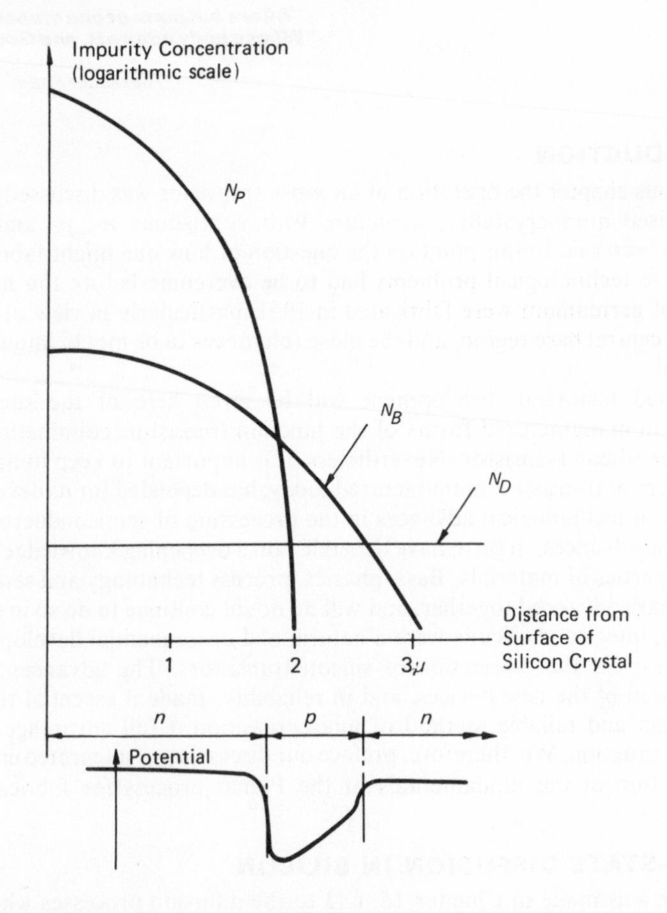

FIG. 17.1 Concentration Profiles of Boron and Phosphorus Atoms after Diffusion Processing, and Distribution of Potential.

If a second diffusion step is now carried out, but this time in an atmosphere of phosphorous (donor) impurities at a lower temperature, the result is a distribution of donors given by curve N_P in Fig. 17.1. In the new surface layer, of depth approximately 2μ, the donor density is greater than the acceptor density and it is therefore *n*-type. At the lower temperature the distribution N_B of boron atoms remains essentially unaltered.

We have thus produced an *n-p-n* structure in a silicon crystal by two sequential solid-state diffusion processes which enable close control to be achieved on the width of the *p*-type layer, and on the impurity densities. Impurity distributions of the type shown in Fig. 17.1 are a little different from the uniform donor and acceptor densities considered in Chapter 16, but there may actually be distinct advantages in having a non-uniform distribution of acceptor density in the base region, as we shall now see.

From the equilibrium distribution of electrostatic potential in the *n-p-n* structure, shown in Fig. 17.1, it can be seen that there is a gradient in potential, and therefore an electric field, in the base region. Consideration will show that this electric field is in such a direction as to accelerate electrons from left to right. In a uniform-base transistor, the electrons travel from emitter to collector by the relatively slow process of charge-carrier diffusion. If we were to make the surface (left-hand) *n*-region of Fig. 17.1 the emitter of a transistor, however, the electrons would actually be accelerated across the base to the right-hand (collector) region, and their transit times reduced by a factor of 10 or more. Thus the impurity distributions obtained in solid-state diffusion processes, at the surface of a silicon crystal, lend themselves ideally to the fabrication of high-frequency transistors, operable at frequencies up to and above 1*GHz*.

C. OXIDE MASKING

The discussion of solid-state diffusion in the previous section has shown that desirable distributions in depth of donor and acceptor impurities may be achieved in the surface layers of a silicon crystal for the purpose of fabricating a junction transistor. However, there remain two problems; making electrical contacts to the three regions (particularly the *p*-type base region of Fig. 17.1), and limiting the active area of the device to a useful value.

An additional discovery in process technology enabled both these problems to be surmounted simultaneously. It was found that a thin layer of silicon dioxide (SiO_2), formed as a glassy layer in intimate contact with a silicon crystal, can effectively mask the underlying silicon against in-diffusion of some donor and acceptor impurities. The oxide layer is most conveniently formed by oxidising the underlying silicon.

Accordingly, in the fabrication of an *n-p-n* silicon transistor, following the steps as outlined in Fig. 17.2, a layer of SiO_2 of thickness approximately 0.5μ is first formed on the crystal by oxidation in water vapour at temperatures around 1150° C. Subsequently, "windows" are opened in this layer by a selective etching process, and the diffusion of boron is carried out as outlined in Section B. The *p*-type region is now seen to be confined to the lateral dimensions of the window, by the masking action of the original 0.5μ oxide layer.

During the boron diffusion process, carried out in an oxidising atmosphere, a layer of borosilicate glass (a mixture of B_2O_3 and SiO_2) is formed. The borosilicate glass reseals the original window-opening in the 0.5μ oxide layer, as shown in the second diagram of Fig. 17.2. At this stage the way is clear for phosphorous to be diffused through a second smaller window opened in the borosilicate glass, as shown in the third and fourth diagrams in Fig. 17.2. Once again the window is resealed during diffusion, this time by a phosphosilicate glass layer.

It is now a relatively simple matter to form metallic contacts to the emitter and base regions at the upper surface of the crystal. The base contact problem has been solved, it should be noted, by restricting the lateral extent of the emitter diffusion to an area somewhat less than that of the base.

The collector contact can be made to the rear surface of the original *n*-type crystal. As a final step in fabrication, the silicon crystal, in which many transistors had been diffused according to the procedures of Fig. 17.2, is divided into small segments each containing one transistor structure.

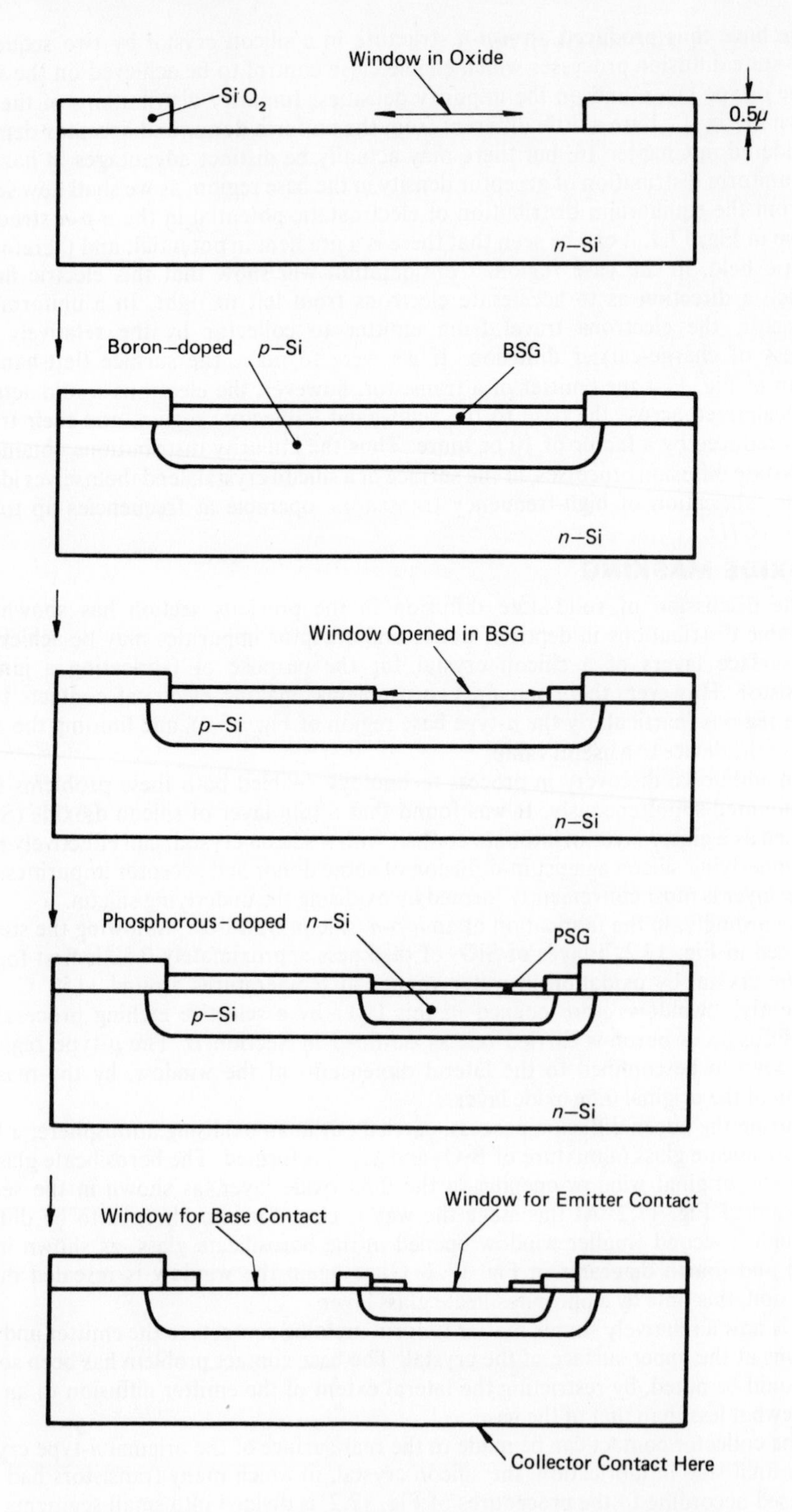

FIG. 17.2 Sequence of Processes in Fabricating Diffused *n-p-n* Silicon Transistor.

D. PLANAR TECHNOLOGY

When a reverse bias voltage is applied to a silicon *p-n* junction, the height of the potential barrier at the junction is increased, and with it the magnitude of the electric field in the depletion layer. As the reverse voltage is increased so also is the electric field. At field strengths around 100 kV cm^{-1} or above there is a dielectric breakdown in the junction: the reverse current increases markedly above the saturation value I_s, as shown in the current-voltage characteristic of Fig. 16.3.

Breakdown occurs because electrons and holes which traverse the depletion layer, contributing to I_s, are so strongly accelerated by the field that they gain sufficient energy to break up some of the valence bonds. There is an avalanching increase of electron-hole pairs, and thus of the current, which is of course carried by the increased number of charge-carriers. Provided the current is externally limited so that the junction temperature does not rise more than approximately 200° C, the breakdown may subsequently be extinguished without damage to the device.

At an external crystal surface intersected by a *p-n* junction there is always some electric field component in the space immediately adjacent to the crystal. Charged ions, originating from the ambient surroundings of the crystal, may become absorbed on the crystal in the vicinity of the junction intersection, and in this situation will be influenced by the electric field associated with the junction. Their effect is to degrade the leakage current I_s, particularly in respect to noise content, and to lower the breakdown voltage below the value which would otherwise be achieved in the bulk of the crystal.

It is clearly desirable to protect the junction intersection from the ambient atmosphere, in order to avoid such degradations of device properties, and the method by which this is achieved is indicated in Fig. 17.3. Illustrated there is an enlarged view of a diffused *p*-type region, produced by diffusing boron through a window in an oxide layer on *n*-type silicon. It will be seen that there is slight lateral diffusion of boron, as well as diffusion normal to the plane of the crystal surface, and that the intersection of the resulting *p-n* junction with the surface is actually protected from the ambient by the masking layer of oxide. To reinforce this protective action, oxygen is added to the surface atmosphere during diffusion, as already mentioned, so that a layer of borosilicate glass is formed on the silicon exposed in the window (shown in Fig. 17.3). A contact window can subsequently be opened in the borosilicate glass layer, somewhat removed from the junction boundary, to obtain a *p-n* junction which exhibits a high degree of stability in its electrical characteristics with respect to variations in the ambient environment.

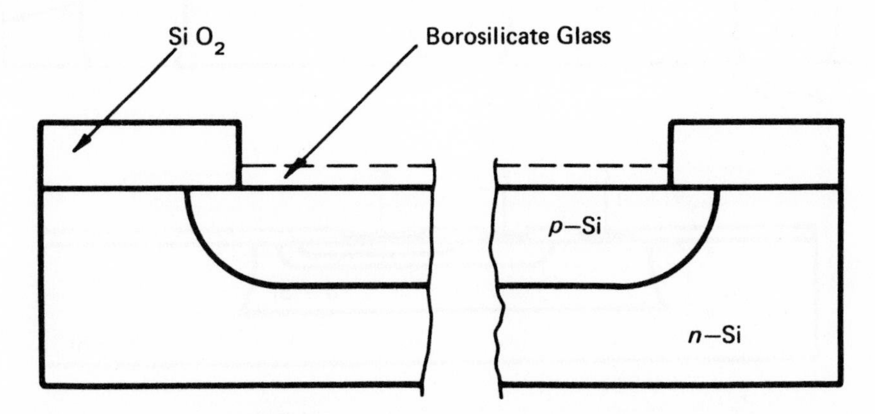

FIG. 17.3 Passivation of Junction-Surface Intersection by Oxide Layer.

E. MONOLITHIC BIPOLAR INTEGRATED CIRCUITS

In order to open up windows in oxide layers for subsequent diffusion operations, extremely precise photolithographic techniques have been developed. The oxidised surface of a silicon crystal is first coated with a photosensitive layer, which has the property of being polymerised and made resistant to acid when it is irradiated with ultra-violet light. When a photographic mask is placed in contact with this layer and exposed to UV radiation, regions of the mask which are blackened prevent polymerisation of the photosensitive layer immediately below them, and these non-polymerised regions can subsequently be removed with a solvent. A solution of hydrofluoric acid can then attack the oxide exposed through the openings in the photosensitive layer ("photoresist"), thus producing windows of identical dimension in the oxide.

Optical techniques have been developed which enable windows of lateral dimensions as small as a few microns to be opened in SiO_2 layers, on a production basis. It is therefore possible to contemplate the manufacture of junction transistors, whose dimensions are very much smaller indeed than the minimum size of silicon crystal chips which it is possible to handle mechanically (approximately 350μ square). Using either or both of the diffusion operations of the *n-p-n* transistor fabrication process outlined in the previous two sections, it is also possible to produce diffused areas which can be used as resistors. One can also fabricate diffused junctions whose capacitance can serve as capacitors in a circuit, when reverse-biased. These possibilities lead, naturally, to the concept of a complete circuit diffused on and into the surface of an *n*-type silicon crystal. Required interconnections of components may be achieved by a pattern of metallisation: a continuous thin film of aluminium is evaporated over the SiO_2 surface, and onto the silicon where contact windows have been opened, and is subsequently etched to the required pattern by a further photoresist operation.

Although the processes outlined above are complex, the costs of integrated circuits produced in this way are low because of the large number (up to several thousand) that can be processed simultaneously on a single slice of silicon.

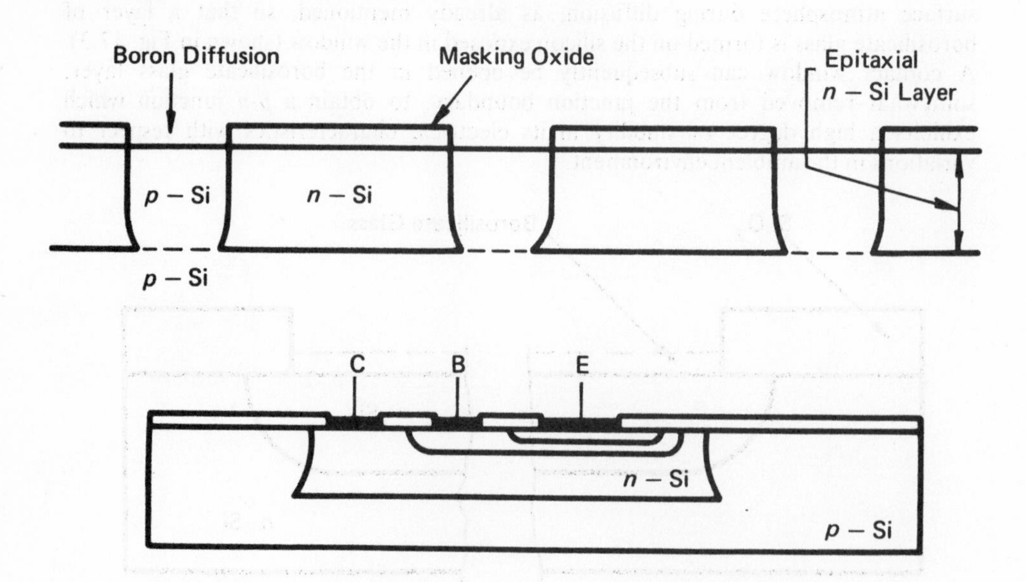

FIG. 17.4 Isolation Procedure, and Isolated Transistor Connections.

There are a number of design problems implicit in the technique outlined above for producing integrated circuits monolithically in a silicon crystal. Capacitances of *p-n* junctions, for example, are voltage-dependent, and there is a parasitic capacitance associated with diffused resistors. A further important problem, however, arises from the fact that all transistors have been diffused into an *n*-type silicon crystal, according to the procedure outlined above, and therefore have their collectors connected electrically within the crystal. This connection problem, which would be a very severe design limitation, can be surmounted in the following way.

With the aid of a so-called epitaxial process it is possible to produce a monocrystalline layer of *n*-type silicon some 10μ thick on an underlying *p*-type substrate. The process is carried out at a temperature below the melting point of the *p*-type crystal, so that donor and acceptor impurities are kept distinct. By means of a preliminary boron diffusion through chosen areas of the epi-layer one is then able to produce "islands" of *n*-type material, as shown in Fig. 17.4, which are electrically isolated from each other by virtue of the low leakage current of reverse-biased silicon *p-n* junctions. Segments of an integrated circuit which it is desired to have non-connected electrically are then diffused in one or more of the islands produced in this way. We note that electrical connection to the collector of *n-p-n* transistors diffused in isolated regions must now be made at the surface of the crystal, as also shown in Fig. 17.4.

F. CONCLUSION

We have described in an elementary way the concept of a complete electronic circuit fabricated monolithically in a silicon crystal. A little thought will indicate that there are many problems, and many possibilities, associated with the design of such circuits, problems which are not encountered in circuit design with more conventional discrete components.

The transistors discussed in relation to integrated circuit fabrication operate, as explained in Chapter 16, by the movement of carriers in excess of equilibrium from emitter to collector. The injected carriers—electrons in the base of an *n-p-n* transistor—are always accompanied by an approximately equal density of carriers of the opposite sign, in order to preserve space-charge neutrality, and for this reason such transistors are referred to as bipolar. Integrated circuits incorporating such transistors are also described as bipolar.

There is another class of transistor, a so-called Field-Effect Transistor (FET), in which there are no injected carriers. Unfortunately there is no opportunity here to discuss the details of their operation, but it may be noted that significant advances in miniaturisation may be achieved in the future with FETs, relative to bipolar integrated circuits.

SUGGESTED READING MATERIAL

HITTINGER, W. C. and SPARKS, M. "Microelectronics", *Scientific American* **213**, No. 5, p. 56, 1965.

The Western Electrical Engineer **11**, No. 4, 1967.

CHAPTER 18

Switching Circuits, Number Systems and Basic Computer Organisation

M. W. Allen

"It has been pointed out by A. M. Turing in 1937 and by W. S. McCulloch and W. Pitts in 1943 that effectively constructive logics, that is intuitionistic logics, can be best studied in terms of automata. Thus logical propositions can be represented as electrical networks or (idealised) nervous systems. Whereas logical propositions are built up by combining certain primitive symbols, networks are formed by connecting basic components, such as relays in electrical circuits and neurons in the nervous system. A logical proposition is then represented as a "black box" which has a finite number of inputs (wires or nerve bundles) and a finite number of outputs. The operation performed by the box is determined by the rules defining which inputs, when stimulated, cause responses in which outputs, just as a propositional function is determined by its values for all possible assignments of values to its variables. There is one important difference between ordinary logic and the automata which represent it. Time never occurs in logic, but every network or nervous system has a definite time lag between the input signal and the output response. A definite temporal sequence is always inherent in the operation of such a real system. This is not entirely a disadvantage. For example, it prevents the occurence of various kinds of more or less overt vicious circles (related to "non-constructivity", "impredicativity", and the like) which represent a major class of dangers in modern logical systems. It should be emphasised again, however, that the representative automaton contains more than the content of the logical proposition which it symbolises—to be precise, it embodies a definite time lag."

John Von Neumann
Probabilistic Logics and the Synthesis of Reliable Organisms from Unreliable Components, 1956

In a physical sense a computer largely consists of organised assemblies of binary elements, i.e., elements that can only take on two distinct values. An understanding of machine organization therefore begins with a study of how binary elements may be arranged in groups, which carry out more complex functions such as binary addition and number storage. These form the building blocks for the next step, which consists of organizing logic groups into larger assemblies which can execute arithmetic and other operations on numbers represented in any convenient base; this unit is called the arithmetic unit or in Babbage's terms the "mill". To build an automatic computer we need to include a store for holding numbers and instructions, an input and output unit, and a control, which determines the harmonious interplay of all units and the detailed steps involved in executing arithmetic operations. The purpose of this chapter is to outline key ideas involved in this synthesis of a computer commencing with logic elements themselves.

A. SWITCHING CIRCUITS AND LOGIC

1. Logic

Logic as a subject originated as a branch of philosophy in the fourth century (B.C.). However, progress was relatively slow until 1847 when George Boole and Augustus de Morgan independently developed algebras for symbolically representing and manipulating logical expressions. The same principles of logic are particularly valuable in the analysis and synthesis of switching circuits. Shannon (ca. 1938) was responsible for initial development of methods of symbolic analysis of relay and switching circuits based on Boolean algebra.

A logical variable, sometimes called a Boolean or binary variable is one that can take on only two distinct values. The value assigned thus represents a choice between two alternatives which can be denoted by the symbols 1 and 0. Alternative symbols sometimes used are T (true) and F (false) or X true and $\overline{X}$ (not X).

2. Truth Tables

A function of one or more binary variables can be completely specified by a table, listing all possible argument values and the corresponding function value for each. Such a table is called a truth table.

TABLE 18.1

Argument		Function			
x	y	AND	OR	NAND	NOR
0	0	0	0	1	1
0	1	0	1	1	0
1	0	0	1	1	0
1	1	1	1	0	0

Table 18.1 is a truth table which defines the functions AND, OR, NAND, NOR. The AND function is true only when x AND y are true, the OR is true when x OR y is true. The NAND is the negate of AND, and NOR the negate of OR. All are capable of simple realisation in relay or electronic circuits. The set AND, OR, NOT are primitives and complete in the sense that all other functions can be expressed in terms of them. Figure 18.1 shows the standard symbols used in logic diagrams, and a selection of different circuit configurations used. Further information may be obtained from manufacturers' data sheets.

A truth table of n input variables has 2^n rows, which are usually listed as if they were consecutive binary numbers. On the right of each row a 0 or 1 is entered depending on the output value required for the combination of inputs represented by the row. Table 18.2 shows a truth table for 3 input variables. The example is that of binary addition in which there are 3 inputs, x, y and c (carry in) and two outputs, sum and carry. (It is worth noting that for 3 input variables there are 2^8 or 256 possible output functions—in general for n input variables there are 2^n rows and therefore 2^{2^n} possible truth tables.) Sometimes a particular input combination is known to be impossible or an output combination can be allowed as 0 or 1. In such cases an x may be used to assign a "don't care" condition, such conditions can often lead to circuit simplifications.

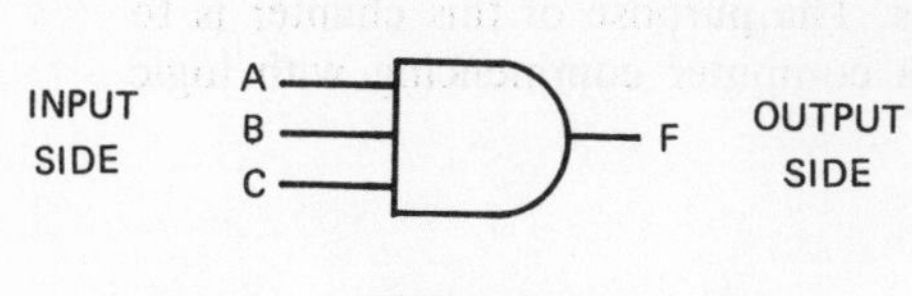

"AND GATE"

The AND output is Active if and only if all the inputs are Active.

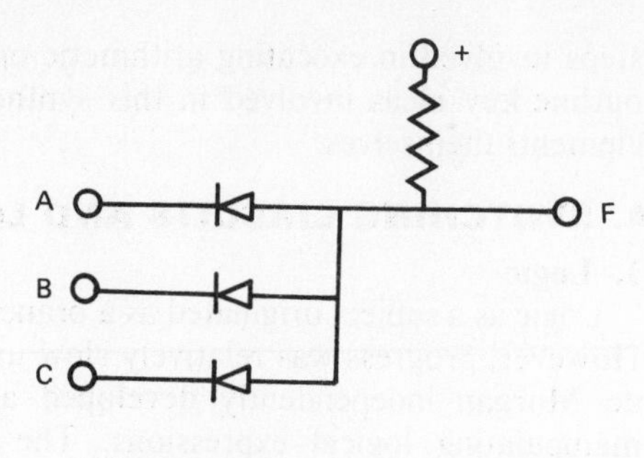

DIODE LOGIC (DL)

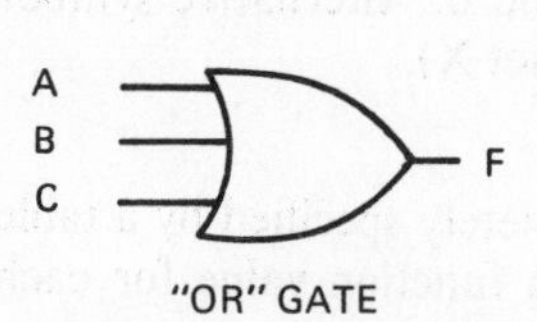

"OR" GATE

The OR output is Active if and only if one or more of the inputs is Active.

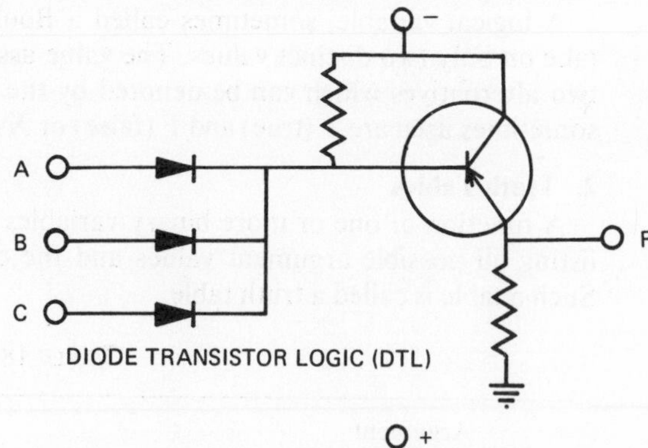

DIODE TRANSISTOR LOGIC (DTL)

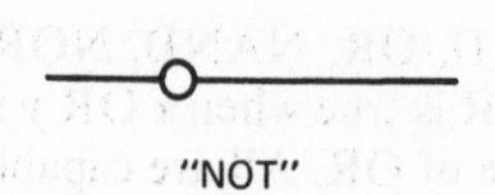

"NOT"

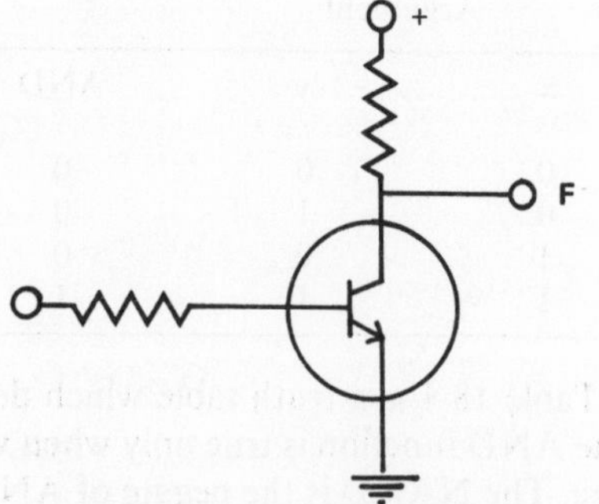

RESISTOR – TRANSISTOR LOGIC (RTL)

A
B
C

"NAND" GATE

The output is low if and only if all the inputs are high.

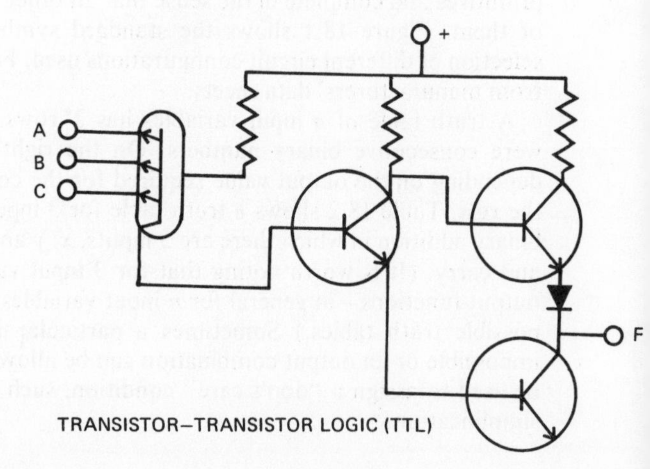

TRANSISTOR–TRANSISTOR LOGIC (TTL)

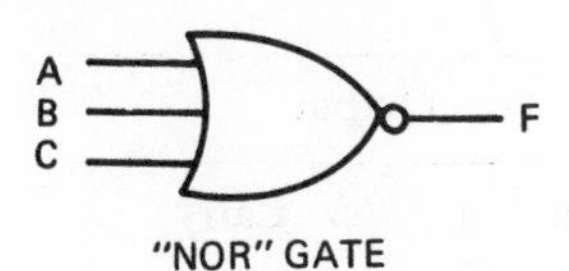

"NOR" GATE

The output is low if any one or more inputs are high.

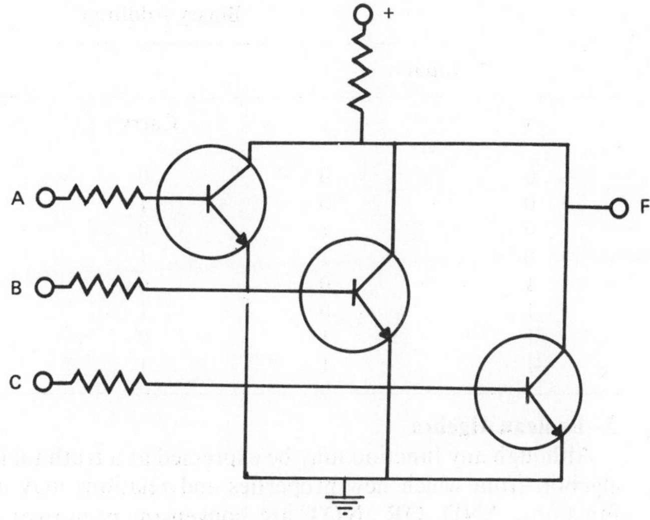

RESISTOR-TRANSISTOR LOGIC (RTL)

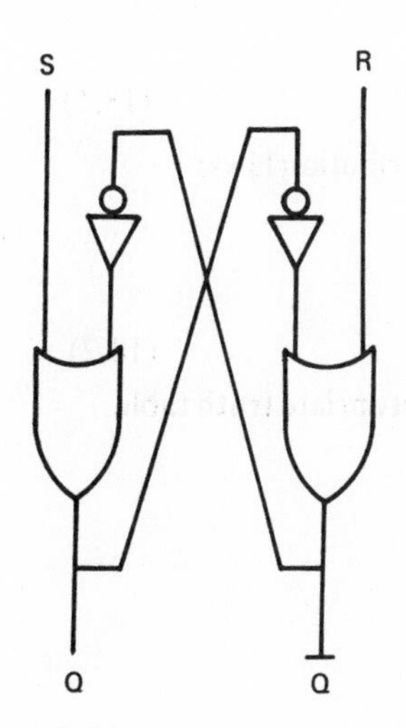

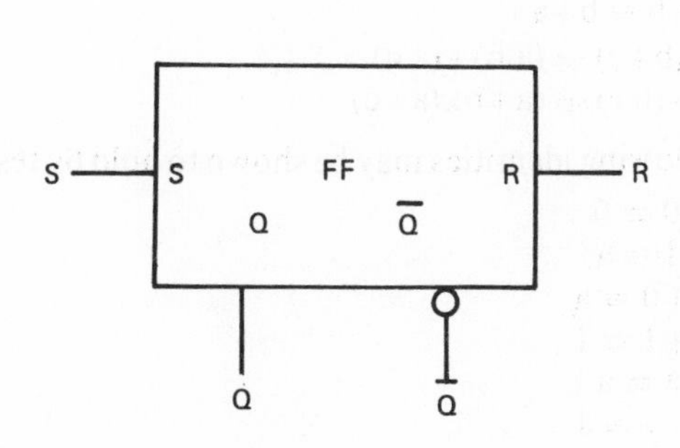

EQUIVALENT DIAGRAMS FOR THE RS FLIP-FLOP

FIG. 18.1 Logic Functions, Symbols and Circuits. The symbol is shown in the left-hand column and a representative circuit for each function in the right-hand column. The flip-flop, however, is shown as a logical configuration and as a symbol.

TABLE 18.2

Binary Addition

Input			Output	
x	y	Carry	Sum	Carry
0	0	0	0	0
0	0	1	1	0
0	1	0	1	0
0	1	1	0	1
1	0	0	1	0
1	0	1	0	1
1	1	0	0	1
1	1	1	1	1

3. Boolean Algebra

Although any function may be expressed as a truth table it is convenient to have an algebra, from which new properties and relations may also be deduced. The set of functions, AND, OR, NOT are convenient primitives and can be used in a way somewhat analogous to multiplication and addition in ordinary algebra. The following points summarise key features. AND is written . and OR is written +. Alternative symbols are $\wedge$ for AND and $\vee$ for OR.

The operators AND and OR obey the association law:

$$a.b.c. \equiv (a.b).c \equiv a.(b.c)$$
$$a+b+c \equiv (a+b)+c \equiv a+(b+c) \qquad (18.1)$$

The operators AND and OR also obey the commutation and distribution laws:

$$a.b \equiv b.a$$
$$a+b \equiv b+a$$
$$a.(b+c) \equiv (a.b)+(a.c)$$
$$a+(b.c) \equiv (a+b).(a+c) \qquad (18.2)$$

The following identities may be shown to hold by testing the appropriate truth table.

$$a.0 \equiv 0$$
$$a.1 \equiv a$$
$$a+0 \equiv a$$
$$a+1 \equiv 1$$
$$a.a \equiv a$$
$$a+a \equiv a$$
$$a+\bar{a} \equiv 1$$
$$a.\bar{a} \equiv 0 \qquad (18.3)$$
$$a \equiv \bar{\bar{a}}$$

$$a+(a.b) \equiv a$$
$$a+(\bar{a}.b) \equiv a+b \qquad (18.4)$$
$$a.(a+b) \equiv a$$

$$\overline{a.b} = \bar{a}+\bar{b}$$
$$a.b = \overline{\bar{a}+\bar{b}}$$
$$\overline{a+b} = \bar{a}.\bar{b}$$
$$a+b = \overline{\bar{a}.\bar{b}} \qquad (18.5)$$

Equations (18.5) are forms of De Morgan's law and are especially useful in analysis of logic circuits. These laws are embodied in a number of methods of simplification, which are outside the scope of this course. Probably the two most useful methods are the Karnaugh map and the Quine-McCluskey simplification algorithm (Marcus, 1967).

Any truth table may be alternately written as an algebraic expression (and vice versa). There are two methods yielding different but equivalent algebraic expressions—the first yields an AND-OR form of circuit and the second an OR-AND form.

METHOD 1 For each case where the output function is 1 provide an AND gate to detect the state of the input variables. Combine the AND gates in an OR gate. Figure 18.2 shows the result obtained from Table 18.2.

METHOD 2 For each case where the output function is 0 provide an OR gate for the complement of the corresponding input variables. Combine the OR outputs in an AND circuit.

The techniques of micro-electronics now enable hundreds of gates to be developed on a single chip of silicon. This is having a significant effect on the cost of logic and on the design methods used to build computer systems.

Before dealing with logic organisation to carry out arithmetic operations we need to review some of the principles of number systems.

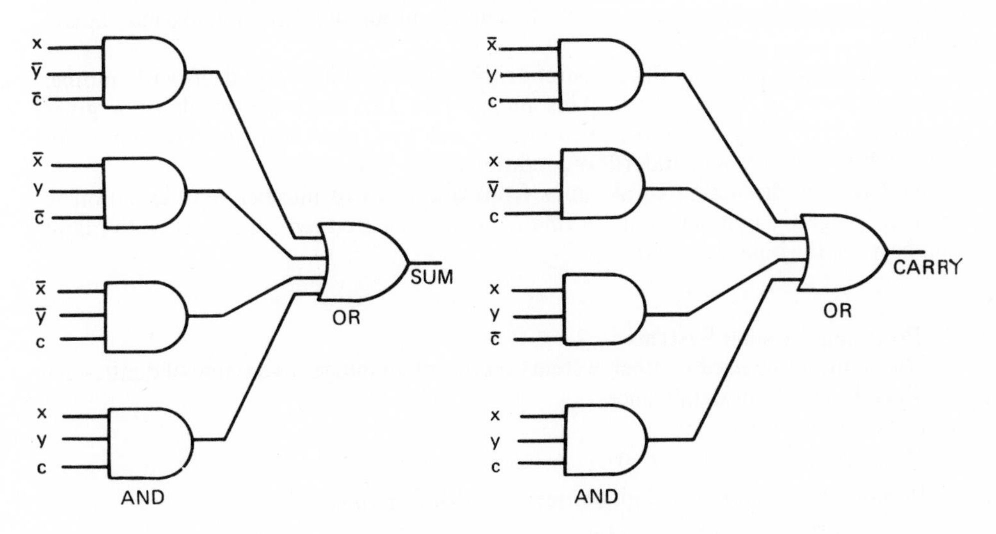

FIG. 18.2 Binary Addition.

B. NUMBER REPRESENTATIONS AND ARITHMETIC OPERATIONS

The historical development of number representations is a fascinating story, since it parallels the development of civilisation itself. The earliest forms of number representations are generally based on groups of fingers or piles of stones, with special conventions about replacing a larger pile by one object of a special kind, or in a special place. Generally, such representations are quite inconvenient for performing arithmetic.

Fixed point positional notation was apparently first developed by the Maya Indians in Central America about 2,000 years ago, but this system had no important influence on the rest of the world. It was a mixed radix system and not very suitable for arithmetic operations.

Our decimal notation was first developed in India among the Hindu people in approximately 600 A.D. The principles were brought to Persia in about 750 A.D. It is interesting to note that the term algorithm is derived from the name of the author of an Arabic text on arithmetic, published during this period (the author was al-Khowarizmi).

The first known appearance of binary notation was about 1600, in manuscripts of Thomas Harriot (1560-1624). However, an article by G. W. Leibnitz in 1703 is usually referred to as the birth of radix 2 arithmetic. He illustrated all of the principles of arithmetic but was more concerned with its use in number theory. Charles XII of Sweden foresaw the advantages of radix 8 arithmetic in 1717. He considered radix 8 or 64 would be more convenient for calculation than the decimal system. Unfortunately for designers of computers, he died in battle before he was able to institute such a change.

The first high speed electronic computer (ENIAC) used decimal arithmetic. However, the advantages of binary arithmetic were realised early in the design of the first stored program general-purpose computer and since then most computers have remained binary.

Numerical or other types of information are usually represented internally in computer systems by finite strings of two-valued signals (binary digits or bits) denoted by 0 or 1. The choice of two valued elements is due to the simplicity and reliability of their hardware realisation. This does not represent any serious restriction in choice of number system because it is possible to encode other number systems into binary form.

A fundamental property of computer arithmetic is that it always deals with number representations of finite length. This means that the hardware result of a given operation may differ from the theoretical, and that provision must be made for detection of overflow and underflow conditions.

We begin by discussing conversions from one form of number representation to another. Negative numbers and arithmetic operations are then considered with major emphasis on the binary system.

1. Positional Number Systems

Most commonly used number systems represent an integer as a string of digits, each of which has position significance.

Thus in radix or base 10

$$789_{10} \equiv 7.10^2 + 8.10^1 + 9.10^0$$

In general terms for an m digit number N_r with digits d in radix r

$$N_r = d_o r^{m-1} + d_1 r^{m-2} \ldots d_{m-1} \qquad (18.6)$$

Thus

$$678_9 = 6.9^2 + 7.9^1 + 8.9^0$$

or

$$356_8 = 3.8^2 + 5.8^1 + 6.8^0$$

A polynomial form of representation is sometimes more convenient for evaluation.

$$367_{10} = ((3.10 + 6)\,10 + 7) \quad \text{or}$$

$$456_8 = ((4.8 + 5)\,8 + 6)$$

in general form.

$$N_r = (\ldots((d_0 r + d_1)r + d_2)r \ldots + d_{m-1}) \qquad (18.7)$$

2. Radix Conversion

Conversion of integers from one radix representation to another is achieved by division or by multiplying out the polynomial. Thus to convert an integer from radix 10 to radix 8 we first divide by 8; the remainder is the least significant digit. The operation is then repeated. Alternatively, the polynomial may be multiplied out. Our choice of method for manual work is generally that which allows the arithmetic to be done in base 10.

Example:

Convert 367_{10} to base 8

$8_{10})367_{10}$

$8_{10})\underline{45}$ and remainder 7

5 and remainder 5

Thus $367_{10} \equiv 557_8$.

Non-integer numbers are conveniently represented in two parts, an integer part and a fraction part. For example:

$62.159 = 6.10^1 + 2.10^0 + 1.10^{-1} + 5.10^{-2} + 9.10^{-3}$.

It is convenient to treat fractions separately and to represent a fraction of i places as:

$$F_r = f_0 r^{-1} + f_1 r^{-2} + f_2 r^{-3} + f_3 r^{-4} + \ldots f_{i-1} r^{-i} \tag{18.8}$$

or in nested form as:

$$F_r = r^{-1}(f_0 + r^{-1}(f_1 \ldots r^{-1}(f)\ldots)). \tag{18.9}$$

The problem is to find f_0 then f_1 and so on. This is achieved by multiplication by r which yields f_0 as the integer part and a fraction part which is again multiplied by r to derive f_1 and so on. Continuation of the process will yield the elements of f as far as desired. Alternatively one may evaluate the polynomial directly.

Example:

Convert 0.367_{10} to 4 places in base 8

0.367
× 8
$\underline{2}$.936
× 8
$\underline{7}$.488
× 8
$\underline{3}$.904
× 8
$\underline{7}$.232

Thus to 4 places $0.367_{10} \equiv 0.2737_8$.

It is important to note that exact conversion of fractions may not be possible.

3. Representation of Negative Numbers

The representation of negative numbers may take various forms the choice of which may be determined by considerations such as the simplicity of hardware realisation. For example, it may be desirable to avoid building a complete subtraction mechanism, as well as an addition mechanism. Practical mechanisms are also finite in length. Thus, for example, a mechanism capable of working to two decimal digits can be represented as a circle (Fig. 18.3) from which it is clear that the maximum and minimum numbers are adjacent.

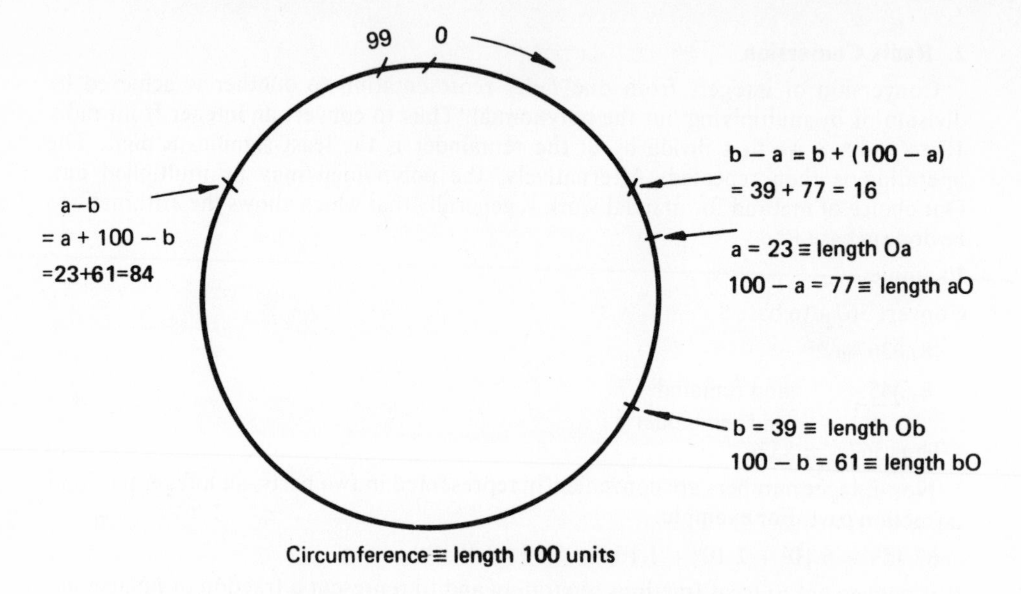

FIG. 18.3 Representation of Negative Numbers in Complement Form.

A given number is then represented as a portion of the circumference, and addition corresponds to clockwise movement. Thus, if subtraction or anti-clockwise motion is not allowed, it is apparent that we can always get to the correct result position by addition of the appropriate quantity. Thus, to form a-b we can instead do the addition

$a + (100 - b)$.

This is the basis of the complement system of representing negative numbers and leads to the general rule that addition of the complement is equivalent to subtraction. Formation of the complement is simply achieved by subtracting each digit from 9 and adding one into the least significant position. This method fits the mechanisms used, since subtraction of a number from 99 . . . 9 cannot involve a borrow digit and the 1 to be added can be done in the add mechanism.

Examples:

If $a = 39, b = 23$

$$a - b = a + 100 - b = 39 + 77 = 16$$
$$b - a = 23 + 61 = 84$$

The result 84 is the representation in complement form of −16. The sign digit may be automatically included by assigning a binary digit to the sign position thus writing a positive number as $0, d_1, d_2 \ldots$ and negative as $1, d_1, d_2 \ldots$ and allowing the rules of binary addition to apply to the sign position. Thus in repeating the examples, note that $39 + 77 = 116$ involves an extra or carry digit for the sign while $23 + 61 = 84$ does not:

$$a - b = 0{,}39 + 1{,}77 = 0{,}16$$
$$b - a = 0{,}23 + 1{,}61 = 1{,}84.$$

This form for complement is called the radix complement—in this case the tens complement. An alternative sometimes is the radix minus one, or nines complement. Thus the nines complement of 23 is $99 - 23 = 76$. However, it is also apparent from the illustration that the number zero now has two alternative forms, 99 and 00. It is also apparent that the correct result for $b - a$ requires a correction digit to be added: $b - a = 39 - 23 = 16$. This may be formed by $39 + (99 - 23) + 1$.

The correction digit is automatically supplied if the carry from the most significant place is allowed to enter the least significant digit position. This is known as end-around carry and is necessary to all radix-1 systems.

```
   0,39
   1,76
(1)0,15
      1   End-around carry
   0,16
```

Although our discussion of negative representations has concentrated on the decimal system the same reasoning is applicable to numbers in other bases. We now give more detailed discussion of binary arithmetic since this is the basis of most computer design practice.

4. Binary Arithmetic

The rules of binary arithmetic are greatly simplified by the fact that only two symbols are possible, 0 and 1, and were demonstrated in the section covering switching circuits and logic. The rules are summarised below:

Addition. The input digits are the operands x and y and input carry c. Table 18.2 is the truth table for binary addition.

Negative Numbers. The three main alternative representations of negative numbers correspond to those of the decimal system; sign and magnitude, two's complement (equivalent to tens) and one's complement (equivalent to nines). In the following examples the sign digit is at the left separated by a comma. We show the three alternative forms of negative representation:

Number	Sign Magnitude	Ones	Twos
– 13	1,01101	1,10010	1,10011
– 27	1,11011	1,00100	1,00101
– 15	1,01111	1,10000	1,10001

Two's complement

```
+ 27 – 15 ≡ 0,11011
          + 1,10001
            0,01100  ≡  + 12

– 13 + 27 ≡ 1,10011
          + 0,11011
            0,01110  ≡  + 14

– 27 + 15 ≡ 1,00101
          + 0,01111
            1,10100  ≡  – 12
```

One's complement

```
+ 27 – 15 ≡ 0,11011
          + 1,10000
            0,01011
                  1  End-around carry
            0,01100  ≡  + 12
```

$$
\begin{array}{rl}
-13 + 27 \equiv & 1{,}10010 \\
& +0{,}11011 \\ \hline
& 0{,}01101 \\
& \quad\quad 1 \text{ End-around carry} \\ \hline
& 0{,}01110 \quad \equiv \quad +14
\end{array}
$$

$$
\begin{array}{rl}
-27 + 15 \equiv & 1{,}00{,}100 \\
& +0{,}01111 \\ \hline
& 1{,}10011 \quad \equiv \quad -12
\end{array}
$$

Multiplication

Basic rule $0 \times 0 = 0$
$0 \times 1 = 0$
$1 \times 1 = 1$

This leads to the simple rule that in step by step multiplication the multiplicand is copied for a multiplier digit of 1.

Example: $13 \times 27 \equiv$ 01101 multiplicand

$$
\begin{array}{r}
01101 \text{ multiplicand} \\
\times\, 11011 \text{ multiplier} \\ \hline
01101 \\
01101 \\
00000 \\
01101 \\
01101 \\ \hline
101011111
\end{array}
$$

This may be checked as equivalent to $351_{10} = (13\times 27)_{10}$

Division

The steps are analogous to those used in decimal arithmetic. A subtraction is allowed if the remainder is positive, and a quotient of 1 is recorded; otherwise a quotient of 0 is recorded and subtraction attempted again after shifting the divisor relative to the remainder.

Example: $450 \div 15 = 30$

$$
\begin{array}{r}
01111)111000010(11110 \\
01111 \\ \hline
011010 \\
01111 \\ \hline
010110 \\
01111 \\ \hline
001111 \\
01111 \\ \hline
000000
\end{array}
$$

Octal representation is a convenient way of working with binary numbers. Numbers form groups of 3 bits which are given octal symbols 0 to 7. The chances of human error are reduced in this way. For example, the binary number 111,110,101,100,011,010,001, 000 would be written in octal as $7\,6\,5\,4\,3\,2\,1\,0_8$.

The hexadecimal system uses groups of 4 bits and the symbol set 0 to 9, A, B, C, D, E, F. Thus a binary computer with a word length of 36 bits could alternatively be represented by 12 octal digits or nine hexadecimal digits. This is commonly done to give a more readable form than pure binary. For example the binary number 1111, 1000,1101,0001 would be written as F8D1.

Exercise: Repeat the previous examples of binary arithmetic in octal and hexadecimal arithmetic.

C. BASIC COMPUTER ORGANISATION

1. Registers

The circuits that store information in a computer can be divided into two classes: registers and memory cells. Register circuits are combined with logic circuits to build up the arithmetic, control and other information processing parts of the computer. The term "memory" is commonly reserved for those parts of a computer that make possible the general storage of information such as the instructions of a program, the information fed into the program, and the results of computation. Our procedure will be, firstly to discuss arithmetic logic, then memory devices and systems and finally the organisation of a simple computer.

Registers are usually made up of one-bit storage circuits called flip-flops. A typical flip-flop circuit has two input terminals, set and reset, and two output terminals to provide both signal and complement. A flip-flop is usually named and may be represented symbolically as shown in Fig. 18.1. Thus, if a signal pulse is applied to the set terminal, Q becomes 1 and remains so until a pulse is applied to the reset terminal. Application of simultaneous pulses to both set and reset is not allowed and the circuit can be regarded as remembering its most recent input.

2. Arithmetic Logic

Logic circuits and registers can be combined to perform elementary arithmetic operations. Table 18.2 is a truth table for one digit binary addition in which the inputs to the adder are the binary digits X and Y combined with the "input carry". The outputs are the sum digit and the "carry out". Figure 18.2 is an implementation of this truth table based on AND, OR circuits.

In a computer employing binary arithmetic the basic number length could be say 32 bits. Numbers of such length can be added in two general ways. One way is to use an adder for each bit assembled as shown in Fig. 18.4. The inputs to this adder are the two four bit numbers: $X_3X_2X_1X_0$ and $Y_3Y_2Y_1Y_0$ and the output a five bit sum: $S_4S_3S_2S_1S_0$. The inputs to the adder can be provided by two four bit registers and the sum can be set into a five bit register following reset of the register to zero. This can be continued to make an adder of any length and is known as parallel addition.

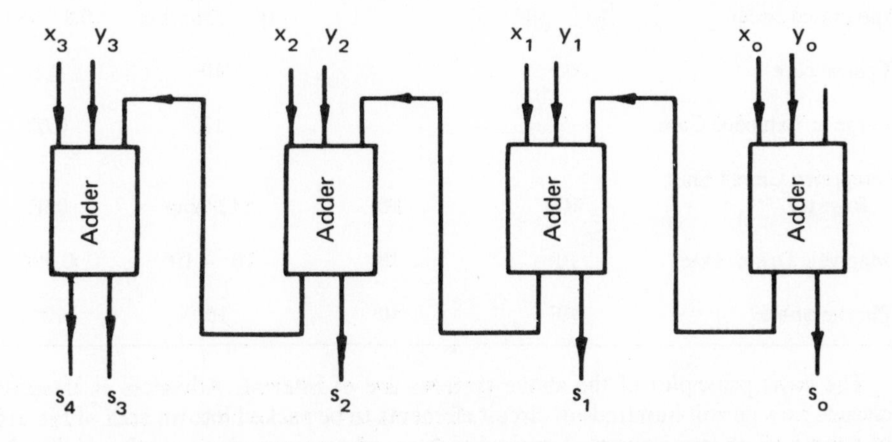

FIG. 18.4 Four Bit Parallel Binary Adder.

The second method of addition is serial which requires presentation of the appropriate bit pairs in correct sequence. One way of achieving this is to use special registers called shift registers, which have the ability to shift information from one stage to the next on application of an input to the shift terminal. The essential difference is that only one binary adder is used and that the "carry out" digit is used to set the state of the flip-flop which then forms the carry in for the next pair.

The reader is referred to references at the end of this chapter for a more complete account.

3. Memory Systems

The speed and cost of a modern computer are largely determined by the speed and capacity of the storage or memory unit. As a result a great deal of effort has been devoted to the development of memory systems and to the improvement of memory devices.

Functionally a memory unit is a device which can read or write data into a location identified by an address. In the common situation each location holds a single word of data and the address is given by a number. An alternative form of address scheme is "content addressing" in which access is determined by the content of the word being sought. Although this form of addressing has many advantages it is more costly to mechanise and will not be considered further.

The important characteristics of a memory are access time, information transfer rate, capacity and cost. Table 18.3 is a comparison summary of different systems based on these characteristics. The range in each of the characteristics should be noted, in access time, capacity, and in cost per bit. Some of the most difficult problems in computer design are connected with proper use of storage. The store should appear to the user to have large capacity and rapid access but for economic reasons this must be realised by a hierarchy of storage devices.

TABLE 18.3
Comparison of Memory Systems

Type of Memory	Typical Access Time (microseconds)	Typical Information Transfer Rate (bits/second)	Capacity (bits)	Cost (Dollars/bit)
Integrated circuit	10^{-2} – 10^{-1}		16 – 256/chip	0.4 – 0.1
Typical core	0.5		10^6	0.1
Large or Extended Core	5		10^7	0.02
Integrated Circuit Shift Register	100	10^6	512/chip	0.01
Magnetic Drum/Disc	10^4	10^7	10^7 – 10^8	0.001
Photographic	10^7	10^6	10^{12}	10^{-6}

The basic principles of the above systems are of interest. Advances in integrated circuits now permit hundreds of circuit elements to be packed into an area of the order of a tenth of an inch square. A typical high speed memory chip contains 16 flip-flops with an access time of less than 7.10^{-9} seconds (7 nanoseconds). A second technique

(MOS) is capable of much higher component density but results in slower circuits (0.1 microseconds). There is little doubt that these components will have an important role in future computers, and that they will become the dominant technique in high speed systems.

Magnetic-core memories are the most widely used random access memory systems. A normal system consists of a three dimensional array of about a million tiny magnetic cores or rings, each of which can store one bit of information. The core is made of ferrite with a rectangular hysteresis loop so that a 1 or a 0 can be stored as a clockwise or anticlockwise magnetic state. The rectangular characteristic also means that the direction of magnetism cannot be changed by a current below the threshold (say 200 ma) but can be safely changed by double this current (400 ma). This threshold characteristic enables lower cost coincident current techniques to be used for addressing and writing into a memory stack. Figure 18.5 and Figure 18.6 illustrate in further detail the principles involved.

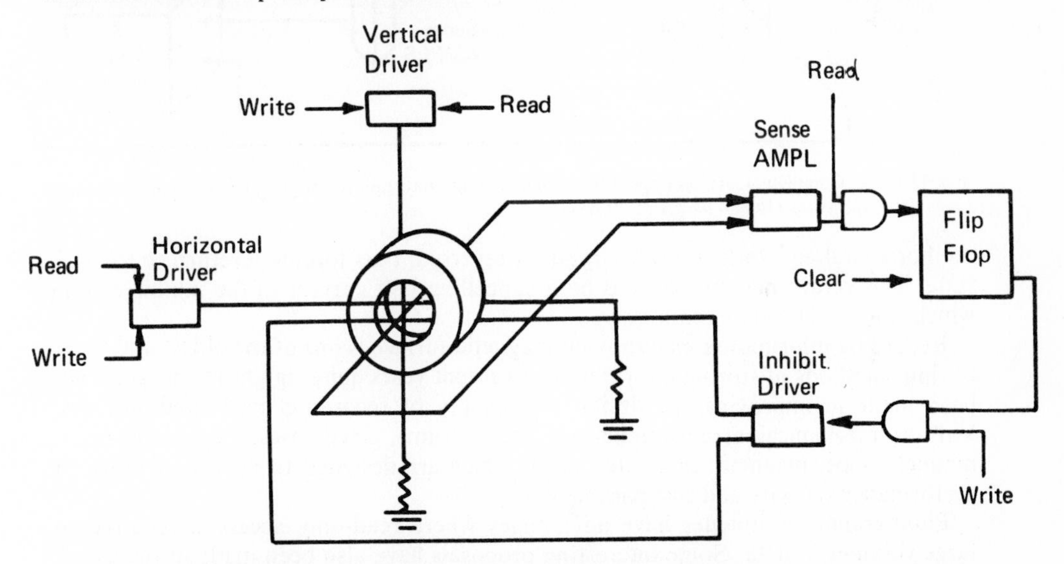

FIG. 18.5 A Single Memory Core with Associated Logic. The cores are made of ferrite with a rectangular hysteresis loop. Operation involves switching the direction of magnetisation between clockwise ('1' state) and anticlockwise ('0' state). Reading and writing signals each carry half the current needed to switch. During read, coincidence of the vertical and horizontal half currents causes switch of a core in the '1' state and an output in the sense amplifier, while the '0' state produces no output. During write the horizontal and vertical currents are reversed which writes a '1' unless an opposing half current is in the inhibit wire in which case '0' state remains.

Figure 18.5 shows the wiring details of a single memory core. In practice cores are strung according to this wiring principle in arrays or planes and a memory stack is assembled using one plane per bit of the memory word. Thus a unit consisting of a stack of 8 planes each containing an array of 32 × 32 cores would have a capacity of 1024 words of 8-bits. A word would be accessed by providing an address number of 10-bits ($2^{10} = 1024$) which would be interpreted as a horizontal selector (5 bits for 1 of 32) and vertical selector (5 bits). Figure 18.6 shows a schematic of a larger system. The principle of operation involves selective switching of one core in each plane. Thus during read coincidence of horizontal and vertical half currents causes the core at the intersection, to be forced into the "0" state. If the core was in the "1" state the reversal of magnetisation causes a signal in the sense amplifier but if the core was already in the "0" state no signal is created. During write, the direction of currents in

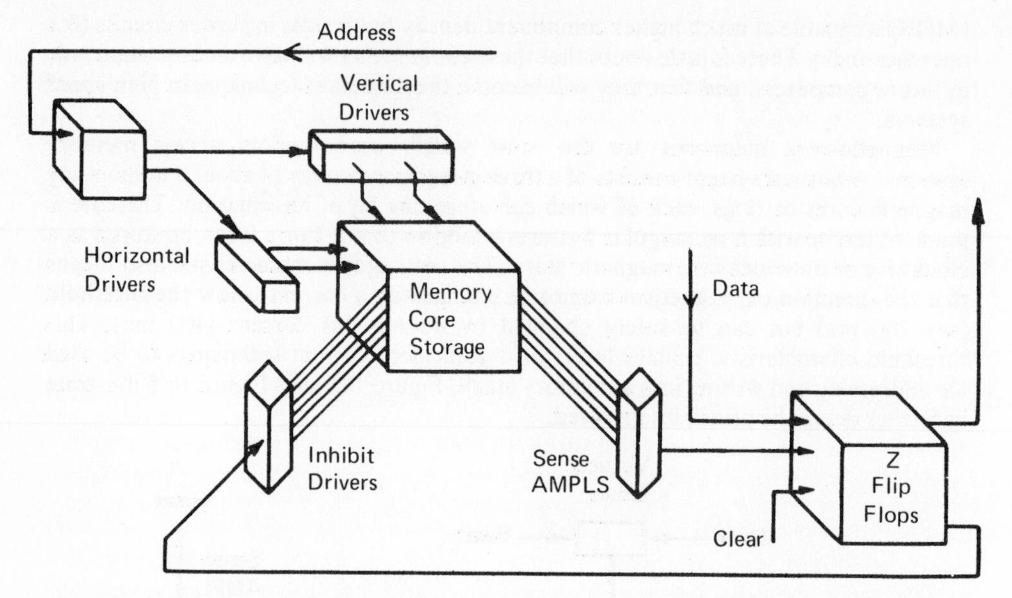

FIG. 18.6 A Complete Core Storage Unit. Such a unit may consist of 50 planes, each containing 128 × 128 cores (16,384 words of 50 bits).

the horizontal and vertical drive circuits is reversed, thus forcing a return to the "1" state. This return may be blocked by a cancelling half current in the inhibit wire in which case the "0" state remains.

Recording information on a moving magnetic surface is one of the oldest and more obvious methods of storing information. In recent years quite significant progress has been made in improving the design and quality of recording heads and surfaces. Various mechanical arrangements are used; drums, fixed discs, removable discs, magnetic tape, magnetic card, etc., all of which are designed to provide a range of performance capacity and cost parameters.

Photographic techniques have advantages where read-only access is required to large volumes of data. Some interesting proposals have also been made in the use of lasers with photographic recording.

4. Arithmetic Unit

The arithmetic unit is responsible for execution of the logical and arithmetic operations. We will consider the organisation of a simple unit capable of fixed point addition, subtraction, multiplication and division. This demonstrates most of the important principles involved and is of historical interest since it is similar in organisation to the early machines from which modern machines have evolved. Figure 18.7 shows a structure for an arithmetic unit using 4 bit words (a practical system may use 32 bits).

Data is transferred from the upper to the lower registers directly or via the adder unit and returned from the lower to upper with a right shift of one place. The example illustrates the individual steps involved in forming the product 1001 × 1101. The bits of the multiplier are sensed successively commencing with the least significant bit, and used to determine whether the multiplicand should be added (m = 1) to form a new partial product. Shifting takes place on return from the lower to upper registers. It will be seen that the growing partial product successively displaces the individual bits of the multiplier thus achieving economy in register accommodation. The essential differences between this machine method and the normal hand method are that partial products are formed for each step rather than leaving the addition to the end, and that the position at which addition takes place is fixed so that the partial product is moved instead to give the same relative shift. Figure 18.7 should be studied from which the essential principles will become apparent.

If we allow either the register value or the bitwise complement to enter the adder, then subtracting is also possible. For example, if two's complement is used then A – D may be done by using the bitwise complement of D and injecting a one into the carry input at the least significant bit position.

Division is possible if we include a path for shift left. In this case the process is one of trial subtraction at the end of which a left shift is done and the corresponding quotient digit placed in the least significant bit portion of Q. The process is illustrated in Fig. 18.8. Each step involves a trial subtraction and if the remainder is positive the subtraction is allowed and quotient of 1 recorded at the right hand end of Q. The partial result is then shifted left. If the subtraction shows a negative remainder it is not performed so that a 0 is recorded and left shift allowed. The process is repeated.

The arithmetic structure so far developed will allow addition, subtraction, multiplication, division, and with minor additions shift right, shift left and the logical operations AND, OR, etc. It is sufficient, therefore, to execute an adequate set of operations for a general-purpose fixed point binary computer. However, generation of the sequence of transfer operations which comprise an arithmetic operation such as multiplication requires a further unit called a control unit. In the next section we outline the organisation of a simple computer based on these concepts.

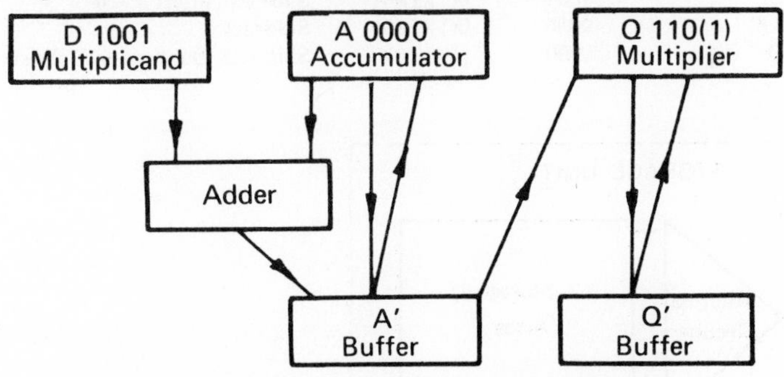

FIG. 18.7 Binary Multiplication in a Simple Arithmetic Unit.

LOCATION	CONTENTS		OPERATION
AQ	0000	110(1)	Sense m = 1
A'Q'	1001	110(1)	Add multiplicand
AQ·	0100	111(0)	Shift right 1. Sense m = 0
A'Q'	0100	111(0)	Copy
AQ	0010	011(1)	Shift right. Sense m = 1
A'Q'	1011	011(1)	Add multiplicand
AQ	0101	101(1)	Shift right. Sense m = 1
A'Q'	1110	101(1)	Add multiplicand
AQ	0111	0101	Shift right (Result)

5. Organisation of a Computer

Figure 18.9 shows the organisation of a simple general-purpose computer.

The storage unit holds both instructions and data; it is accessed by supplying the address to the address register and then passing data to or from the store via a buffer register known as the data register. Data and instructions are initially loaded into the store by the input unit and results taken from the store by the output unit. This allows the arithmetic unit to derive program instructions and data from the store and thus to avoid being bound to peripheral unit speed. The basic operation cycle is in two phases as follows:

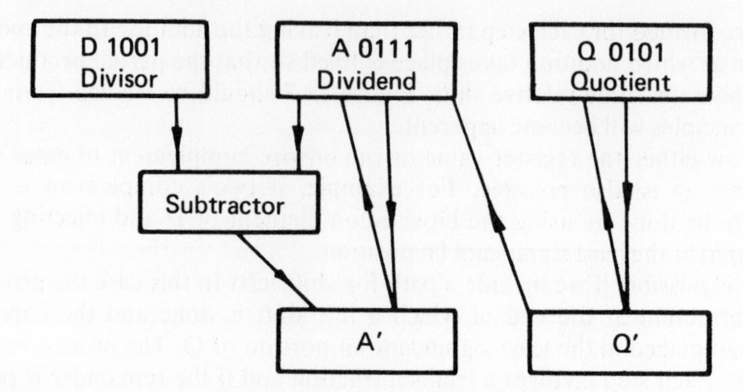

FIG. 18.8 Binary Division in a Simple Arithmetic Unit.

LOCATION	CONTENTS		OPERATION
AQ	0111	0101	
A'Q'	0111	0101	Transfer
AQ	1110	1010	Shift left
A'Q'	0101	1010	Subtract divisor
AQ	1011	010(1)	Shift left, insert quotient bit = 2
A'Q'	0010	010(1)	Subtract divisor
AQ	0100	10(1)(1)	Shift left, insert quotient bit = 1
A'Q'	0100	10(1)(1)	Transfer (subtraction goes neg.)
AQ	1001	0(1)(1)(0)	Shift left, insert q = 0
A'Q'	0000	0(1)(1)(0)	Subtract
AQ	0000	(1)(1)(0)(1)	Shift left. Quotient is Q. (Result)

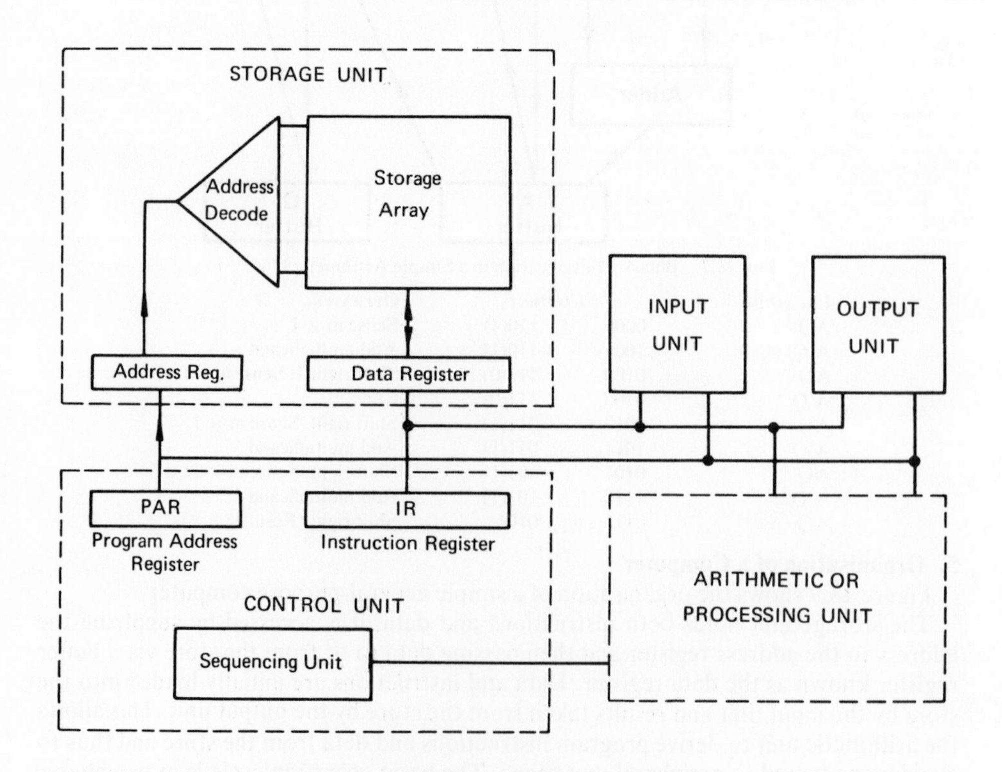

FIG. 18.9 Organization of a Simple Computer.

(1) Read next instruction. The location of the next instruction is the address held in the program address register. This address is transferred to the store address register and the contents of the location passed via the store data register to the instruction register where it sets up decoding circuits.
(2) Execute instruction. The instruction held in the instruction register is decoded and initiates the appropriate control sequence for execution. In general this will require another access to the store and in this case the store address is derived from the instruction register and the data word is passed to or from the arithmetic unit. Before returning to the "read next instruction" phase the address in the program address register is updated (in the usual case this consists of adding unity to the last value).

In the next chapter these concepts will be illustrated by specifying a machine and instruction set to be used in programming examples.

D. EXERCISES

1. Sketch a logic circuit to realise the carry output in a binary adder. Can you suggest more than one solution?
2. Parity checks are often used for error checking. The scheme involves addition of an extra bit to cause the number of ones to be even (even parity) or odd (odd parity). For example odd parity on the following 3-bit numbers would appear as:

Digits	Parity
000	1
001	0
010	0
011	1
100	0
101	1
110	1
111	0

   Devise a circuit to check whether parity is correct in a given group. Comment on the effectiveness of this method of error detection.
3. Convert the following:
   157_{10} to binary
   133_9 to octal
   567_8 to base 9
   110110_2 to decimal
4. Carry out the operations of addition, subtraction and multiplication on the following number pairs:
   1011_2 1011_2
   56_8 35_8
   23_9 38_9
5. Using the number pairs in 4 repeat the operation of subtraction using the radix complement and radix − 1 complement.

SUGGESTED READING MATERIAL

MARCUS, M. P. *Switching Circuits for Engineers*. Prentice-Hall, N.J., 1967.
HELLERMAN, H. *Digital Computer System Principles*. McGraw-Hill, New York, 1967.
Scientific American. September, 1966. (This was a special computer issue.)
CHU, Y. *Introduction to Computer Organisation*. Prentice-Hall, N.J., 1970.

CHAPTER 19

Introduction to Computer Programming

A. A. Thompson

"No, THE FUTURE offers very little hope for those who expect that our new mechanical slaves will offer us a world in which we may rest from thinking. Help us they may, but at the cost of supreme demands upon our honesty and our intelligence. The world of the future will be an even more demanding struggle against the limitations of our intelligence, not a comfortable hammock in which we can lie down to be waited upon by our robot slaves."

Norbert Wiener
God & Golem, Inc., *1964.*

A. INTRODUCTION

1. General

The purpose of this chapter is to provide you with a brief introduction to digital computer, machine language programming. The pertinent details of computer component parts, structure and operating characteristics are briefly described because we need to understand these details before we can program the computer. If you understand the material in this chapter you will be in a position to write simple machine language programs for the IBM/360 computer.

The discussion is kept as general as possible and many simplifications have been made for the sake of clarity and conciseness. The structure of a "typical computer" is described for example—however in practice computer structures take on a multitude of forms. Even so, if you understand this "typical" computer, it will be the first step to an understanding of present day digital computers.

2. The Nature of Computers

A digital computer is a machine that processes or transforms information that is represented in digital form. Computers are directed to do useful work by performing ordered sets of transformations on ordered sets of information items, each particular transformation is defined by specifying the item or items of information to be transformed and the transformation function.

The set of transformation functions invariably include the arithmetic functions. The addition function, for example, is a procedure which takes the specified function input values (the addends) and forms their sum, which is the function output or result. The conventional *arithmetical instruction* or statement $3 + 7$ is a transformation specification, it specifies the transformation function and the items to be operated on by that function: these items are variously known as function *operands*, function *arguments* or function *input-data*.

In an analogous way the transformations to be performed by a computer are specified by means of *computer instructions*, which specify the function to be performed together with its operands. Computers are designed to interpret these instructions and perform the transformations specified by them.

The essential features of a computer are: A *storage unit*, in which sets of instructions and data are stored, a set of *executable functions*, and the facility to *interpret* each instruction and *execute* the transformation that it specifies. The set of executable functions are called *hardware* functions, because they are executed by the computer hardware which consists of specially designed electronic circuits. Specific hardware functions are activated by specific instructions, to perform specific transformations on specific items of data. The computer is directed to do useful work by performing an ordered set of transformations on an ordered set of data as specified by an ordered set of instructions. The set of instructions and the set of data items are held in the computer storage unit.

In the currently available types of digital computers the hardware functions are quite primitive compared with the type of function that we want the computer to perform, such as matrix inversion, averaging a set of numbers, computing trigonometric functions, and so on. On the other hand, however, the hardware functions are relatively complex compared with the digital logic circuit functions with which the machine is built (such as AND, OR and NAND functions); the effect of this is that, the more complex the hardware function then the more complex the logic circuits, the more it costs, and the more difficult it is to design and maintain in operation. These conflicting factors result in a compromise between the type of hardware function suitable to the users' needs and the complexity and cost of building these functions into the computer. A typical set of hardware functions is listed in Table 19.4. The programmer uses these hardware functions to program more elaborate functions such as matrix inversion, trigonometric functions, and so on, which more closely meet his needs. These functions, called software (i.e., programmed) functions, are used in turn to construct programs which fulfil the users' needs, such as programs for statistical analysis, engineering design, scientific calculations, processing of business data, and so on.

Function	Operands	Instruction
SET	O & SUM	SET SUM TO O
ADD	A & SUM	ADD A TO SUM
ADD	B & SUM	ADD B TO SUM
ADD	C & SUM	ADD C TO SUM
SET	AVE & SUM	SET AVE TO SUM
DIVIDE	AVE & 3	DIVIDE AVE BY 3

FIG. 19.1 Schematic Illustration of Computer Program.

3. Computer Programs

The computer-user specifies what he wants the computer to do by preparing a program of computer instructions, such as that schematically illustrated in Fig. 19.1, which is designed to compute the sum and average value of the three numbers denoted A, B, and C. Programs of a more complex nature such as those for statistical calculations or engineering design calculations, would have the same general characteristics but consist of a much larger number of instructions. In addition to the type of instruction shown in Fig. 19.1 the user would also specify instructions to control the operation of computer input and output devices, and hence specify control over the input of data and the output of results.

Program execution commences after the program is fed into the computer and stored in the storage unit. During the execution of a typical program it will read its input-data, which will usually be set up in an input device such as a card reader unit, it will process this data and produce results on an output device, usually in printed form on a printer unit.

4. The Binary Nature of Computers

Digital computers are constructed in the main from *binary devices* that is, electronic, mechanical, magnetic, optical, etc., devices which have only two stable states. Since about 1958 the majority of the binary devices used for building computers have been constructed using one of three major types of physical components, they are: magnetic ferrite cores; semiconductor devices such as transistors, diodes and integrated circuits; and magnetic surfaces in the form of magnetic drums or cylinders, magnetic disks and magnetic tapes.

The main reason for using binary devices is their high reliability; two other important reasons are the relative simplicity and the low cost of the basic binary devices made from the components mentioned above. When one considers that a computer may consist of hundreds of thousands of binary elements it is apparent that the operational reliability of these elements is crucial to the reliable operation of the computer as a whole.

The inherent reliability of binary devices stems from the fact that they have only two stable states (and hence are commonly called *bistable devices*). For example, a transistor gate circuit can be either open or shut, a semiconductor flip-flop circuit can be either on or off, a magnetic ferrite core can be magnetised in either one direction or the other, a spot on a magnetic surface can be either magnetised or not magnetised, and so on. This inherent reliability of binary mechanisms is employed in a multitude of forms in computer systems—with punched cards for example information is conveyed by the fact that a hole is either punched or not punched in a specific position on the card.

Because computers are constructed from binary devices it follows that they can only store and process information that is represented in binary form. Hence all numerical data must be either in the form of a binary number (for example decimal 43 is represented as binary 101011) or in some binary coded form, the most common of which is Binary Coded Decimal (BCD) in which a code consisting of four binary digits is used to represent each of the decimal digits as indicated in Table 19.1.

The Binary Coded Decimal representation for the decimal number 43 is for example, 0100 0011. Alphabetic data is also stored and processed in binary coded form, a commonly used code employs eight binary digits to represent each alphabetic character as shown in Table 19.2.

TABLE 19.1
Binary Coded Decimal Representation.

Decimal Digit	Binary Coded Decimal Representation	Decimal Digit	Binary Coded Decimal Representation
0	0000	5	0101
1	0001	6	0110
2	0010	7	0111
3	0011	8	1000
4	0100	9	1001

TABLE 19.2
Binary Coded Character Representation

Alphabetic Character	Binary Coded Character Representation	Alphabetic Character	Binary Coded Character Representation
A	1100 0001	—	— —
B	1100 0010	—	— —
C	1100 0011	U	1110 0100
D	1100 0100	V	1110 0101
E	1100 0101	W	1110 0110
—	— —	X	1110 0111
—	— —	Y	1110 1000

B. GENERAL STRUCTURE OF A COMPUTER SYSTEM

1. Central Processing Unit

(a) GENERAL. A typical digital computer system consists of three major parts: an input unit, a Central Processing Unit and an output unit, as shown in Fig. 19.2. The central processing unit (CPU) can be further subdivided into the following subunits; a storage unit (which is usually called the main or primary storage unit), a control unit, a function unit and a register unit. The interrelationship between these units is also depicted in Fig. 19.2, where the solid lines indicate the source and direction of control signals within the system and the dashed lines indicate the direction of information flow within the system.

(b) MAIN STORAGE UNIT. This unit consists of a set of storage cells, in which items of information can be stored. Each cell can be used to store a fixed amount of information in binary coded form. The size or storage capacity of these cells varies from one type of computer to another, typical sizes include 8-bits, 16-bits, 32-bits, 48-bits and 64-bits. In an 8-bit cell, for example, we can store one alphabetic character, represented by an 8-bit code, or two decimal digits represented in Binary Coded Decimal form, or an 8-bit pure binary number. Alternatively, with computers having 32-bit storage cells we can store four 8-bit alphabetic character codes, and so on. Once the information in a particular cell is no longer required, that cell can be used to store some other different item of information.

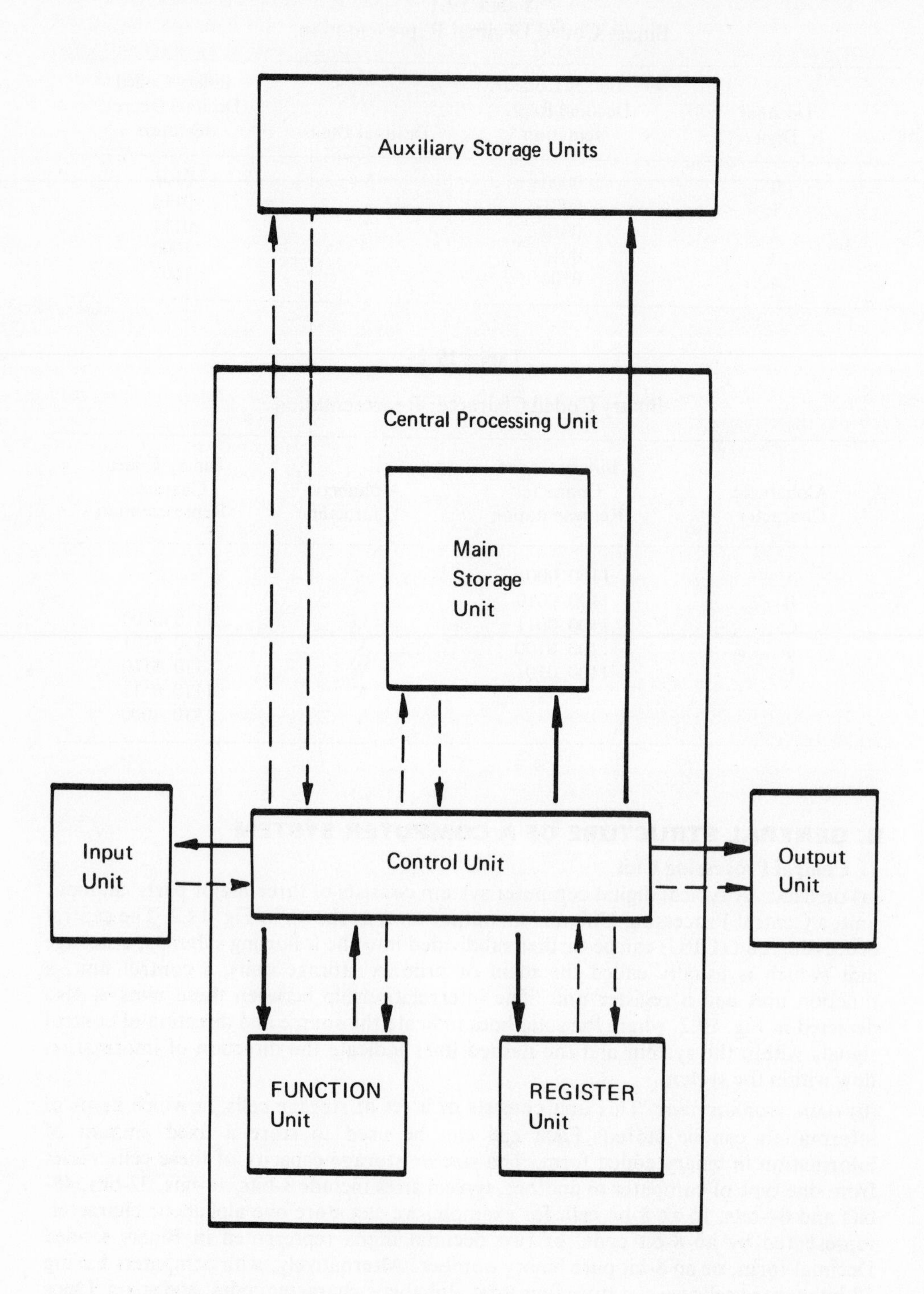

Fig. 19.2 Schematic of Computer System showing Functional Units and the Direction of Control (—) and Information (- - - -) Flow.

Each storage cell is identified by a numeral address (in the actual machine this will be a binary number address, however, we will use decimal numbers for simplicity). When we want either to obtain information that is stored in a particular cell, or to store information in a particular cell we specify the address of that particular cell in an appropriate computer instruction. The storage unit electronic circuitry is designed to use the specified address to locate the storage cell that we want to reference. The programmer keeps track of the items of information held in the storage unit by keeping track of the address of the storage cell in which each particular item of information is stored.

The most common type of main storage device is the magnetic core store unit. The size of these units vary from computer to computer, ranging from about 4000 cells for small computers to 250,000 cells or more for very large computers, with medium sized computers typically having 32,000 to 64,000 storage cells. The most important feature of core store units is their operating speeds—a cell of information can be stored in, or retrieved from, a magnetic core storage unit at rates ranging from one cell every 750 nanoseconds to one cell every two or three microseconds, depending on the operating speed of the particular storage unit.

The operating speed of the main storage unit is of major importance because the computer's operating speed, that is the rate at which it can process information is directly related to the rate at which it can store information in, and access information from its main storage unit.

In most computers all information that flows through the computer system passes through the main storage unit. For example, input information is fed into main store from whence it may either be processed immediately or it may be transferred to an auxiliary storage unit from whence it can be fetched and processed at a later time. Similarly output information can only be fed out (e.g., on a line printer unit) from main storage, for example, if the information required to be output is held in an auxiliary storage unit it must first be transferred to main storage and then fed to the output unit from there.

(c) FUNCTION UNIT. This unit contains the electronic circuits which perform the set of hardware functions built into the computer, these functions include arithmetic functions such as add, subtract, multiply and divide; and Boolean-logic functions such as AND, OR, Exclusive-OR, and so on. The speed of operation of an arithmetic unit varies with the type of computer and the type of function to be performed; some typical performance figures are as follows:

Addition, Subtraction:	0.1 to 3.0 microseconds
Multiplication, Division:	0.5 to 30 microseconds
AND, OR:	0.1 to 3.0 microseconds.

As a matter of interest, if the operating speed of the main storage unit is assumed to be two microseconds for each access/storage operation and the function unit takes 250 nanoseconds to perform an addition function then the total time taken to perform the addition of two operands from core store will be 6.25 microseconds as indicated below.

Fetch 1st Operand		Fetch 2nd Operand		Add		Store Result		
2.0	+	2.0	+	0.25	+	2.0	=	6.25

(d) REGISTER UNIT. Many computers contain a limited number of very high speed storage cells known as *registers* in which information can be stored or accessed at speeds ranging, typically, from 100 to 250 nanoseconds, depending upon the make of computer. The programmer can use these register cells (by specifying the appropriate instructions) for the high speed storage of such things as operands that are to be used repeatedly in a set of arithmetic operations, or for intermediate arithmetic results that are to be used more or less immediately in following arithmetic operations.

To illustrate, assume that we want to add a constant value, CV say, to a set of ten other values, V1, V2 . . . V10; now if CV had to be fetched from a main storage cell for each add operation, then each such operation would take a total of 6.25 microseconds. This time can be reduced by storing CV in a register and by proceeding as follows (where the register fetch/store speed is assumed to be 0.25 microseconds).

Fetch CV from main storage 2.0	+	Store CV at register 0.25					=	2.25 microseconds
Fetch V1 from main storage 2.0	+	Fetch CV from register 0.25	+	Add 0.25	+	Store result in main storage 2.0	=	4.5 microseconds
Fetch V2 from main storage 2.0	+	Fetch CV from register 0.25	+	Add 0.25	+	Store result in main storage 2.0	=	4.5 microseconds
etc.		etc.						

In this case a total time reduction of approximately 27 per cent is gained through the use of a high speed register for this sequence of operations.

(e) CONTROL UNIT. As its name implies this unit controls and supervises the operation of the computer system, which it does by sending control signals to each of the units in the system specifying the various operations that they are to perform. Control unit operations are specified by the program of instructions prepared by the programmer. The control unit takes instructions, one at a time from main storage, decodes each instruction and sends signals to the various units required to take part in the performance of the specified operation. The control unit then coordinates the activities of the various units to complete the execution of the operation specified by the instruction, it then takes the next instruction from main storage, decodes it, and so on.

When an instruction specifying an arithmetic function is executed, for example, the control unit fetches the function arguments (data items) from the storage cell address specified in the instruction, and transmits them to the function unit. The control unit then sends a signal to the function unit causing it to perform the required function (add, subtract, multiply or divide). When the function unit has completed its operation the control unit takes the result and transmits it back to the storage cell specified in the instruction.

When an instruction specifies the reading of a punched card, for example, the control unit sends a command signal to the card reader unit directing it to read the contents of the next card in its input hopper. The control unit then supervises the transmission of this input information from the card reader to the main storage cell locations specified in the instruction. And so on.

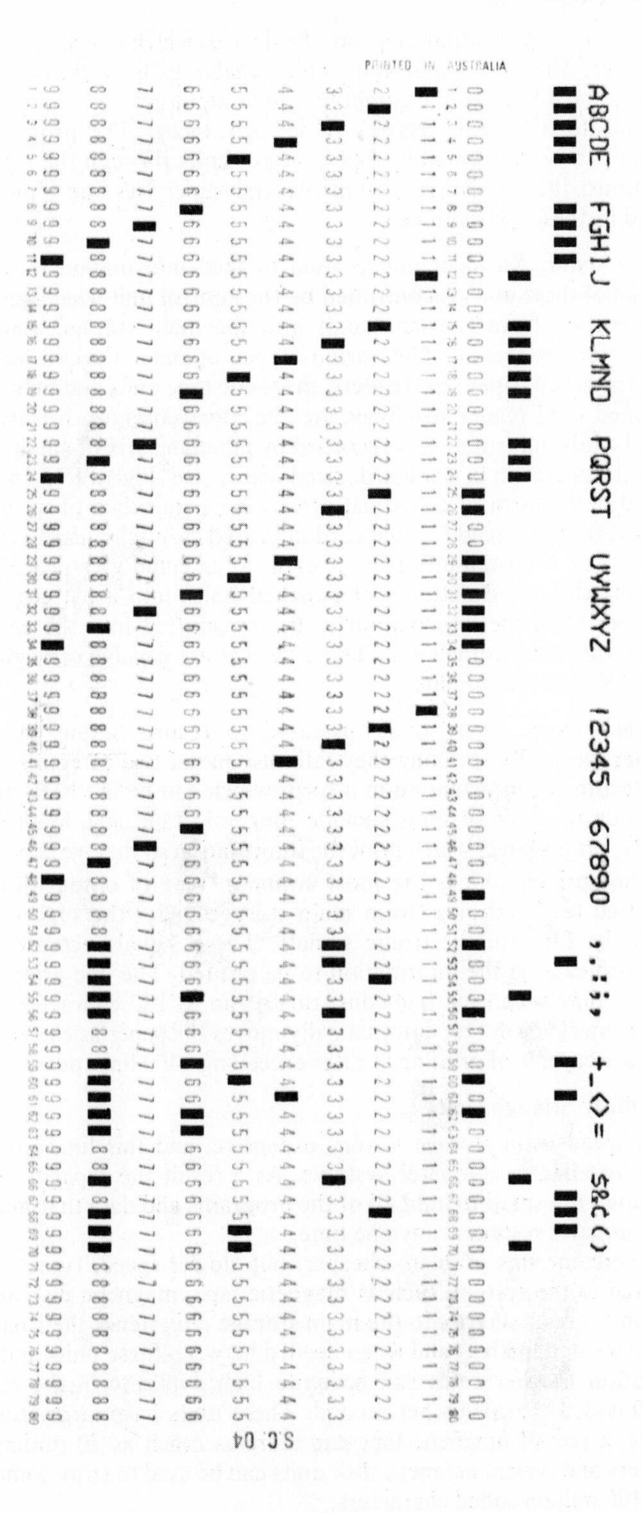

Fig. 19.3 Standard 80-column Punched Card, showing a set of Punched Card Character Codes.

2. Input/Output Units

(a) GENERAL. Input/output units are the devices which enable us to communicate with a computer. There are input units which enable us to feed or send information (i.e., programs and data) into the computer, and output units which enable the computer to communicate to use the results of its processing. The programmer specifies the operations to be performed by input/output units through the use of special machine instructions; the control unit controls and supervises the input/output operations specified by these instructions.

(b) INPUT UNITS. An input unit is used to feed information into the computer; the operation of these units is controlled by the control unit which supervises the transfer of information from the input unit into the main storage unit where it becomes available for processing. The various types of input devices include punched card readers, punched paper tape readers, magnetic tape units and magnetic disk units.

Punched card reader machines are the most common form of input unit. With punched cards, information is recorded by punching sets of small rectangular holes in specific locations in a standard sized card, see Fig. 19.3. Information which is recorded in the form of coded patterns of holes punched in specific locations on the card, is sensed or "read" as the card is moved through a card reading machine. The "reading" of information from a card is essentially a process of automatically converting the coded patterns of punched holes into a pattern of electronic pulses thereby enabling the information to be transmitted into the computer system and stored in the main storage unit. These devices are capable of reading at rates of up to about 1000 cards per minute.

(c) OUTPUT UNITS. These units are used to record output information from the computer, generally speaking they fall into one of two categories; firstly, those units which record the information in a form which can be fed back into the computer at some later time, these units include magnetic tape and magnetic disk units, and secondly, printer units which print the information so that we can read it.

The line printer unit is the most common type of output device, information is transmitted to this device (from main storage under the supervision of the control unit) in the form of electronic signals. These signals actuate the unit's printing mechanism causing the information to be printed. The line printer prints one line of print at a time with each line consisting of up to 132 characters and it has a paper transport mechanism that automatically moves the paper as each line is printed. These devices are capable of printing at rates exceeding 1400 lines per minute.

3. Auxiliary Storage Units

High speed main storage is very expensive, and this limits the amount of main storage installed in computer systems. As a result the capacity of the main storage unit is often too small to hold all of the programs and data that may need to be stored in the computer system at any one time.

To overcome this problem cheaper, but slower speed, types of storage units are connected to the system, such as magnetic tape, magnetic disk and magnetic drum. These units are auxiliaries to the main storage unit, hence their name. Information is passed back and forth as and when needed between these units and main storage. The information transfer rates can be quite high, typically in the range of 150,000 to 300,000 coded characters per second. These units have large storage capacities, for example, a reel of magnetic tape can store as much as 20 million coded alphabetic characters and typical magnetic disk units can be used to store somewhere in the range of 5 to 100 million coded characters.

C. COMPUTER INSTRUCTIONS AND PROGRAMS

1. Machine Instructions

Digital computers can only store and process information that is represented in binary form, this also applies, of course, to computer instructions and hence these instructions consist of binary coded information. To illustrate, an instruction specifying the transfer of the contents of a main storage cell to a register cell might have the form illustrated in Fig. 19.4 where it can be seen that instructions consist of several predefined bit fields, each of which contains either binary coded information or binary numbers relating to some specific thing. For example, the first eight bits of the example instruction constitutes a function code field, in this case the function code 01011000 specifies the function "transfer the contents of a storage cell to a register cell". The second field consisting of bits 8 to 11 is a register address field, in this case the register address is register number 9 (binary 1001). The contents of the third and last field specifies the address of the storage cell the contents of which are to be transferred to register number 9; in this case it is storage cell number 3075 (binary 1100 0000 0011).

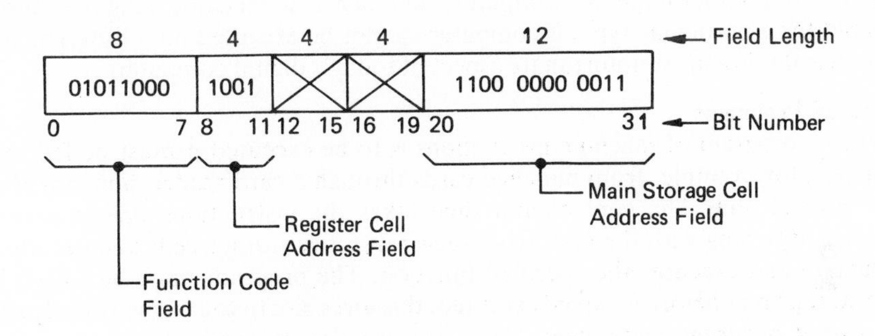

FIG. 19.4 Typical Store-to-Register Instruction Format.

To take another example, an instruction specifying the addition of the contents of register six to the contents of register ten might have the format shown in Fig. 19.5. The result obtained is returned to register ten thus replacing the value that was in register ten immediately before execution of the instruction.

Note that in computer instructions the function arguments are not specified directly, the arguments are identified by specifying the address of either the storage cell or the register cell in which the arguments are stored.

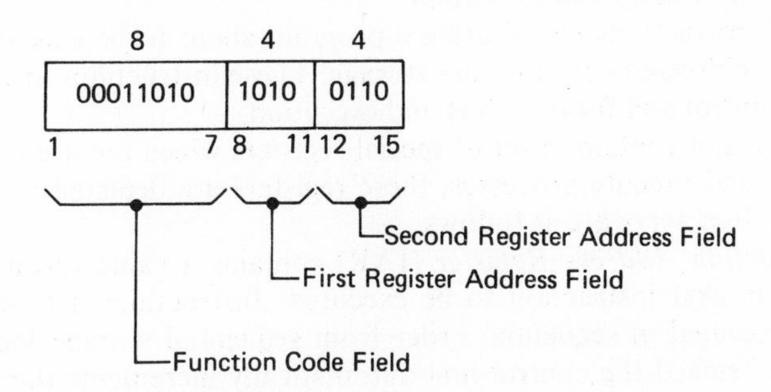

FIG. 19.5 Typical Register-to-Register Instruction Format.

We call these instructions, *machine instructions* or *hardware instructions*, because the operations specified by them are directly executed by the machine hardware. (They are also called binary or machine code instructions, however, these terms are somewhat misleading because there are other sorts of instructions which of course have to be represented in binary, i.e., machine code form, but which cannot be directly executed by hardware.)

The set of hardware or machine instructions that a computer is designed to interpret and execute constitutes the machine language for that particular computer—it is the language that the programmer uses to specify what he wants the machine to do.

A typical digital computer might have from about 20 to 100 or so different types of machine instructions depending upon its type and make. Whilst the set of hardware functions (e.g., add, subtract, multiply, AND, OR, etc.) specified by machine instructions, and hence performed by their execution, are much the same from one type of computer to another, the format or structure of these machine instructions and their field codes are different for different types of computers. This means that the machine instructions designed for one type of computer cannot be interpreted and executed by another type of computer, and hence a machine language program prepared for a particular type of computer cannot be executed on a different type of computer; this is a most unfortunate aspect of today's digital computers.

2. Stored Programs

When a program of machine instructions is to be executed it must be fed into the computer, for example, from punched cards through a card reader, and stored in the main storage unit. The control unit then takes the instructions one at a time, in sequence beginning with the first, from their respective storage cells and decodes each instruction and executes the specified function. The program must be stored in the main storage unit before it can be executed, this gives rise to the terms *stored program* and *stored program computers*. Also because the program is stored within the computer they are sometimes referred to as *internally programmed computers*.

The stored program is the most significant feature of modern day digital computers. The significance of this feature lies in the fact that once the program of instructions is stored inside the computer the instructions can be fetched, decoded and executed at very high speeds. Each instruction will usually occupy one storage cell and hence can be accessed for execution in a few microseconds or a fraction of a microsecond, depending upon the operating speed of the storage unit. The decoding and execution of each instruction will take a few more microseconds. Hence the time taken to execute a program consisting of a hundred or so instructions may be only a few hundred or so microseconds.

3. Instruction Decoding and Execution

The set of instructions constituting a program about to be executed are stored in contiguously addressed cells in main storage. These instructions are fetched one at a time by the control unit for decoding and execution.

The control unit contains a set of special registers which are used in the instruction fetch decode and execute processes, these registers are depicted in Fig. 19.6(b), the purposes that they serve are as follows.

The Instruction Address Register (IAR) contains a value specifying the storage address of the next instruction to be executed. Instructions are fetched (i.e., read from) and executed in sequential order from sequential storage locations. As each instruction is fetched the control unit automatically increments the address value in the Instruction Address Register to the address of the cell containing the next instruction to be executed.

The Instruction Execution Register (IER), instructions are fetched one at a time from storage and loaded into this register and held there whilst they are decoded and executed. The instruction held in this register at any given time is called the current instruction, i.e., the instruction currently being executed.

The Operand Address Register (OAR), also known as the Data Address Register, holds the address of the store location to be referenced by the current instruction. The control unit determines the operand address, if any, when decoding the current instruction and loads that address into the Operand Address Register.

The Store Buffer Register (SBR) is used to buffer (i.e., store temporarily) information being sent to and from the storage unit. When a word of information is to be put into storage the control unit loads the word into the Store Buffer Register, it then loads the address of the storage cell in which the word is to be stored into the Operand Address Register. The control unit then transmits the contents of these two registers to the storage unit together with a "store information" command signal. The storage unit then proceeds to store the transmitted word of information at the specified store address. Similarly when a word of information is to be read from storage the control unit transmits a "read information" command signal, in this case the contents of the storage cell having the address specified in the Operand Address Register is read out and transmitted into the Store Buffer Register where it becomes available for processing.

Determination of the actual address of that storage cell, if any, to be referenced by the current instruction may involve a fairly complex *hardware address computation procedure*. In order to describe this procedure more clearly we will define the instruction execution cycle in more detail, it consists of the following three phases.

(i) *Instruction Fetch phase*. During this phase the control unit fetches the next instruction to be executed (from the main storage cell specified by the address value in the Instruction Address Register) and puts it into the Instruction Execution Register, it then becomes the current instruction, i.e., the instruction currently being executed.

(ii) *Address Computation phase*. During this phase the control unit decodes the current instruction and uses the information in its address field(s) to determine the address of the main storage cells (if any) to be referenced by the current instruction. This address is then loaded into the Operand Address Register for use in the next phase.

(iii) *Instruction Execution phase*. During this phase the control unit: (1) increments the address value in the instruction address register to the address of the next instruction to be executed; (2) then executes the current instruction by performing the function specified by its function code; and (3) returns to the Instruction Fetch phase and repeats the cycle for the next instruction to be executed.

The address of the main storage cell (if any) to be referenced by the current instruction is put in to the Operand Address Register. If this instruction specifies the use of an operand held in a main storage cell (such as an instruction which specifies the transfer of an operand from a main storage cell to a register cell) then the control unit fetches the operand from the main store address specified by the current contents of the Operand Address Register. If the current instruction specifies the storing of an operand in a main storage cell (for example, an instruction which specifies the transfer of an operand from a register cell to a main storage cell), then the control unit will supervise the storing of the operand at the address specified by the current contents of the Operand Address Register.

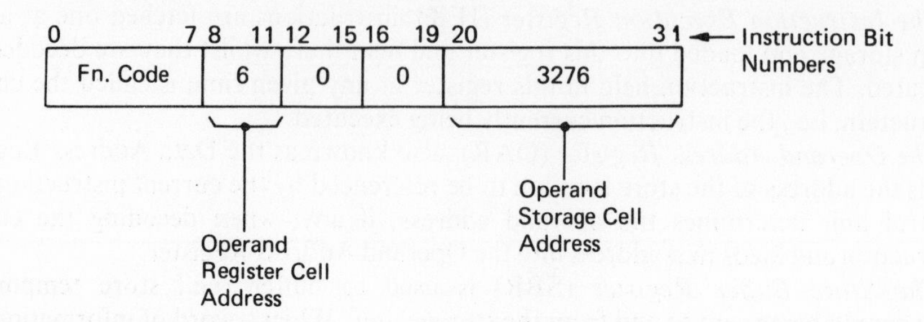

FIG. 19.6(a) Instruction which uses Direct Addressing to Specify the Transfer of the Contents of Main Storage Cell 3276 to Register Cell 6.

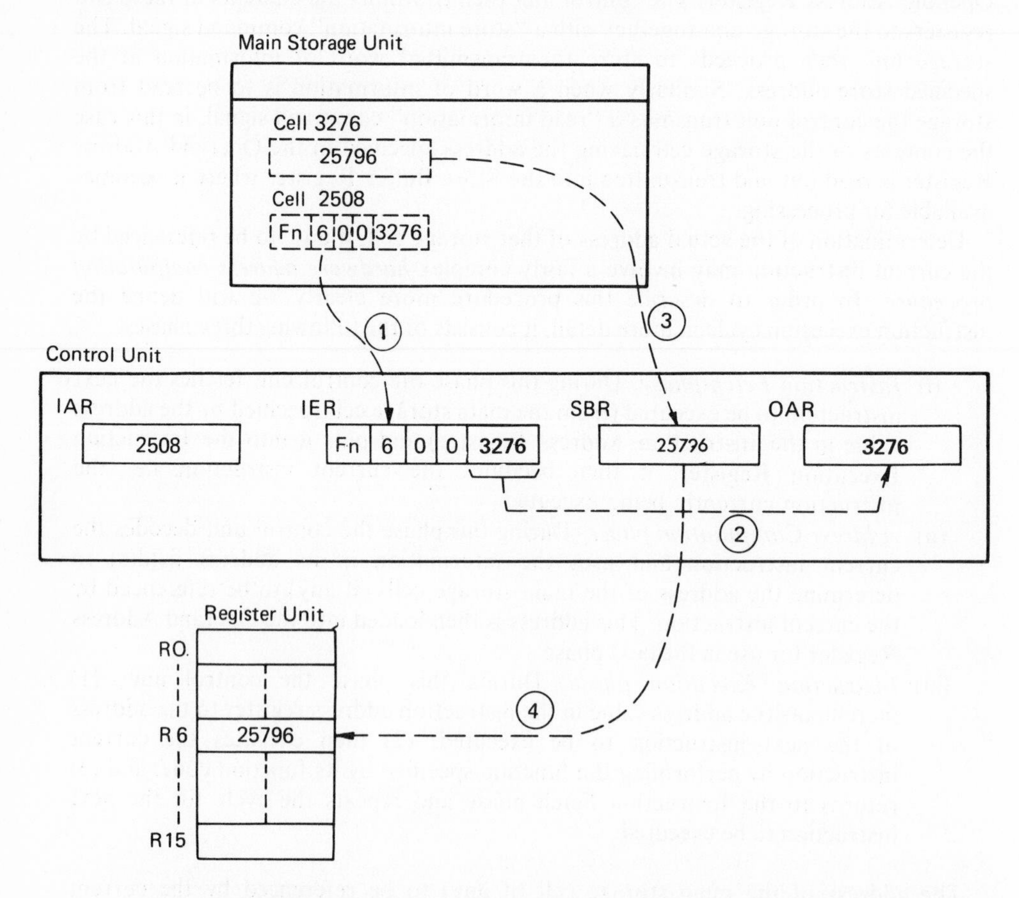

FIG. 19.6(b) Schematic Illustration of Direct-Addressing Process Performed during the Execution of the Instruction Shown in Fig. 19.6(a).

4. Addressing

(a) DIRECT ADDRESSING. With this mode of addressing the actual address of the storage cell to be referenced is transferred direct, without modification, from the storage address field of the current instruction and put into the Operand Address Register. The process is illustrated in Fig. 19.6(b), for the instruction shown in Fig. 19.6(a), which specifies the transfer of the contents of storage cell 3276 to register number 6.

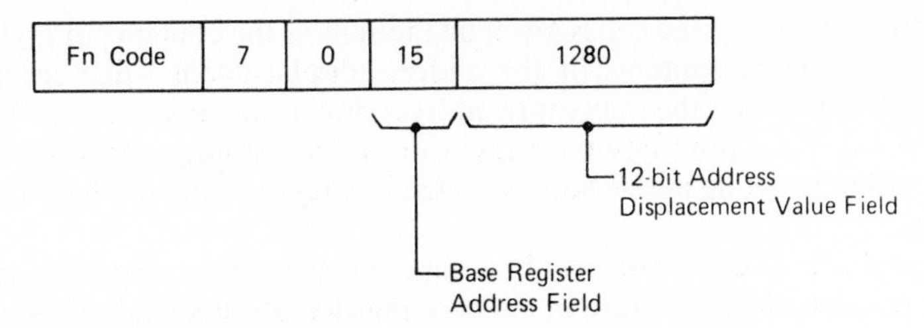

FIG. 19.7(a) Instruction which uses Base-Displacement Addressing to Specify the Transfer of the Contents of Storage Cell 3280 to Register Cell 7.

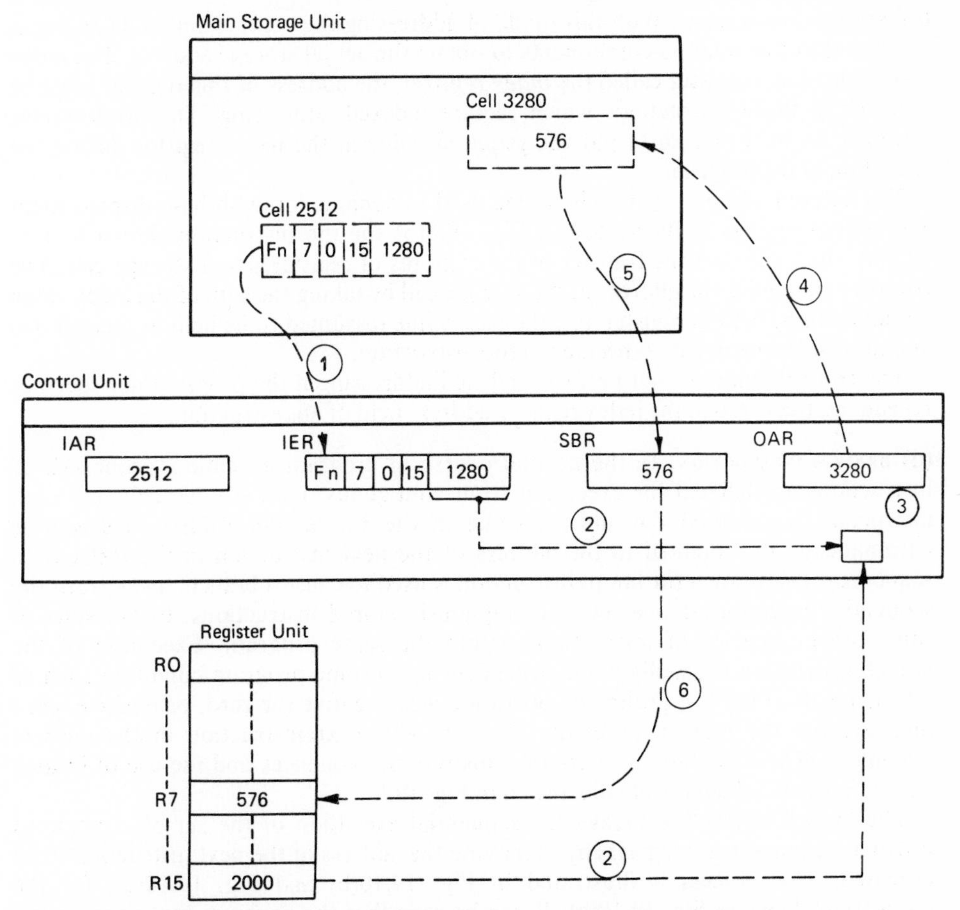

FIG. 19.7(b) Schematic Illustration of Base-Displacement Addressing Process performed during the Execution of the Instruction shown in Fig. 19.7(a).

(b) BASE-DISPLACEMENT ADDRESSING. With this mode of addressing the control unit obtains the actual address of the storage cell to be referenced by adding an *address base value* and an *address displacement value*. The base address value is held in a register called the *base address register*; the address of this register must be specified in those instructions which require base-displacement addressing. The address displacement value is specified in the instruction as indicated in Fig. 19.7(a).

The addressing process is illustrated in Fig. 19.7(b) for the instruction shown in Fig. 19.7(a) which specifies the transfer of the contents of a storage cell to register number

7. The address of the storage cell is given by the sum of the contents of register 15 (the base register) and the contents of the address displacement value specified in the instruction. Observe that the maximum address displacement value is $(2^{12} - 1)$ which is 4095 decimal, if the programmer wants to reference storage addresses higher than 4095 then he has to set up a base address value in a register and use base-displacement addressing.

The control unit does not perform base-displacement addressing if the programmer specifies a register address of zero in the base register address field of his instructions: in this case the control unit treats the value specified in the instruction's address displacement field as a direct address.

(c) INDEXED ADDRESSING. With this mode of addressing the control unit adds an *index value* to the other address components to obtain the actual storage address. The index value is held in a register called the *index register*, the address of this register must be specified in those instructions which require indexed addressing. The programmer arranges for his program to put the required value in the index register during the execution of the program.

The indexed addressing (which can be used in conjunction with base-displacement addressing) process is illustrated in Fig. 19.8(b) for the instruction shown in Fig. 19.8(a) which specifies the transfer of the contents of register|8| to a storage cell. The control unit obtains the address of the storage cell by taking the sum of the index value (assumed to be held in register 10), the base value (assumed to be held in register 15) and the displacement value specified in the instruction.

The control unit does not perform indexed addressing if the programmer specifies register address zero in the index register address field of an instruction.

(d) BRANCH INSTRUCTION. In the normal course of program execution, sequences of instructions are fetched for execution from contiguous main storage cells (as each instruction is executed the address value in the Instruction Address Register is automatically incremented to the address of the next instruction in the instruction sequence). The Branch (or jump) instruction is used to cause a branch, away from the sequential execution of one set of contiguously stored instructions, to the start of some other sequence of instructions within the same program. Execution of the branch instruction is usually made dependent upon some program condition such as "if the result of the last arithmetic operation was negative (or zero, or positive, etc.) then execute the branch, otherwise proceed with next instruction in the current sequence". The execution of alternative instruction sequences and the use of branch instructions are schematically illustrated in Fig. 19.9.

The branch instruction breaks the sequential execution of the set of contiguous instructions, in which it appears by specifying the address of the next instruction to be executed. This process is illustrated in Fig. 19.10(b) and Fig. 19.10(c), for the instruction shown in Fig. 19.10(a). It can be seen that the program branch is simply achieved by loading the address specified in the branch instruction into the Instruction Address Register. The conditional execution of branch instructions, is dealt with in more detail in the section on IBM/360 Machine Instructions.

The ability to program the conditional execution of alternative program procedures (or sets of instructions) is a fundamentally important feature of modern day digital computers. This feature enables us to program the computer to make decisions such as "if the result of the last arithmetic operation was zero (e.g., $A - B = 0$ or $A - 3.141 = 0$, etc.) then branch to and execute an alternative procedure, otherwise continue on executing the current procedure". The ability to program a computer to make decisions and follow alternative processing paths enables us to write generalised

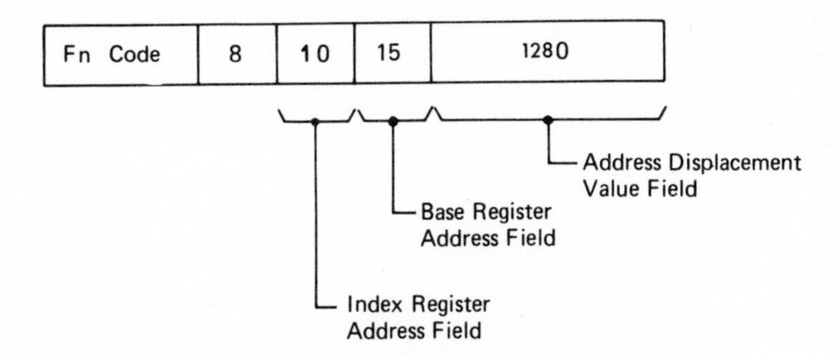

FIG. 19.8(a) Instruction which uses Indexed, Base-Displacement Addressing to Specify the Transfer of the Contents of Register Cell 8 to Storage Cell 3284.

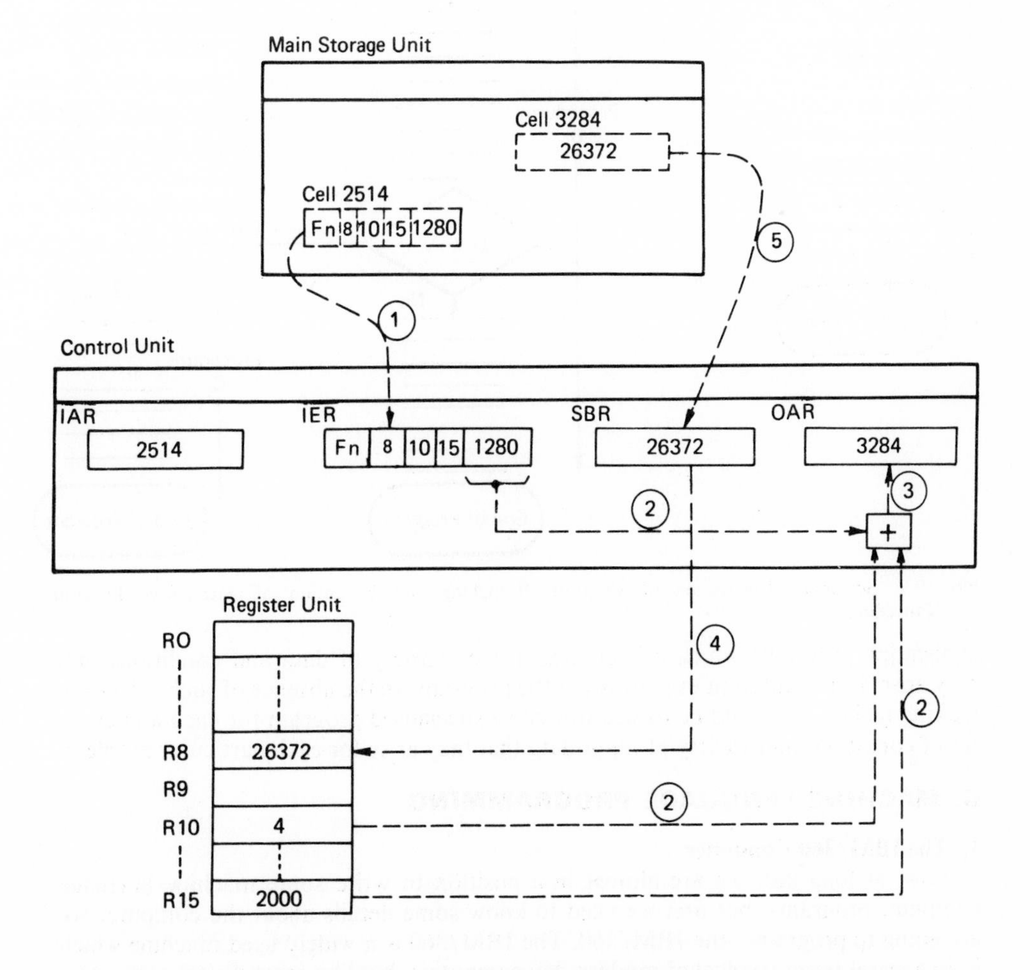

FIG. 19.8(b) Schematic Illustration of Indexed-Base-Displacement Addressing Process performed during the execution of the Instruction Shown in Fig. 19.8(a).

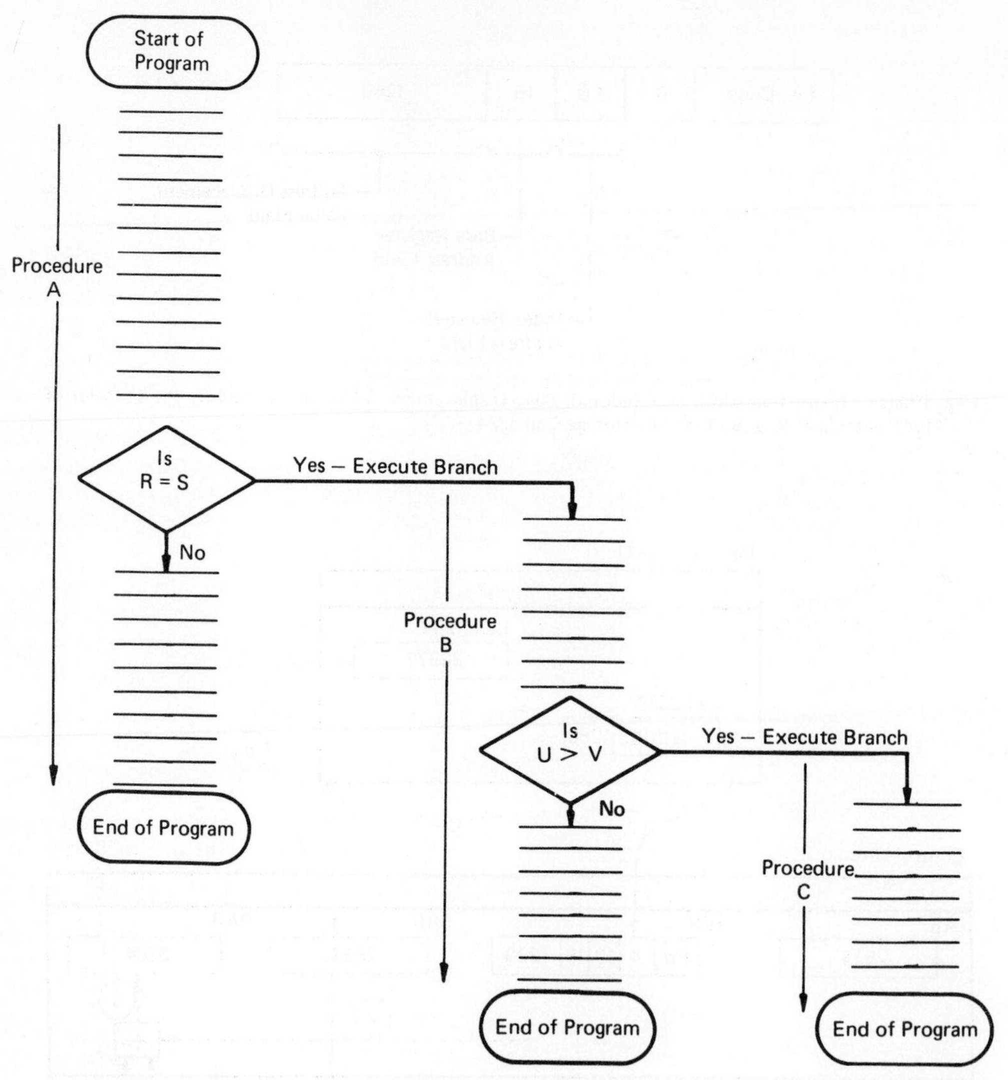

FIG. 19.9 Schematic Illustration of Program Branching and Execution of Alternative Program Procedures.

application programs which can process a wide variety of data and conditions that may arise during different executions of the program. In the absence of such a decision making facility we would be forced to write a specialised program for each and every set of conditions and variety of input data that may arise for each particular problem.

D. MACHINE LANGUAGE PROGRAMMING

1. The IBM/360 Computer

Now, at long last, we are almost in a position to write some machine language computer programs, but first we need to know some details about the computer we are going to program—the IBM/360. The IBM/360 is a widely used machine which is, in general terms, typical of modern day computers, but like most digital computers it has some characteristics peculiar to itself, these are described below.

(a) THE MAIN STORAGE UNIT. The main storage unit of the 360 consists of individually

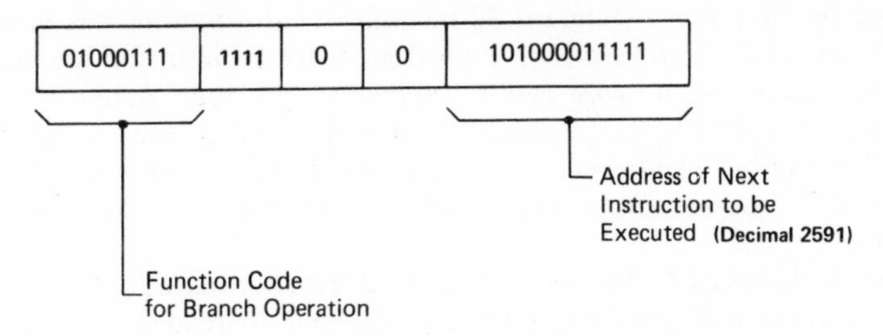

FIG. 19.10(a) Format of a Branch Instruction.

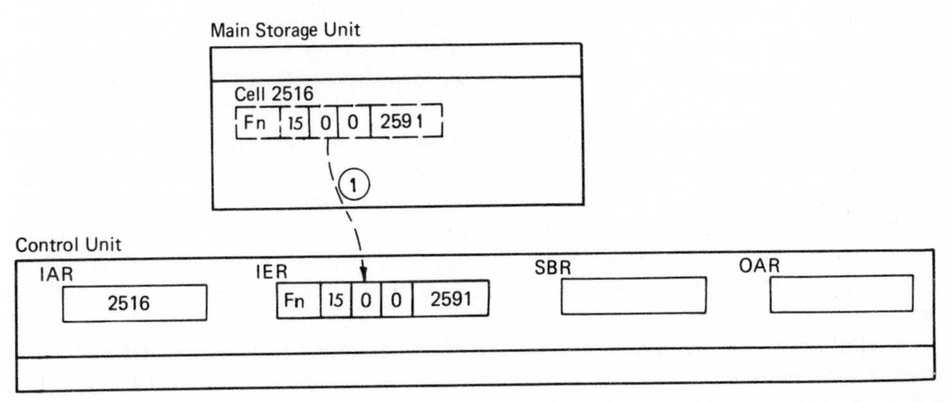

FIG. 19.10(b) The Instruction in Storage Cell 2516 is the next Instruction to be Executed in the Current Sequence of Instructions, it is fetched from storage in the usual way and loaded into the Instruction Execution Register.

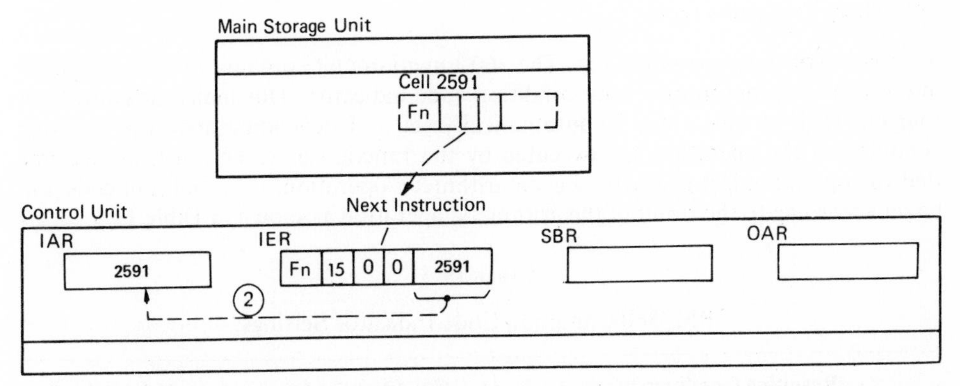

FIG. 19.10(c) The Instruction Cell 2516 happened to be a Branch Instruction, its Execution causes the Program to Branch to the Instruction in Cell 2591 and Continue Program Execution from there.

addressable 8-bit storage cells, referred to as *byte-cells*. The capacity of 360 main storage units range from a few thousand up to several million byte-storage cells. We will assume that the 360 computer we are going to program has 4096 byte-storage cells with addresses ranging from 0 to 4095.

Each byte-storage cell has sufficient capacity to hold one 8-bit binary coded character or one-bit binary number, the maximum number value that it can hold is only $(2^8 - 1)$, equivalent to the decimal value 255, which is too small for practical computing purposes. When operating with numbers in the 360 computer this difficulty is resolved by adopting a standard 32-bit unit of information, this standard unit is called a *word of 360 information*; a word is stored in four contiguously addressed byte-storage cells, so a *word-cell* is equivalent to four *byte-cells*. A word of information may be used to represent a signed 31-bit integer number and hence, provides for number values with magnitudes up to $(2^{31} - 1)$ which is equivalent to 2,187,483,647 decimal (note that, negative integers are represented in two's complement form).

The 360 is designed to handle words of information automatically; for example, when the control unit commands the storage unit to read a word of information starting at, say, byte-cell address 1024 the storage unit will read the contents of byte-cells 1024, 1025, 1026 and 1027, and transmit the 32 bits read from those cells to the control unit. Similarly when the control unit sends a word of information to the storage unit and commands it to store the word beginning at byte-cell address 1024, then the storage unit will store the 32-bit word in byte-cells 1024, 1025, 1026 and 1027.

There are two important factors to be observed when programming the 360 to operate on word-length numbers. Firstly, the storage address of the word-cell must be specified as the address of the first byte-storage-cell used to store the number (for example, byte-cell 1024 in the example discussed above). Secondly, the start address of each word, that is the address of the first byte-cell of a word, must be divisible by 4 (for example, byte addresses, 4, 8, 12, 16, 32 . . . 1024, 1028 . . . 2048, 2052 . . . and so on).

In addition to the *byte* and the *word* the 360 is designed to handle other standard units of information such as 16-bits or *halfword* and 64-bits or *doubleword*. We will, however, restrict our attention to byte and word units.

(b) REGISTER UNIT. This unit contains sixteen 32-bit registers, each of which may be used as a general purpose register for arithmetic operations, as a base address register, or as an address index register.

(c) THE CONDITION CODE INDICATOR. The 360 Function unit contains an item of special interest to the programmer—the condition code indicator. This indicator consists of four bits each of which may be automatically set to 1 depending upon the resulting condition of the operation last executed by the function unit. To illustrate, assume that the operation last executed was an arithmetic operation, the condition code will be set according to the result of the arithmetic operation as shown in Table 19.3.

TABLE 19.3

IBM/360 Condition Code Indicator Settings.

Resulting Condition from Arithmetic Operation	*Condition Code* Setting	*M Specification* Binary	Decimal
Zero	1000	1000	8
Less than zero	0100	0100	4
Greater than zero	0010	0010	2
Magnitude of result $\geq 2^{31}$ (overflow condition)	0001	0001	1

There are many situations where the programmer needs to be able to test the outcome of an arithmetic operation to determine whether it is zero, negative or positive, he is able to do this by testing the setting of the condition code indicator.

(Note the overflow result listed in Table 19.3, this condition occurs when the arithmetic result exceeds, i.e., overflows, the maximum value that can be represented in a 32-bit word.)
(d) THE 360 ADDRESSING SYSTEM. When writing a 360 program we need to be able to make reference to register cells and main storage cells. Particular register cells are addressed using a four-bit register address which provides for addresses from zero (0000) through to fifteen (1111). The addressing of particular main storage cells involves a base address value BV, a displacement address value DV, and may involve an address index value XV. The actual main storage address is given by the sum of these three values, i.e.,

$$\text{Actual Main Storage Address} = BV + DV + XV.$$

The base address value must be stored in one of the general purpose registers, any one of these registers may be used for this purpose, the one that is used is known as the Base Register BR. The address index value must also be stored in one of the general purpose registers, again any one of these registers may be used for this purpose (except the one that is being used as the base register); the one that is used to hold the index value is known as the index register XR. The actual main storage location to be referenced is given by the sum of the contents of the base register (BR), the contents of the index register (XR), and the displacement value DV, i.e.,

$$\text{Actual Main Storage Address} = (BR) + DV + (XR).$$

Because these three values may be used to identify a particular main storage address, those machine instructions that reference main storage must make provision for specifying a base register, an index register and a displacement value. Such an instruction is illustrated by the RX1 type instruction shown in Fig. 19.11, note that it has a four bit index register address field, a four bit base register address field, and a twelve bit displacement value field.

The displacement value can be up to 12 bits long and hence can range from 0 to 4095, i.e., $(2^{12} - 1)$, this means that by varying the displacement value, and keeping the base value and index value (if any) constant, we can address a range of 4096 byte-storage cells, or 1024 word storage cells. The use of an address index value is optional, it is only required in special situations, in most cases the base address value and the displacement address value are all that are required.

2. IBM/360 Machine Instructions

(a) GENERAL. The 360 has a large set of machine instructions which may be used to specify quite a wide variety of machine functions such as integer and floating point arithmetic, character data manipulation, and so on. In order to simplify things we will use only a small sub-set of these instructions and restrict our attention to integer arithmetic type machine functions. The formats for the instructions that we will be using are shown in Fig. 19.11 (note that all the instructions have an 8-bit function code field).

(b) RR TYPE INSTRUCTIONS. These instructions are used to specify operations on operands held in registers. The RR1 format has two 4-bit register address fields, denoted RA and RB, each of which is used to specify the address of one of the sixteen general purpose registers. The specified registers contain the operands for the operation specified by the function code.

In the RR2 format the RA specification is replaced by an M specification, when the programmer wants to test the Condition-Code-Indicator to determine the result of a previous operation he specifies the condition, that he is interested in, in the M field in an RR2 type instruction. For example when testing the result of an arithmetic

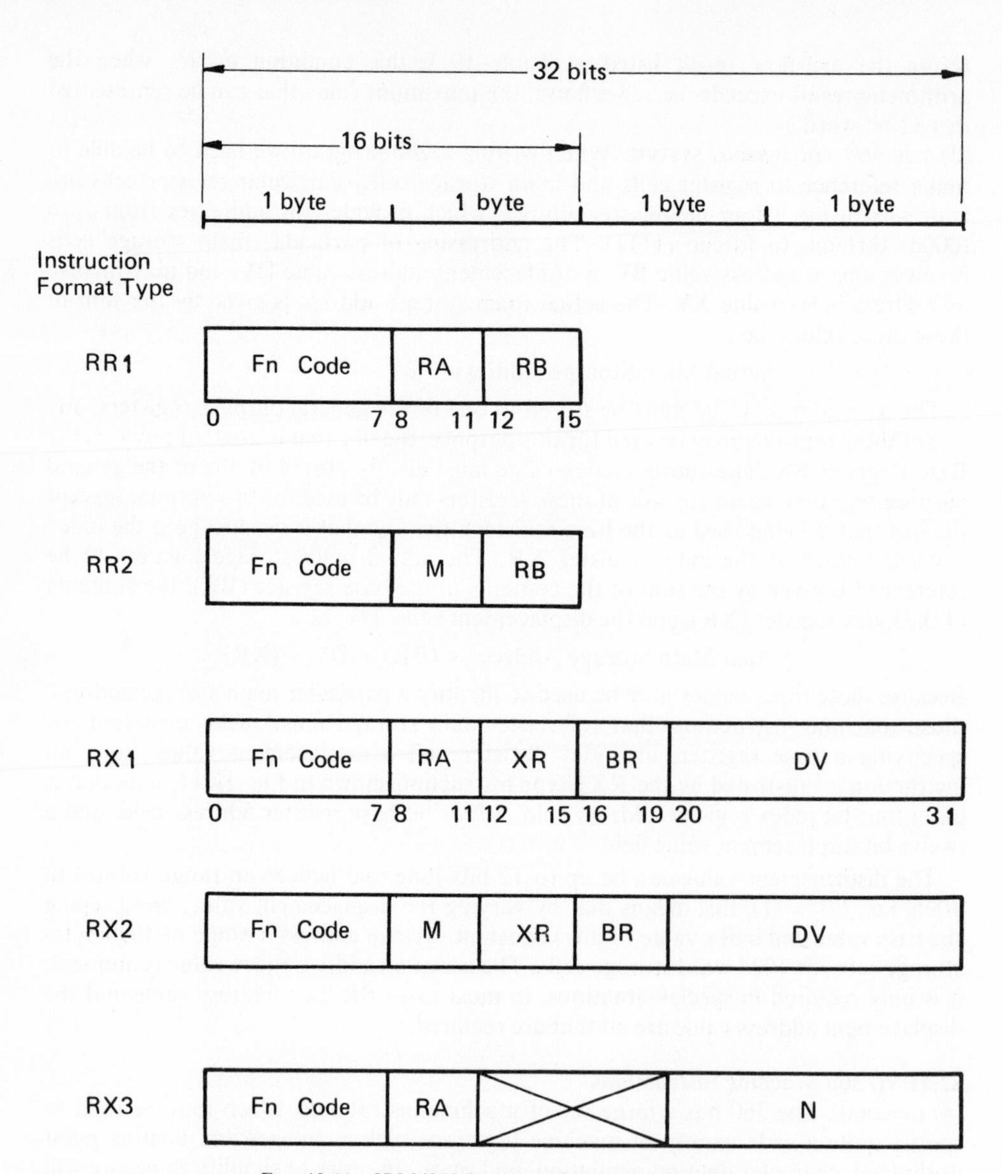

FIG. 19.11 IBM/360 Computer Instruction Formats.

operation (see Table 19.3) M is specified as 1000 when testing for a zero result; M is specified as 0100 when testing for a negative result; and so on. Further details are given below in the discussion on Branch Instructions.

(c) RX TYPE INSTRUCTIONS. The RX1 type instruction is used to transfer a word (32-bits) of information between a particular register and a particular main storage cell. The address of the particular register involved in the transfer is specified in the register address field, labelled RA. The address of the particular word of main storage involved in the transfer is specified, using the BR (base register), and DV (displacement value) fields and when required, the XR (index register) field.

In the RX2 type instruction the RA specification is replaced by an M specification which is used to test the setting of the Condition-Code in the same way as the M specification in the RR2 type instruction.

The RX3 type instruction is used to set up a positive integer number value in the general purpose register specified in the RA field. The integer number value is specified in the N field which occupies the last twelve bits of the instruction, hence this value can range from zero to 4095, i.e., $(2^{12} - 1)$.

(d) BRANCH INSTRUCTIONS. The Branch-on-Condition instructions enable us to program the conditional execution of alternative program procedures. The branch condition is specified by setting the value in the M field of the appropriate instruction to correspond to that Condition-Code-Indicator setting for which branching is to take place. When this instruction is executed the control unit checks the M specification against the current Condition-Code-Setting, if they correspond the control unit will load the branch address from the branch instruction into the Instruction Address Register thus causing a program branch to a new instruction sequence, if they do not correspond branching does not take place and the control unit proceeds to fetch and execute the next instruction in the current sequence. To illustrate, the various M values that may be used to test the setting of the Condition-Code-Indicator resulting from an arithmetic operation are listed in Table 19.3, note that the M specifications are listed in both binary form and decimal form; when writing binary coded instructions the binary form is used, when writing symbolically coded instructions the decimal form is used.

So far we have been discussing *conditional-branch instructions*, that is, instructions which result in branching to the start address of an alternative set of instructions if, and only if, some specified condition exists at the time that the branch instruction is executed. Another important and useful operation is an *unconditional branch*, that is one which is performed regardless of what conditions exist at the time of its execution. In the 360 computer an unconditional branch is specified by setting the M value of a branch instruction to 15, i.e., M = 1111.

(e) SYMBOLIC INSTRUCTION CODES. Details of the sub-set of the IBM/360 instructions that we will be using are set out in Table 19.4. These machine instructions must be represented in the machine in binary coded form. We could write the instructions in binary coded form, in which case the written instructions represent the form that the instructions take in the machine and the process required to convert the instructions from their external to internal form is a simple one-to-one conversion process. If we were to write the instructions in binary form they would appear as shown in Fig. 19.12. However, writing instructions in binary form is a cumbersome task, and so to simplify things we will write instructions using symbolic (alphabetic) operation codes and symbolic (decimal) address notations as shown in Fig. 19.12. A special computer program is used to translate programs of these symbolically coded instructions into executable programs of binary coded instructions. The symbolic form of each of the IBM/360 instructions that we will be using are set out in Table 19.4.

(f) SPECIAL INSTRUCTIONS. The set of special instructions listed in Table 19.4 are not machine instructions, that is they are not directly executed by the machine hardware. They are known as macro-instructions and are executed by means of small programs known as macro-routines that are permanently stored in the machine. Each macro-instruction function code identifies the particular macro-routine to be executed, and the macro-instruction operands constitute the input data to that macro-routine. For all intents and purposes we can treat these macro-instructions as symbolic machine instructions, even though some of them appear as very long instructions.

(g) INTEGER NUMBER REPRESENTATION. In order to simplify the discussion we have restricted our attention to integer number functions and instructions and therefore to programs which process integer numbers only. This being the case we should know how integer numbers are represented within the 360 computer. Each integer number

TABLE 19.4
Sub-set of IBM/360 Machine Instructions and Macro-Instructions.

NOTE 1. The contents of a register or storage cell are denoted by the use of parenthesis, for example (RA), (RB). The following expression, (RA) + (RB)→ RA, denotes that the contents of register B are to be added to the contents of register A; the arrow denotes that the result is to be stored in the register specified on the right hand side, in this case RA.

NOTE 2. The programmer can only use registers 0 to 9 because registers 10 to 15 are reserved for special purposes. Register 15, for example, is used as the base address register.

Instruction	Instruction Format and Length in Bytes (n)	Function Code (Binary and Mnemonic)	Symbolically written Form of Instruction	Effect on Condition code Indicator: Result of Operation	Effect on Condition code Indicator: C.C. Setting
REGISTER—REGISTER INSTRUCTIONS					
R-R Add: (RA) + (RB)→ RA Add the second operand (contents of register RB) to the first operand (contents of register RA) and store the result in register RA.	RR1 (2)	AR 00011010	AR RA, RB RA = 0, 1,——9 RB = 0, 1,——9	Result = 0 <0 >0 $<-2^{31}$ $>2^{31}-1$	8 4 2 2 1
R-R Subtraction: (RA) – (RB)→ RA Subtract the second operand (RB) from the first operand (RA) and store the result in register RA.	RR1 (2)	SR 00011011	SR RA,RB	Result = 0 <0 >0 $<-2^{31}$ $>2_{31}-1$	8 4 2 1 1
R-R Transfer: (RB)→ RA The contents of register RB are copied into register RA.	RR1 (2)	LR 00011000	LR RA,RB	No effect	
R-R Comparison: The first operand (RA) is algebraically compared with the second operand (RB) and the result of the comparison sets the value of the condition core indicator. (RA) and (RB) remain unchanged.	RR1	CR 00011001	CR RA,RB	(RA) = (RB) (RA) < (RB) (RA) > (RB)	8 4 2

R-R AND: The logical product (AND) of the 32-bits of the first operand (RA) and the 32-bits of the second operand (RB) is stored in register RA.	RR1 (2)	NR 00010100	NR RA,RB	Result=0 Result≠0	8 4
R-R OR: The logical sum (OR) of the 32-bits of the first operand (RA) and the 32-bits of the second operand (RB) is stored in in register RA.	RR1 (2)	OR 00010110	OR RA,RB	Result=0 Result≠0	8 4
R-R Exclusive OR: The modulo-two sum (Exclusive OR) of the 32 bits of the first operand (RA) and the 32 bits of the second operand (RB) is stored in register RA.	RR1 (2)	XR 00010111	XR RA, RB	Result=0 Result≠0	8 4
STORE—REGISTER INSTRUCTIONS					
Store Register: The 32-bit long first operand (RA) is stored in the four main storage byte-cells starting with the byte-cell having the address given by [DV + (BR) + (XR)].	RX1 (4)	ST 01010000	ST RA, DV/(XR,BR)	No effect	
Load Register: The 32-bit second operand—which is made up of the contents of the four main storage byte-cells starting with the byte-cell having the address given by [DV + (BR) + (XR)]—is loaded into register RA.	RX1 (4)	L 01011000	L RA,DV(XR,BR)	No effect	
LOAD CONSTANT INSTRUCTION					
Load Constant: The value, C, specified in bits 20 to 31 of this instruction is loaded into Register RA and replaces the previous contents of RA.	RX3 (4)	LA 01000001	LA RA, C	No effect	
BRANCH INSTRUCTION					
Branch on Condition: If the value of the condition code indicator corresponds with the value specified in the M field of the instruction then the address of the next instruction to be executed is given by the 32-bit second operand which is in main store, starting at the byte-cell address given by [DV + (BR) = +(XR)]. The 32-bit value stored at this address is automatically loaded into the Instruction Address Register.	RX2 (4)	BC 01000111	BC M,DV(XR,BR)	No effect	

TABLE 19.4 (contd.)
Subset of IBM/360 machine instructions and macro-instructions.

Instruction	Instruction Format and Length in Bytes (n)	Function Code (Binary and Mnemonic)	Symbolically written Form of Instruction	Effect on Condition code Indicator: Result of Operation	Effect on Condition code Indicator: C.C. Setting
SPECIAL (MACRO) INSTRUCTION					
ENTRM: The first instruction in each program must always be an ENTRM instruction. It automatically loads the base address value for the program into register 15.	0 bytes	ENTRM	ENTRM	No effect	
EXITM: The last instruction in each program must always be an EXITM instruction, it terminates program execution.	24 bytes	EXITM	EXITM	No effect	
GETB: Read the N digit decimal integer number from the next input card, convert it into a 32-bit pure binary integer number and store it in the four main storage cells beginning at cell address [DV, (BR)].	20 bytes	GETB	GETB DV(BR)	No effect	
PUTBS: Start printing on a new line on the output printer. Take the 32-bit binary integer numbers beginning at the main storage cell addresses DV_1(BR), DV_2(BR), DV_3(BR),— and print their decimal equivalent in the next line of print on the output printer. Each of the decimal numbers occupies a ten character field width, since a printed line is 132 characters wide, a maximum of 13 numbers can be printed on one line (any more than this are automatically printed on the next line.)	20 bytes	PUTBS	PUTBS DV_1(BR) DV_2(BR), DV_3(BR)	No effect	
PUTB: Is the same as PUTBS except that it does not start a new line of printing, rather the values specified in this instruction are printed on the current line. PUTB output values are cumulative—i.e., PUTB DV_1 (BR), PUTB DV_2(BR) is equivalent to PUTB DV_1(BR), DV_2(BR) and are printed in sequential positions across the output line.	20 bytes	PUTB	PUTB DV_1(BR), DV_2(BR), DV_3(BR)	No effect	

PUTMS: Start printing on a new line. Print the string of characters specified as an operand, enclosed in quote marks, on the next line of print on the output printer. The string must not exceed 56 characters.	70 bytes	PUTMS	PUTMS "character string"	No effect	
PUTM: Is the same as PUTMS except that it does not start a new line of printing, rather the specified string is printed as a continuation of the current line of print. PUTM output messages are cumulative—i.e., PUTM 'TODAY', PUTM 'IS', PUTM 'MONDAY' is equivalent to PUTM 'TODAY IS MONDAY'. Up to 132 characters can be printed on one line in this way. The maximum number of output characters specified in an individual PUTM instruction must not exceed 56.	70 bytes	PUTM	PUTM "character string"	No effect	
MPR: Multiply register contents. Multiply the contents of register A by the contents of register B and store the truncated 32-bit, most significant portion of the result, in register RA.	10 bytes	MPR	MPR RA,RB	Result $=0$ <0 >0 $>2^{31}-1$ $<-2^{31}$	C.Code 8 4 2 1 1
DVR: Divide register contents. Divide the contents of register A by the contents of register B and store the quotient in register A, the remainder is put into register 10 and can be examined by transferring it to one of the registers 0 to 9.	10 bytes	DVR	DVR RA,RB	Result $=0$ <0 >0 $>2^{31}-1$ $<-2^{31}$	C.Code 8 4 2 1 1
END INSTRUCTION					
END: Every program that is fed into the 360 must end with an END statement; this enables the computer system to detect when a program has been completely read in, the system then proceeds to execute the program.			END	No effect	

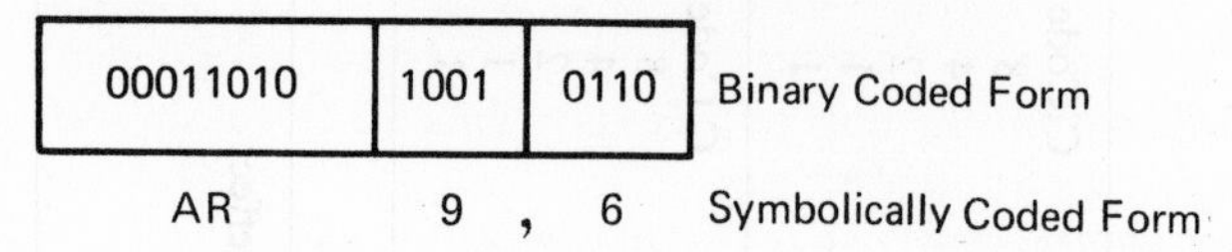

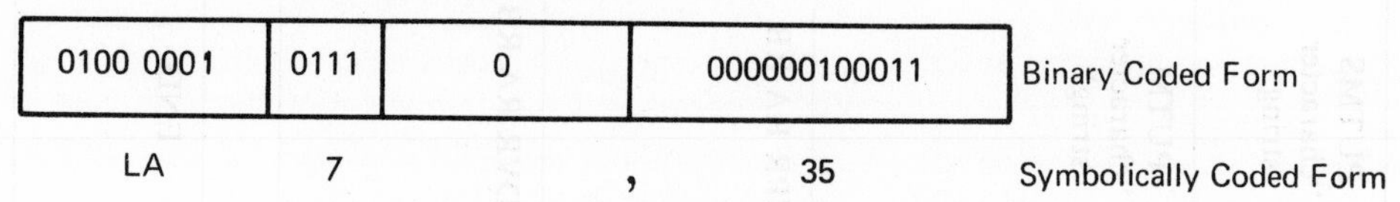

FIG. 19.12 Binary Coded and Symbolic Coded Instructions.

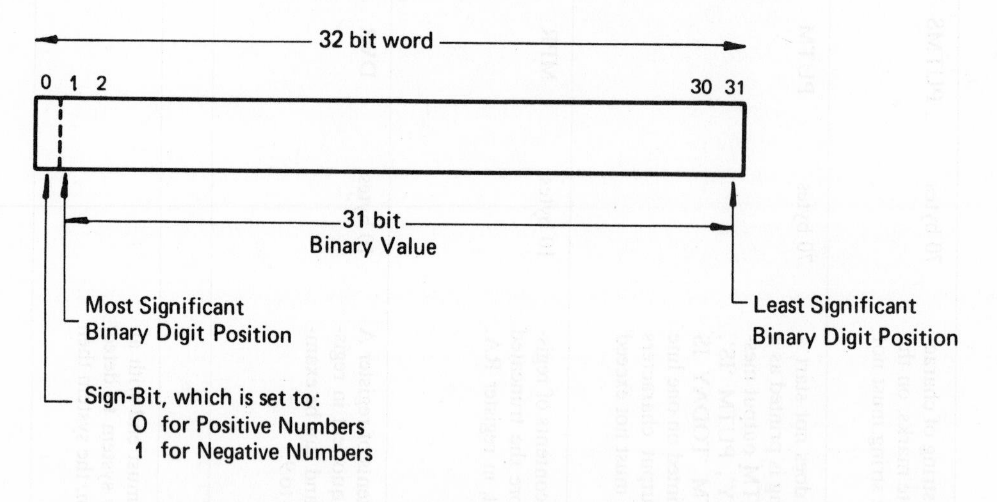

FIG. 19.13 Format of IBM/360 Integer Number Binary Representation.

occupies one 360 word unit and is represented in the form of a signed 31-bit binary number as indicated in Fig. 19.13. Positive integers are represented as pure binary numbers, whereas negative integers are represented in two's complement form (see Chapter 18).

3. Algorithms and Programs

(a) GENERAL. In this section machine language programming is illustrated by a number of program examples. It should be noted that a considerable amount of planning is needed when preparing a computer program for a specific problem. In particular, a complete, correct, unambiguous statement of the problem solution procedure is needed, in order to prepare a correct program, such a procedure is called an *algorithm*. A computer program is simply an algorithm specified in programming language form. The term *algorithm* derives from an ancient Arabic text, "Kitub aljabra w'al-mugabula" by Al-Khouarizmi, which describes the Arabic system of counting using numerals instead of a counting frame (such as an abacus). The algorithms which interest us in computing are those which are capable of implementation as computer programs—such algorithms should have the following features:

Finiteness: The algorithm must always terminate after a finite number of steps.
Unambiguous: Each step in the algorithm must be precisely defined, rigorously and unambiguously for all possible cases.
Efficient: The algorithm should lead to the efficient production of results.
Input and Output: In general an algorithm will require input data of specified type and will produce an output which is the desired transformation of the input.

(b) FLOWCHARTS are a graphical method of representing algorithms and are a very useful and universal way of expressing such procedures. It is good practice to flowchart an algorithm before we attempt to program it because this helps us to prove and improve its correctness and effectiveness, and also because flowcharts provide a universal and relatively easily understood method of communicating the description and definition of the algorithm. Some example flowcharts are shown in Figs 19.14, 19.15 and 19.16 where, for example, the following symbolic conventions are adopted.

Rectangular symbols with circular ends identify the starting and terminating points of the algorithm.

Plain rectangles represent either a single machine function or a small set of related machine functions.

Angular ends are used to represent branch decision points which lead to alternate paths of action. And so on.

When a suitable algorithm has been worked out and expressed in flowchart form the problem is then ready for programming. This involves the coding of the procedure embodied in the flowchart as a specific set of machine instructions that will direct the computer to execute that procedure.

(c) PROGRAM EXAMPLE NO. 1. This problem involves the reading of three integer numbers from punched cards, one number per card, and determining the largest of the three numbers. A flowchart of a procedure (algorithm) for solving this problem is shown in Fig. 19.14(a). A symbolic machine language program for solving the problem is shown in Fig. 19.14(b).

As shown, the program is written on a specially designed form called a program coding sheet. This sheet has eighty columns corresponding to the eighty columns of a standard punched card, each line written on the coding sheet is recorded column for column on a punched card. When the complete symbolic program has been recorded on punched cards it is fed into the computer and translated into binary machine code form by a special translator program and then it is executed.

There are a number of specific points to note about the coding form. Instruction function codes must be written in columns 10 to 14. Instruction operands may be written starting at column 16, a comma must be used to separate individual operands, and there must be no blank characters (i.e., spaces) appearing in the operand specifications. A comment can be written alongside each instruction to annotate the program, each comment must be separated from the instruction operands by one or more blank characters. A whole line can be used for a comment by recording an asterisk(*) in the first column of the line as shown. The "length" field in columns 73 to 76 is provided to assist the programmer, who uses this field to record "the length" (in bytes) of the instruction appearing on the same line (Note that comments are not stored with the program and hence have no effect on the spacing of instructions when they are stored in the computer storage unit). The "Address" field in columns 77 to 80 is used to record the start address of the instruction or data item appearing on the same line. This address is calculated relative to the start address of the program and is

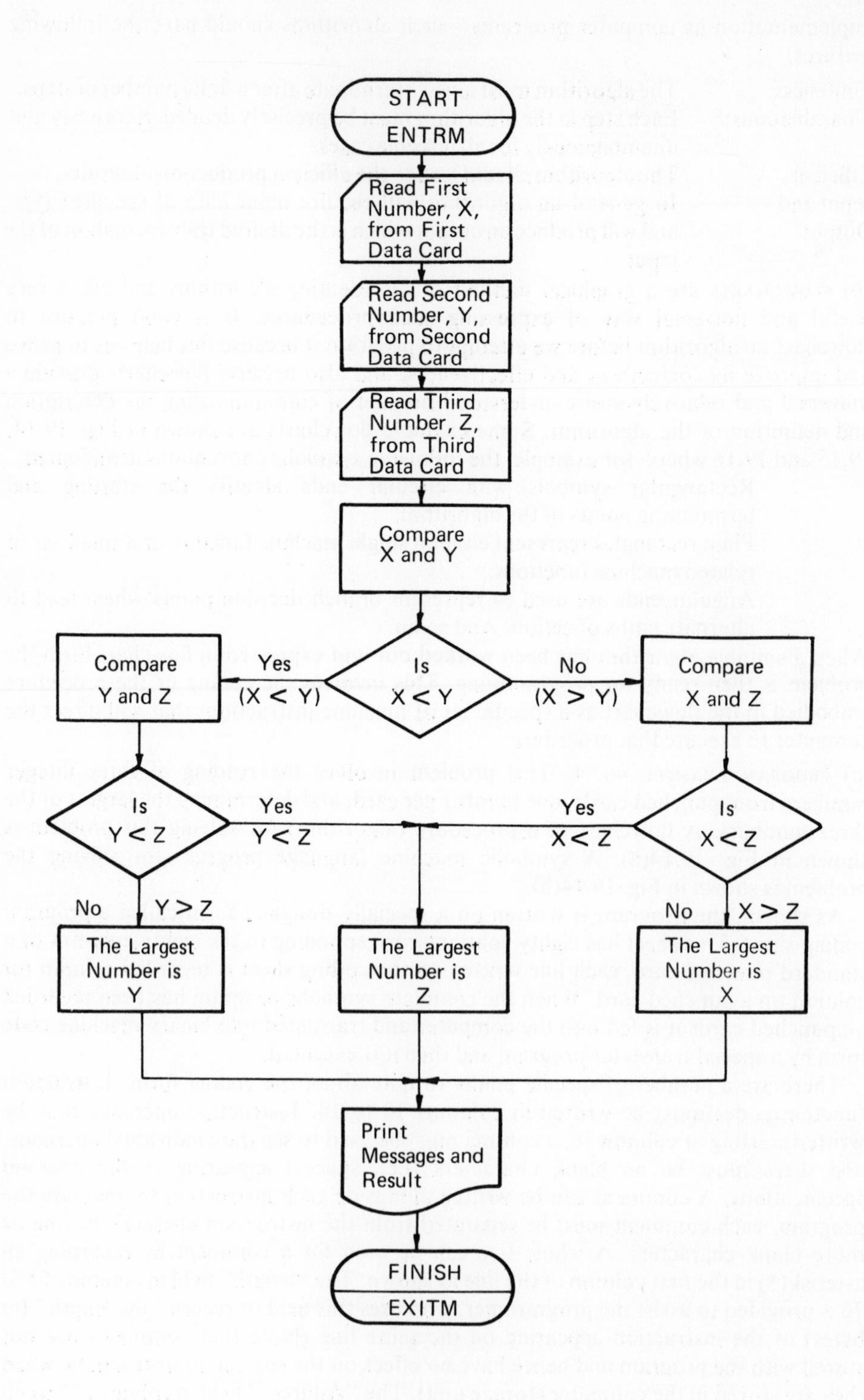

FIG. 19.14(a) Flowcharted Algorithm for Example Program No. 1.

	Function	Operands	Comments	Length	Address
1 2 3 4 5 6 7 8 12345678901234567890123456789012345678901234567890123456789012345678901234567890					
*		PROGRAM EXAMPLE NO. 1			
	ENTRM		MUST BE FIRST INSTRUCTION IN PROGRAM	0	0
*			IT SETS UP PROGRAM BASE ADDRESS IN REG 15.		
	PUTMS	'PROGRAM TO DETERMINE LARGEST OF THREE NUMBERS'		70	0
	GETB	292(,15)	READ THE VALUES IN FROM CARDS (CALL THEM	20	70
	GETB	296(,15)	X, Y, Z) AND STORE THEM AT STORAGE CELL	20	90
	GETB	300(,15)	LOCATIONS 292(,15), 296(,15), 300(,15).	20	110
	PUTBS	292(,15), 296(,15),300(,15)	PRINT THE THREE NUMBERS.	70	130
	L	5,292(,15)	LOAD FIRST VALUE, X, INTO REG 5.	4	200
	L	4,296(,15)	LOAD SECOND VALUE, Y, INTO REG 4.	4	206
	L	3,300(,15)	LOAD THIRD VALUE, Z, INTO REG 3.	4	208
	CR	5,4	COMPARE (R5) WITH (R4), I.E. X WITH Y.	2	212
	BC	4,230(,15)	IF X< Y BRANCH TO ADDRESS 230(,15)	4	214
*			OTHERWISE CONTINUE.		
	CR	5,3	COMPARE (R5) WITH (R3), IE X WITH Z.	2	218
	BC	4,242(,15)	IF X< Z BRANCH TO ADDRESS 242(,15).	4	220
	LR	2,5	X IS LARGEST VALUE LOAD IT INTO R2 AND	2	224
	BC	15,244(,15)	BRANCH TO 244(,15).	4	226
	CR	4,3	COMPARE (R4) WITH (R3), I.E. Y WITH Z.	2	230
	BC	4,242(,15)	IF Y< Z BRANCH TO 242(,15)	4	232
	LR	2,4	Y IS LARGEST VALUE LOAD IT INTO REG 2	2	236
	BC	15,244(,15)	AND BRANCH TO 244,(,15).	4	238
	LR	2,3	Z IS LARGEST VALUE LOAD IT INTO REG 2.	2	242
	ST	2,304(,15)	STORE THE LARGEST VALUE AT ADDRESS 304(,15).	4	244
	PUTBS	304(,15)	PRINT THE LARGEST VALUE.	20	218
	EXITM		END OF PROGRAM.	24	268
		START OF DATA STORAGE			292
		STORE X VALUE AT STORAGE CELL ADDRESS 292(,15)		4	292
		STORE Y VALUE AT STORAGE CELL ADDRESS 296(,15)		4	296
		STORE Z VALUE AT STORAGE CELL ADDRESS 300(,15)		4	300
		STORE MAX VALUE AT STORAGE CELL ADDRESS 304(,15)		4	304

Fig. 19.14(b) Program Example No. 1.

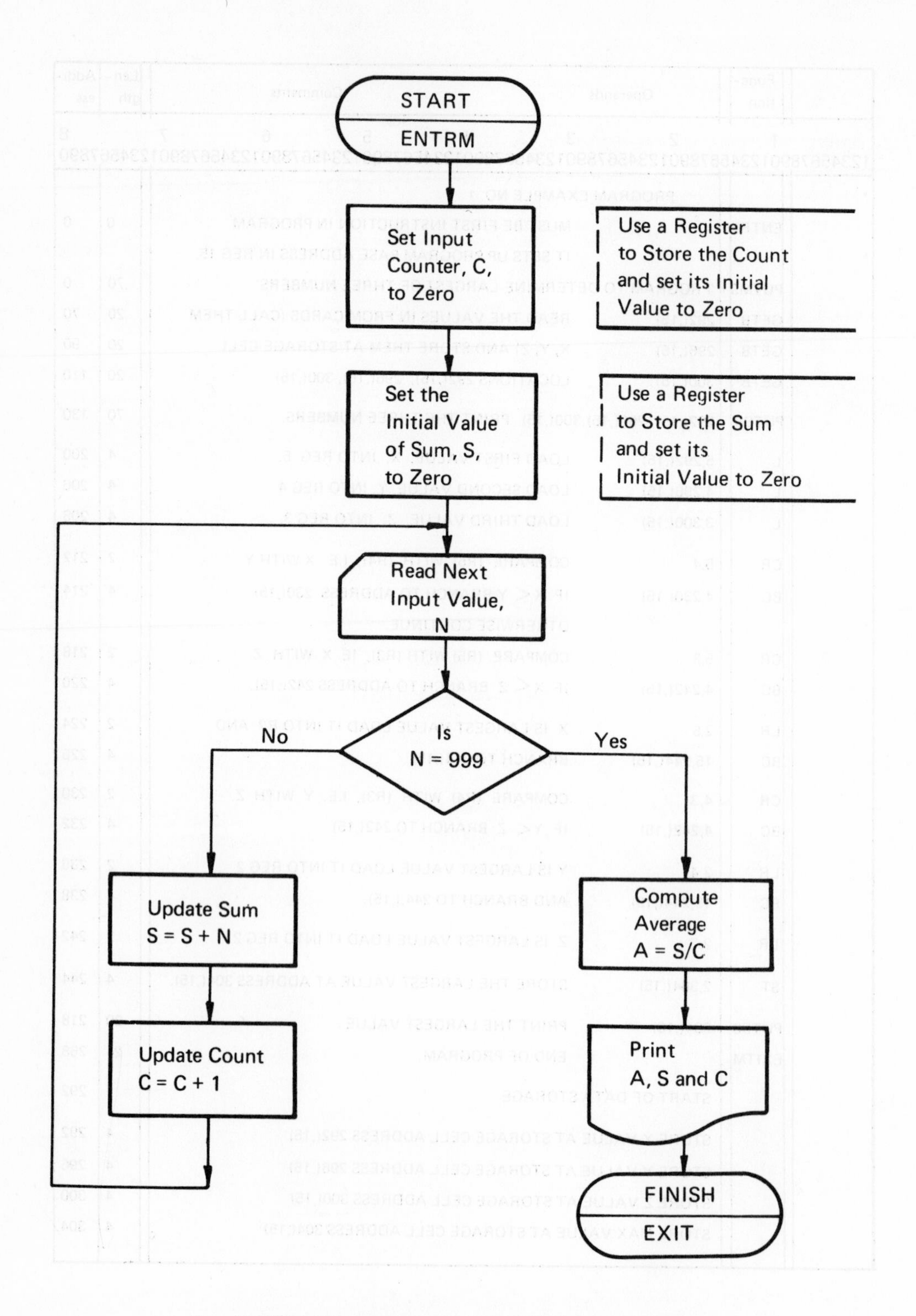

FIG. 19.15(a) Flowcharted Algorithm for Program Example No. 2.

Funct -ion	Operands	Comments	Len-gth	Addr-ess
123456789012345678901234567890123456789012345678901234567890123456789012345678901234567890				
	PROGRAM EXAMPLE NO. 2			
ENTRM		MUST BE FIRST INSTRUCTION IN PROGRAM	0	0
		IT SETS UP PROGRAM BASE ADDRESS IN REG 15		
PUTMS	'PROGRAM TO COMPUTE SUM & AVERAGE OF SET OF INPUT VALUES'		70	0
PUTMS	COUNT SUM AVERAGE'		70	70
LA	5, 0	LOAD THE VALUE O INTO REG 5.	4	140
LA	4, 0	LOAD THE VALUE O INTO REG 4.	4	144
LA	3, 99	LOAD THE VALUE 999 INTO REG 3.	4	148
LA	2, 1	LOAD THE VALUE 1 INTO REG 2.	4	152
GETB	310(,15)	READ NEXT INPUT VALUE INTO STORE CELL.	20	156
L	6,310(,15)	LOAD VALUE INTO REG 6.	4	176
CR	6, 3	TEST FOR LAST CARD.	2	180
BC	8, 194(,15)	BRANCH IF VALUE IS 999.	4	182
AR	5, 6	ADD VALUE TO (R5); SUM IS HELD IN R5,	2	186
AR	4, 2	ADD 1 TO COUNT.	2	188
BC	15, 156(,15)	BRANCH BACK AND READ NEXT INPUT VALUE.	4	190
ST	4, 314(,15)	STORE COUNT AT STORE LOCATION 308(,15).	4	194
ST	5,318(,15)	STORE SUM AT STORE LOCATION 312(,15).	4	198
DVR	5, 4	COMPUTE AVERAGE.	10	202
ST	5, 322(,15)	STORE AVERAGE AT STORE LOCATION 316(,15).	4	212
PUTBS	314(,15), 318(,15), 322(,15)	PRINT COUNT, SUM & AVERAGE.	70	216
EXITM		END OF PROGRAM.	24	286
	START OF DATA STORAGE.			310
	STORE INPUT VALUE AT STORAGE CELL ADDRESS 310(,15).		4	310
	STORE COUNT AT STORAGE CELL ADDRESS 314(,15).		4	314
	STORE SUM AT STORAGE CELL ADDRESS 318(,15).		4	318
	STORE AVERAGE AT STORAGE CELL ADDRESS 322(,15).		4	322

FIG. 19.15(b) Program Example No. 2.

in fact the displacement address value; the start address is the base address value which is stored in convention in register 15. The "Length" and "Address" fields are specially provided to help the programmer keep track of addresses when writing his program.

Note that, apart from comment cards, the program must start with an ENTRM macro-instruction. The ENTRM macro-instruction sets up an appropriate base address value in register 15, so that the programmer does not have to go to the trouble of doing this.

Note also that the programmer can only use registers 0 to 9, the remaining registers, 10 to 15, have been reserved for special purposes.

It can be seen in Fig. 19.14(b) that the program of instructions occupies the first portion of storage from displacement address zero onwards, and that data items may occupy any storage locations (up to displacement address value 4095) following on from the end of the program. It should be remembered that 32-bit integer number data items must be stored starting at addresses that are divisible by 4 (e.g., 292, 296, 300, 304, etc.).

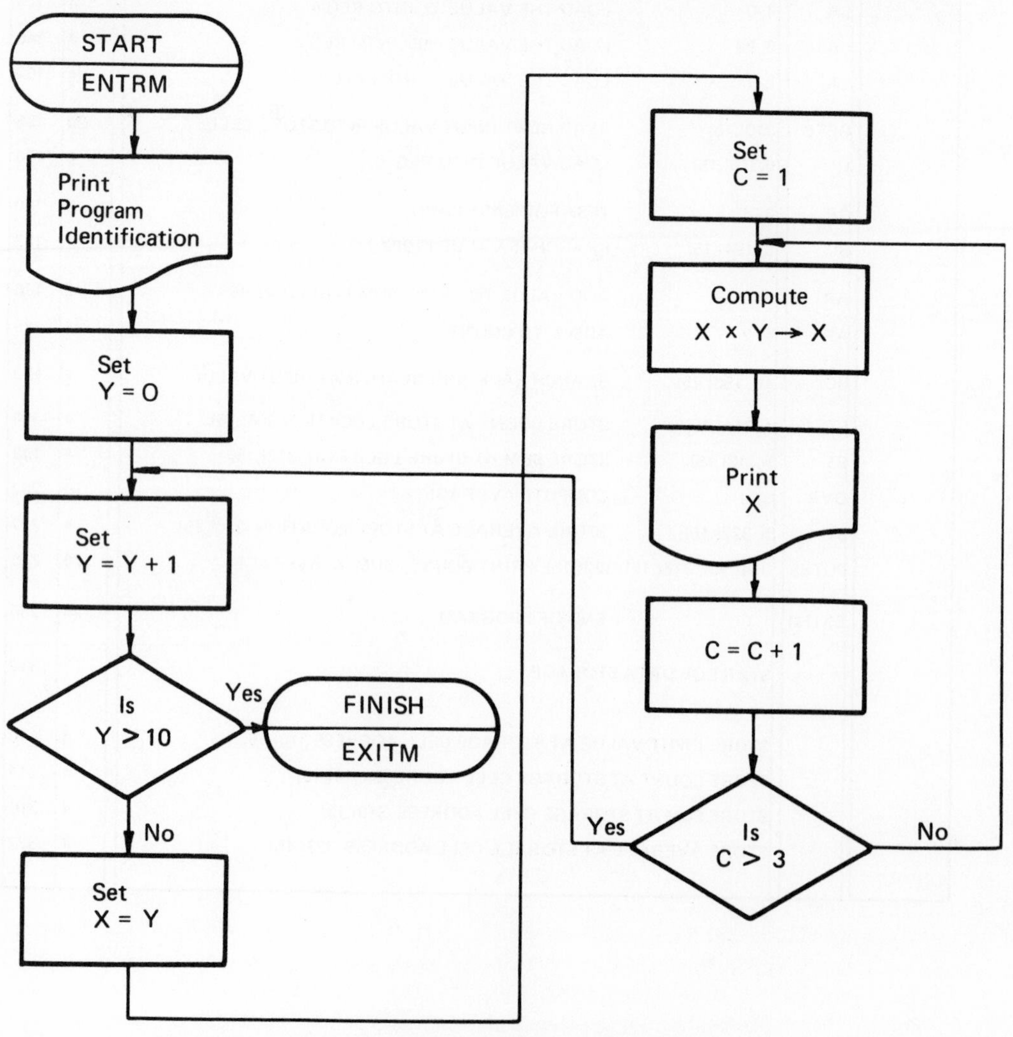

FIG. 19.16(a) Flowcharted Algorithm for Program Problem No. 3.

	Function	Operands	Comments	Length	Address
1234567890123456789012345678901234567890123456789012345678901234567890123456789 0 (columns 1–8)					
*			PROGRAM EXAMPLE NO. 3		
	ENTRM		MUST BE FIRST INSTRUCTION IN PROGRAM	0	0
*			IT SETS UP PROGRAM BASE ADDRESS IN REG 15		
	PUTMS	'PROGRAM TO COMPUTE TABLE OF SQUARES AND CUBES'		70	0
	PUTMS	'VALUE SQUARED CUBED'		70	70
	LA	5,0	STORE VALUE O IN REG 5 (INITIALISE Y).	4	140
	LA	4,1	STORE VALUE 1 IN REG 4.	4	144
	LA	3, 10	STORE VALUE 10 IN REG 3.	4	148
	AR	5,4	ADD (R4) TO (R5) I.E. Y=Y+1.	2	152
	CR	5,3	COMPARE (R5) WITH (R3)	2	154
	BC	2,268(,15)	IF (R5) > (R3), I.E. > 10, BRANCH TO EXITM	4	156
*			INSTRUCTION, OTHERWISE CONTINUE.		
	ST	5,292(,15)	STORE (R5), I.E. Y, AT STORE CELL 292(,15)	4	160
	LR	6,5	LOAD A COPY OF Y INTO REG 6.	2	164
	MPR	6,5	COMPUTE Y SQUARED.	10	166
	ST	6, 296(,15)	STORE Y SQUARED AT STORE CELL 296(,15).	4	176
	MPR	6,5	COMPUTE Y CUBED.	10	180
	ST	6, 300(,15)	STORE Y CUBED AT STORE CELL 300(,15)	4	190
	PUTBS	292(,15),296(,15),300(,15)	PRINT Y , Y SQUARED AND	70	194
*			Y CUBED.		
	BC	15,152(,15)	BRANCH BACK TO 152(,15), INCREMENT Y	4	264
*			AND COMPUTE ITS SQUARE AND CUBE.		
	EXITM			24	268
	START	OF DATA STORAGE AREA			292
	STORE	Y AT STORAGE CELL ADDRESS 292(,15).		4	292
	STORE	Y SQUARED AT STORAGE CELL ADDRESS 296(,15)		4	296
	STORE	Y CUBED AT STORAGE CELL ADDRESS 300(,15)		4	300

FIG. 19.16(b) Program Example No. 3.

(d) PROGRAM EXAMPLE NO. 2. See Fig. 19.15. The problem is to read in an unknown number of integer decimal values from punched cards and compute their sum and average value. The decimal values are recorded one per punched card; the maximum input decimal value will not exceed one hundred; the last input card will have the value 999 recorded in it, this enables us to determine when the complete set of input values has been read in by testing each value to see if it equals 999.

Note that because we are using integer arithmetic machine functions the computed average value will be automatically rounded to the nearest integer by the computer;

the average could be obtained to greater precision in decimal point form however, this would require the use of floating point arithmetic machine functions, which are available on the IBM/360, but which we have not had time to study.

(e) PROGRAM EXAMPLE NO. 3. See Fig. 19.16. The problem is to compute and print a table of the squared and cubed powers of the integers 1 through to 10.

E. EXERCISES

1. Write symbolic machine program* (for the IBM/360/50 Computer) to read two numbers from punched cards, find their highest common factor, and print the HCF together with the two numbers. Use the Euclidean algorithm to find the HCF, which is as follows: subtract the smaller number from the larger, L; compare the new set of two numbers S and (L-S), if they are not equal subtract the smaller from the larger; compare the new set of two numbers if they are not equal subtract the smaller from the larger; continue in this way until the two numbers are equal at which stage either one of the two numbers is equal to the HCF of the original two numbers. Before writing the program draw up a flow chart of the algorithm.
2. Write a sympolic machine language program to generate and print a set of multiplication tables.
3. Write a symbolic machine language program to find the accumulative sum and average of the sequence of numbers between 1 and 100, and print the result.
4. Write a symbolic machine language program to compute and print a compound interest table. The table is to show the approximate compound interest earned by $100.00 for periods of one to ten years, at interest rates ranging from 1 per cent to 10 per cent per annum, in steps of 1 per cent. Note that in order to avoid excessive errors due to integer-division-truncation errors, the numbers that you work with have to be scaled up by a suitable scaling factor. Be sure that you flow-chart the solution before you program it.
5. Write a symbolic machine language program for the IBM/360 computer to generate and print the Fibonacci series (1,2,3,5,8,13 . . .). Note that each element of the series is equal to the sum of its preceding two terms.

SUGGESTED READING MATERIAL

Student Text: *Introduction to IBM Data Processing Systems*. 2nd ed. International Business Machines Corporation, New York, 1968.
This is a very good introductory manual for the beginner—it describes and illustrates computer hardware components and units, input/output media, computer programs and their execution and programming and operating systems.

RICHARDS, R. K., *Electronic Digital Systems*. John Wiley & Sons, New York, 1966.
Richards discusses the history, theory, construction and operation of digital computers. His book is very comprehensive, interesting and readable and is recommended to students with a strong interest in computers.

Student Text: *A Programmer's Introduction to the IBM System/360 Architecture, Instructions, and Assembler Language*. International Business Machines Corporation, White Plains, N.Y. 1965. This manual is recommended for those students interested in learning more about the IBM/360 computer. (See also the books on the IBM/360 assembler language listed at the end of Chapter 20).

* Note: a IBM/360 assembler language system having the features described in this chapter has been developed by Mr K. Robinson at the University of N.S.W. The translator is designed for student use, it remains core resident throughout the processing of a batch of student programs which are assembled and executed in the one step.

CHAPTER 20

Introduction to Computer Software

A. A. Thompson

"There is only one condition in which we can imagine managers not needing subordinates, and masters not needing slaves. This condition would be that each (inanimate) instrument could do its own work, at the word of command or by intelligent anticipation, like the statues of Daedalus or the tripods made by Hephaestus, of which Homer relates that 'Of their own motion they entered the conclave of Gods on Olympus' as if a shuttle should weave of itself, and a plectrum should do its own harp playing."

Aristotle
The Politics, *ca 350 B.C.*

A. INTRODUCTION

1. General

In general, computer systems consist of two complementary parts: *hardware* and *software*. The term *hardware* refers to the computing machinery, that collection of electromechanical/electronic units which constitute the computing machine, including such things as the central processing unit, input/output units, storage units and so on. The term *software* refers to the programs which specify and control machine operations, broadly speaking there are two classes of software—*user-software* and *system-software*. User-software denotes the programs written by the various users to perform their own particular computing processes. System-software denotes a set of standard programs which facilitate and simplify the use of the computer system, including such things as translator programs, programs for controlling standard types of input/output operations, and programs to schedule and supervise the execution of batches of user programs. System-software may also include standardised programs for the solution of commonly encountered problems such as bridge design, file sorting, statistical analysis, and so on.

2. System-Software

(a) HIGH LEVEL PROGRAMMING LANGUAGES. The difficulty of using machine languages has resulted in the development of *high level* programming languages, that is, languages which have a close correspondence to more conventional languages such as those of English and mathematics. Programs written in these high level languages have to be translated into machine language before they can be executed, this is done by specially designed translator programs.

The most elementary of the high level languages are the so-called Assembly languages which are simply extensions of symbolic machine languages (see the symbolic machine language programs appearing in the previous chapter). In fact, Assembly languages are so closely related to machine language that they are not referred to as a high level language but rather as *machine oriented languages*.

Fortran was the first major high level language to be developed, it was first released for general use in 1957 and required some twenty-five man-years of design and development work, Fortran (an abbreviation of Formula Translation) corresponds to the conventional arithmetic and algebraic languages of mathematics, it is designed to simplify the writing of mathematical type programs, and hence is widely used by engineers, mathematicians, scientists, surveyors, statisticians and similar groups of people.

Another significant development in high level programming languages was that of Algol (Algorithmic language), which resulted from the activities of an international committee which endeavoured to define a universal algorithmic, algebraic, computer language. The first version of Algol was released in 1960 and today this language is available for use on a variety of different computers; its major role, however, is as an international language for the specification and definition of computer algorithms.

Developments comparable to Fortran and Algol have also taken place with programming languages in the field of commercial data processing; where the currently accepted standard language is Cobol (Commercial Business Oriented Language), the first version of which was released in 1962. Cobol is a computer language which is as close to the conventionally written language of commerce as practicable; and is designed to facilitate the programming of business and administrative data processing activities.

Languages such as Fortran, Algol and Cobol are generally classified as procedure oriented languages, they are generalised languages which are used to specify the processes to be performed in terms of the procedures to be executed and the data to be processed. Such languages must be able to specify in detail the mathematical, logical and other types of procedures that the user wants performed, and should do so in a form corresponding to the accepted procedures and terminology of the particular problem (e.g., mathematical, engineering, business) being programmed. The language must also have facilities for describing the details of the data to be processed by each program, this is done by defining data types, data dimensions, allowable data magnitudes, data formats, and so on. For example, with numeric data items, the programmer must specify in his program whether they are integer numbers, floating point numbers, complex numbers, etc., and where arrays or matrices are involved, their dimensions must also be declared.

One of the interesting developments in recent years has been with programming languages for processing non-numeric data, such as the Snobol language which is designed to facilitate the processing of data in the form of character strings. The need for these languages has arisen from the application of computers to such areas as information retrieval systems, automatic abstracting and indexing systems, library cataloguing system, and so on.

(b) PROBLEM-ORIENTED LANGUAGES. A second group of languages are the so-called *problem-oriented* languages. They differ from procedure-oriented languages in that they are designed to facilitate the solution of specific types of problems, they consist of sets of standardised specification statements to standardised problem programs. Such languages are a relatively recent development and will undoubtedly play a very significant role in the future, particularly in engineering design. One of the major achievements in this area is the ICES (Integrated Civil Engineering System) designed and developed at the Massachusetts Institute of Technology, this is a comprehensive system which simplifies the application of computers to many areas of civil engineering, such as road, highway, bridge and structural engineering design. Several systems are also available for circuit analysis in electrical and electronic engineering, of which ECAP (Electronic Circuit Analysis Program) is probably the best known. Such systems enable the practising engineer to thoroughly investigate alternative designs rapidly and economically, thus enabling him to prove and optimise his designs before they leave the "drawing board".

(c) SUPERVISOR PROGRAMS. It is not unreasonable to draw an analogy between a computer system and a factory production line. Perhaps the most outstanding difference is that apart from input and output a computer does not deal with physical quantities but rather with the processing of information; in other respects, however, there is a great deal of similarity. A computer system consists of a set of processing resources including hardware resources such as input/output units, storage units and function unit, and software resources such as language translators, problem programs and standardised input/output programs. The system work load consists of one or more input streams of user programs, commonly called jobs, that are fed into the system through one or more input devices such as a card reader.

Special supervisor or executive programs are provided to automate and supervise the flow of work passing through the system. These supervisor programs automatically sequence the processing of incoming jobs, allocate the system resources required by each job (such as input/output units, storage space and language translators) and take care of any user program errors that may arise (such as the wrong kind of input data)—in which case the supervisor may decide, depending upon the type of error, to terminate execution of the error program and proceed on to the processing of the next job in the input stream.

B. SYMBOLIC PROGRAMMING LANGUAGES

1. Assembly Languages

Binary machine language computer programs are difficult to write and understand because we are not used to writing and reading information in binary form (the simple example in Fig. 20.1(a) illustrates the point), also, it can be seen that coding errors may be easily made and are difficult to find. In short, writing programs in binary machine language is a difficult and tedious task requiring a lot of time and effort.

Ever since the inception of digital computers efforts have been directed at simplifying the writing of programs. One of the earliest methods used was to write programs in decimal notation and use a translator program to translate the decimal notation into binary coded instructions. Using this scheme the equivalent of the binary program in Fig. 20.1(a) would have the form shown in Fig. 20.1(b), at a glance you can see that this version is easier to read, and write. An extension of the decimal notation scheme led to the use of alphabetic mnemonic symbols to represent instruction function codes as illustrated in Fig. 20.1(c). Further developments led to the use of alphabetic/numeric symbols to represent storage cell addresses in place of the actual numeric addresses, as illustrated in Fig. 20.1(d).

These developments made programs much easier to write and read and understand, as a result the incidence of programming errors was reduced, errors became easier to find and correct, and so on. In this way the use of computers has been simplified, thus reducing the effort and cost involved in the preparation of computer programs; this is a most significant factor when one considers that even a medium sized program may take many man-weeks to prepare.

Probably the most important thing to note is that the computer itself is used to translate symbolic programs into binary machine language programs. A special translator program performs all the tedious, time consuming, detailed work needed to produce the machine language program; therefore, the *computer itself is the key factor in the development and viability of symbolic programming languages.* The complexity of these translator programs increases as additional symbolic programming facilities such as symbolic function codes and symbolic addresses are

INSTRUCTIONS					COMMENTS	ADDRESS
FN.CODE	RA	RB				
FN.CODE	RA	XR	BR	DA		
FN.CODE	M	XR	BR	DA		
FN.CODE	RA			N		
1234567890123456789012345678901234567890123456789012345678901234567890123456789O (columns 1–8)						
					INITIALISATION PROCEDURE:	
01000001	0110	0000	0000	000000000000	STORE VALUE ZERO IN REG 6.	00000
01000001	0101	0000	0000	000000000000	STORE VALUE ZERO IN REG 5.	00100
01000001	0100	0000	0000	000000000001	STORE VALUE ONE IN REG 4.	01000
01000001	0011	0000	0000	000000001001	STORE VALUE NINE IN REG 3.	01100
					SUMMATION PROCEDURE:	
00011010	0101	0100			ADD ONE TO (R5):I.E. (R4) + (R5) to R5.	010000
00011010	0110	0101			ADD (R5) TO (R6): I.E. ACCUMULATE SUM IN REG 6.	010010
00011001	0101	0011			COMPARE (R5) TO (R3).	010100
01000111	0100	0000	1111	000000010000	IF (R5) < (R3) BRANCH BACK TO START OF SUMMATION PROCEDURE.	010110
					IF (R5) = (R3) THEN END OF PROCEDURE, THE REQUIRED SUM IS IN REG 6.	

FIG. 20.1(a) Example Binary Coded Program.

added to the programming language, these translator programs are amongst the most complex programs that have been developed.

Upon examining the example programs in Fig. 20.1, you will note that the symbolic instructions have a one-to-one relationship with the binary machine instructions. For example, the mnemonic function code LA has a one-to-one correspondence with the binary function code (01000001); the decimal register address (6) corresponds to the binary register address (0110); and the symbolic address (SUMPR) corresponds to the binary address (000000010000). Symbolic languages having this one-to-one relationship with machine language are called Assembly languages because each symbolic instruction can be assembled directly into a binary machine instruction. Each type of computer has its own particular Assembly language corresponding to its own particular machine language.

Apart from the symbolic machine instructions, additional language elements are included in Assembly language systems to facilitate the writing of programs. Two of the most useful of these language elements are: the Define Constant (DC) statement, and the Define (or Reserve) Storage (DS) statement.

The Define Constant statement provides a convenient means for establishing data constants; for example, consider the following statement:

CONST1 DC F'21'.

When this statement is encountered by the translator program, instead of assembling a binary machine instruction, it assembles a binary number value equal to the specified decimal valued constant '21'. The symbolic address of this constant value is specified by the programmer to be CONST1, the programmer can proceed to use this symbolic

INSTRUCTIONS					COMMENTS	ADDRESS
FN.CODE	RA	RB				
FN.CODE	RA	XB	BR	DA		
FN.CODE	M	XR	BR	DA		
FN.CODE	RA			N		
1 2 3 4 5 6 7 8						
12345678901234567890123456789012345678901234567890123456789012345678901234567890						
					INITIALISATION PROCEDURE:	
65	6	0	0	0	STORE VALUE ZERO IN REG 6.	0
65	5	0	0	0	STORE VALUE ZERO IN REG 5.	4
65	4	0	0	1	STORE VALUE ONE IN REG 4.	8
65	3	0	0	9	STORE VALUE NINE IN REG 3.	12
					SUMMATION PROCEDURE :	
26	5	4			ADD ONE TO (R5).	16
26	6	5			ADD (R5) TO (R6).	18
25	5	3			COMPARE (R5) TO (R3).	20
71	4	0	15	16	IF (R5) < (R3) BRANCH BACK TO START OF SUMMATION PROCEDURE.	22
					IF (R5)=(R3) THEN END OF PROCEDURE, THE REQUIRED SUM IS IN REG 6.	

FIG. 20.1(b) Program Example Coded in Decimal Notation.

name in his program to refer to the defined constant, the translator program will recognise this name as the symbolic address of the defined constant, and will translate this symbolic address into the actual address of the storage cell in which the constant will be stored when the translated program is executed.

The Define (Reserve) Storage statement is used to reserve one or more contiguously addressed storage cells and to assign a symbolic address name to the first of these cells, thus enabling the programmer to use the name to address the cells wherever he needs to in his program. For example, the statement:

NUMBS DS 100F

causes the translator program to set aside 100 fullwords of main storage (this amounts to one hundred 32-bit word storage cells, or four hundred 8-bit storage cells in the IBM/360 computers), also, as a result of this statement the translator recognises the symbol NUMBS as the start address of the set of storage cells.

Another useful Assembly language feature is relative addressing; to illustrate, the symbolic addresses NUMBS + 1 is the address of the second of the 8-bit storage cells reserved by the Define Storage statement written above; the symbolic address NUMBS + 3 is the start address of the fourth 8-bit cell, which corresponds to the start of the second 32-bit fullword, and so on—the address of each storage cell is specified relative to some reference address such as NUMBS (the start address of the set of storage cells). The use of these language facilities is illustrated in the example program in Fig. 20.1(e).

2. High Level Languages

(a) GENERAL. Symbolic programming languages have been developed beyond the Assembly language level to provide languages that simplify programming even

INSTRUCTIONS					COMMENTS	ADDRESS
FN.CODE	RA	RB				
FN.CODE	RA	XR	BR	DA		
FN.CODE	M	XR	BR	DA		
FN.CODE	RA			N		
12345678901234567890123456789012345678901234567890123456789012345678901234567890						
					INITIALISATION PROCEDURE:	
LA	6	0	0	0	STORE VALUE ZERO IN REG 6.	
LA	5	0	0	0	STORE VALUE ZERO IN REG 5.	
LA	4	0	0	1	STORE VALUE ONE IN REG 4.	
LA	3	0	0	9	STORE VALUE NINE IN REG. 3.	
					SUMMATION PROCEDURE:	
AR	5	4			ADD ONE TO (R5).	
AR	6	5			ADD (R5) TO (R6).	
CR	5	3			COMPARE (R5) WITH (R3).	
BC	4	0	15	16	IF (R5) $<$ (R3) BRANCH BACK TO START OF SUMMATION PROCEDURE.	
					IF (R5) $\geqslant$(R3 THEN END OF PROCEDURE, THE REQUIRED SUM IS IN REG 6.	

FIG. 20.1(c) Program Example Coded with Mnemonic Function Codes and Decimal Notation.

Symbolic Address	Function	Operands	Comments
12345678901234567890123456789012345678901234567890123456789012345678901234567890			
			INITIALISATION PROCEDURE:
	LA	5, 0	STORE VALUE ZERO IN REG 6.
	LA	5, 0	STORE VALUE ZERO IN REG 5.
	LA	4, 1	STORE VALUE ONE IN REG 4.
	LA	3, 9	STORE VALUE NINE IN REG 3.
			SUMMATION PROCEDURE:
SUMPR	AR	5, 4	ADD (R4) TO (R5).
	AR	6, 5	ADD (R5) TO (R6).
	CR	5, 3	COMPARE (R5) WITH (R3).
	BC	4, SUMPR	IF (R5) $<$ (R3) BRANCH TO SUMPR.
*			IF (R5) $>$ (R3) THEN END OF PROCEDURE,
*			THE REQUIRED SUM IS IN R6.

FIG. 20.1(d) Program Example Coded using Symbolic Instruction Storage Address (SUMPR).

Symbolic Address	Function	Operands	Comments
1		2 3	4 5 6 7 8
12345678	901234	567890123456789	0123456789012345678901234567890123456789012345678901234567890
*			INITIALISATION PROCEDURE:
	L	6, ZERO	LOAD VALUE IN STORE CELL WITH SYMBOLIC
	L	5, ZERO	ADDRESS 'ZERO' INTO REG 6 AND REG 5.
	L	4, ONE	LOAD VALUE IN STORE CELL 'ONE' INTO REG 4.
	L	3, NINE	LOAD VALUE IN STORE CELL NINE' INTO REG 3.
*			SUMMATION PROCEDURE:
SUMPR	AR	5,4	ADD (R4) TO (R5).
	AR	6,5	ADD (R5) TO (R6).
	CR	5,3	COMPARE (R5) TO (R3).
	BC	4, SUMPR	IF (R5) < (R3) BRANCH TO SUMPR, OTHERWISE
*			CONTINUE ON WITH NEXT INSTRUCTION.
*			OUTPUT PROCEDURE:
	ST	6, SUM	STORE (R6) IN STORE CELL WITH SYMBOLIC
			ADDRESS 'SUM'.
	PUTB	SUM	PRINT THE VALUE IN STORE CELL 'SUM' ON
*			LINE PRINTER; I.E. PRINT THE REQ'D SUM.
*			DATA:
ZERO	DC	F'O'	
ONE	DC	F'1'	
NINE	DC	F'9'	
SUM	DS	F	
	END		

FIG. 20.1(e) Program Example Coded in IBM/360 Assembly Language showing the use of Symbolic Storage Addresses, the Output Macro-Instruction PUTB and the Define Constant and Define Storage Assembly Language Statements.

further. Not only does this mean that professional programmers can write programs more quickly and economically, but what is more important, it has greatly facilitated the widespread application of computers to all sorts of problems by such people as engineers, architects, chemists, medical scientists, business men, and surveyors to mention just a few. There has been a proliferation of these languages, but we shall only consider three of the most widely used languages, Fortran, Cobol and Snobol.

(b) FORTRAN. This is a language that corresponds very closely to the conventional arithmetic and algebraic language of mathematics. Fortran was the first of the high level compiler languages to be developed, and is probably the most widely used programming language. It was developed over the period 1954 to 1957 by J. W. Backus and his associates at the International Business Machine Corporation. It is interesting to note what Backus *et al.* (1967) had to say about the objectives to be met by Fortran: "The goal of the Fortran project was to enable the programmer to specify a numerical procedure using a concise language like that of mathematics and obtain automatically from this specification an efficient 704 [machine language] program to

```
C THIS EXAMPLE FORTRAN PROGRAM COMPUTES THE ROOTS OF THE EQUATION
C AX.X + BX + C = 0
C
C THE FOLLOWING STATEMENT CAUSES VALUES OF THE VARIABLES A, B
C AND C TO BE READ FROM AN INPUT CARD
        READ, A, B, C
C THE FOLLOWING STATEMENTS SPECIFY COMPUTATION OF THE ROOTS OF
C THE EQUATION AND ASSIGN THE VALUES TO THE VARIABLE NAMES
C ROOT1 AND ROOT2. NOTE THAT B**2 – 4*A*C IS ASSUMED TO BE
C POSITIVE
        ROOT1 = (–B+SQRT(B**2–4*A*C))/2*A
        ROOT2 = (–B–SQRT(B**2–4*A*C))/2*A
C THE NEXT STATEMENT CAUSES THE VALUES OF THE VARIABLES ROOT1
C AND ROOT2 TO BE PRINTED ON THE LINE PRINTER
        PRINT, ROOT1, ROOT2
C THE NEXT STATEMENT TERMINATES PROGRAM EXECUTION
        STOP
```

FIG. 20.2(a) Example Fortran Language Program.

```
START           OPEN OUTPUT SALARY-FILE, WRITE SALARY RECORD.
                PERFORM CALCULATIONS
                VARYING MONTHLY-PAY
                FROM  500
                BY     10
                UNTIL MONTHLY-PAY IS GREATER THAN 1000
                CLOSE SALARY-FILE. STOP RUN
CALCULATIONS
                COMPUTER WEEKLY IN SALARIES =3*MONTHLY-PAY/13
                COMPUTER ANNUAL IN SALARIES = 12*MONTHLY-PAY
                MOVE MONTHLY-PAY TO MONTHLY IN SALARIES
                WRITE SALARY-RECORD FROM SALARIES
```

FIG. 20.2(b) Example Cobol Language Program.

carry out the procedure. It was expected that such a system would reduce the [program] coding and debugging task to less than one fifth of the job it had been." The Fortran language is illustrated by the example program shown in Fig. 20.2(a).

(c) COBOL. This is a language designed for business data processing applications where the mathematical aspects of the data processing are comparatively simple. An objective in the development of Cobol has been to devise a language which is usable with computers and which is as close as practicable to the conventional spoken and written language of business and commerce. As a direct consequence of this objective, abbreviated notation (which is "of the essence" in mathematics and Fortran) is avoided as much as possible. A simple addition operation that might be expressed as

TOTINC = OTHINC + WAGES

in Fortran, might appear as follows in Cobol:

ADD OTHER-INCOME TO WAGES GIVING TOTAL INCOME

The Cobol language is illustrated by the example program shown in Fig. 20.2(b). Only part of the program is shown; the program is designed to compute and print a table of annual and weekly salary rates, using a monthly salary rate which is initially set to $500.00 and increased in steps of $10.00. The process continues until a table is computed and printed showing annual and weekly salary rates for monthly salaries ranging from $500.00 to $1000.00 in increments of $10.00.

```
*  THIS EXAMPLE SNOBOL PROGRAM READS TEXT FROM PUNCHED CARDS AND
*  CHANGES THE SPELLING OF THE WORD WIZE TO WISE WHEREVER IT
*  APPEARS THROUGHOUT THE TEXT
*
*  THE FOLLOWING STATEMENT CAUSES THE CONTENTS OF THE NEXT INPUT
*  CARD TO BE READ AND ASSIGNED TO THE VARIABLE NAME TEXT. IF
*  THE INPUT OPERATION FAILS, I.E. THERE ARE NO MORE INPUT CARDS,
*  THEN PROGRAM EXECUTION BRANCHES TO THE STATEMENT LABELLED
*  'COMPLETED' WHICH TERMINATES PROGRAM EXECUTION
   START      TEXT = INPUT        :F (COMPLETED)
*  THE FOLLOWING PATTERN MATCHING STATEMENT CAUSES WIZE TO BE
*  REPLACED BY WISE IF WIZE OCCURS IN THE CONTENTS OF THE CARD
*  JUST READ
              TEXT 'WIZE' = 'WISE'
*  THE FOLLOWING STATEMENTS CAUSE THE CORRECTED VERSION OF THE
*  CONTENTS OF THE CARD JUST READ TO BE PRINTED ON LINE PRINTER
*  AND PUNCHED INTO A NEW CARD
              OUTPUT = TEXT
              PUNCH = TEXT        :S(START)
*  THE PUNCH STATEMENT CAUSES PROGRAM EXECUTION TO BRANCH TO THE
*  STATEMENT LABELLED 'START' WHICH READS THE NEXT INPUT CARD
   COMPLETED  END
```

FIG. 20.2(c) Example Snobol Language Program.

(d) SNOBOL. Snobol is a programming language specially designed to facilitate the processing of strings of symbols (characters) as opposed to Fortran and Cobol which are primarily concerned with processing numerical data and files of alphanumeric data. The basic data element in Snobol is a string of characters. The language has operations for joining and separating strings, for testing their contents, and for detecting and replacing substrings. For example, if we assume a particular string to be an English language sentence, then we can write a Snobol program to split the sentence up into words and phrases, to eliminate certain words and/or phrases, to substitute new or alternative words and phrases, and hence restructure the sentence, according to our requirements. A string value (i.e., a string of characters) can be assigned to a variable identifier by assigning a literal string enclosed in inverted commas, as in the following statements.*

LINE1 = 'THIS IS'
LINE2 = 'A STRING'.

Or by assigning the value of one variable to another, as in the statement

LINE = LINE1

which assigns the value THIS IS to the variable name LINE. Concatenation is the basic operation for combining two strings to form a third, this may be specified as follows:

LINE = LINE1 LINE2.

This statement assigns the string THIS IS A STRING to the variable LINE.

* A variable identifier or name is simply the symbolic address of the storage cell(s) in which the value assigned to the variable is stored. Hence, variable identifier or variable name is synonymous with symbolic address.

The operation of examining strings for the occurrence of substrings is fundamental to the Snobol language, this is achieved through the use of the pattern matching statement which takes the general form:

subject pattern = object.

If the pattern matching operation succeeds, the subject string is modified by replacing the matched substring (as specified by the pattern string) by the object string. For example, the statement,

LINE 'IS A' = ' '

specifies that if the pattern string 'IS A' is a substring of the string named LINE then that substring is to be replaced by the null string ' '. Hence the statement will result in the value of LINE being modified to be THIS STRING. The statement

LINE 'IS A' = 'IS A CHARACTER'

will result in the value of the variable LINE being modified to THIS IS A CHARACTER STRING. The statement

LINE 'ALPHABETIC' = 'NUMERIC'

will not result in any change to LINE because the pattern match will not succeed since the pattern substring ALPHABETIC does not appear in the character string labelled LINE.

The output of information is specified as follows, the statement

OUTPUT = 'THE DATE TODAY IS'

causes the literal data on the right hand side of the statement to be printed as the next line on the output printer. Alternately, the statement,

OUTPUT = LINE

causes the string value of LINE, which is THIS IS A CHARACTER STRING, to be printed as the next line of print on the output printer. The input of information is specified as follows, the statement

NEXTIN = INPUT

causes the next punched card to be read and the 80 character string recorded in it to be assigned to the variable name, in this case NEXTIN, on the left hand side of the statement. We can also write a statement to punch information out on cards, for example, the statement

PUNCH = INPUT

causes the next card to be read (INPUT) and a copy of its contents to be punched into an output card. The statement

PUNCH = LINE

causes the string THIS IS A CHARACTER STRING to be punched into an output card.

A simple example Snobol program is shown in Fig. 20.2(c). The purpose of the program is to read a set of input cards, one at a time, search for occurrences of the word WIZE and replace it with the modern spelling WISE; the program prints and punches a corrected version of the input cards. We could assume, for example, that a short story, article or novel is recorded on the input cards and the program is to modernise the spelling of the word WIZE wherever it appears.

C. FORMAL DEFINITION OF PROGRAMMING LANGUAGES

1. General

High level languages are translated by so-called *compiler programs* into executable machine language programs, Fortran programs are translated by a Fortran compiler,

Cobol programs are translated by a Cobol compiler, and so on. In order to be able to translate a high level language it must be free of ambiguities, otherwise it is impossible to design compiler programs that will give an accurate and consistent translation for all programs written in that high level language.

To avoid ambiguity, encourage precision and clearly specify details of language structure, various methods of formal description are employed. Such methods facilitate complete and unambiguous language descriptions in the sense that they allow no alternative interpretations of elements of the language and describe all of the allowed constructs of the language. We refer to the definition of the structure of a language as the *syntax* of the language.

When defining the syntax of a language the first thing to do is to define a set of *symbols* (e.g., alphabetic, numeric and operator symbols) which constitute the alphabet for the language. We then have to define a set of rules specifying the use of these symbols for constructing a set of allowable *words* for the language. We proceed in this way to define sets of rules for constructing *statements* as an ordered construct of symbols and words, *procedures* as an ordered construct of statements, and *programs* as an ordered construct of procedures. It should be observed that there is a close correspondence here with the structure of English language elements such as "clauses", "sentences", "paragraphs", and so on. Furthermore, English language sentences may be classified as declarative, imperative, interrogative, exclamatory, and so on; in a corresponding way programming language statements may be classified as declarative, imperative, assignment, conditional, control, procedural, and so on.

2. Backus Normal Form of Syntax Definition

One commonly used method of formally defining the syntax of a programming language is the so-called Backus Normal Form (BNF) of syntax definition, this method has come into widespread use since it was devised by J. W. Backus and used by P. Naur *et al.* (1967) to formalise the description of Algol in 1963. The principal features of BNF are set out in Fig. 20.3; its use is illustrated in Fig. 20.4 where it is used to define a sub-set of the syntax of the Fortran language. Note that some of these definitions are recursive, that is, some entities are defined in terms of themselves, as in the definitions for identifier, unsigned integer, and so on. To take an example consider the definition of an identifier, which states that an identifier is a string of letter and digit symbols, the first (the left most) of which must be a letter and that the identifier can have at most six symbols. The application of this definition for the integer

fig 20.3

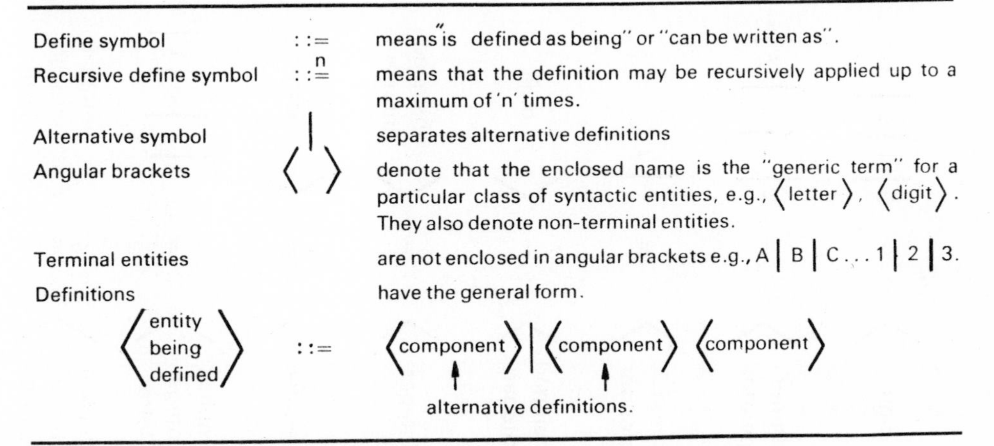

FIG. 20.3 Principal Features of the BNF Notation.

```
<digit> ::= 0 | 1 | 2 | 3 ..... | 9

<letter> ::= A | B | C | D .... | Z

<arithmetic operator> ::= + | − | * | / | **

<sign> ::= + | −

<mult. operator> ::= * | /

<add operator> ::= + | −

<unsigned integer> ::= <digit> | <unsigned integer> <digit>

<integer> ::= <unsigned integer> | <sign> <unsigned integer>

<identifier> ::= <letter> | <identifier> <letter> <identifier> <digit>

<implicit integer letter> ::= I | J | K | L | M | N

<implicit integer identifier> ::= <implicit integer letter> | <implicit integer identifier> <letter> | <implicit integer identifier> <digit>
```

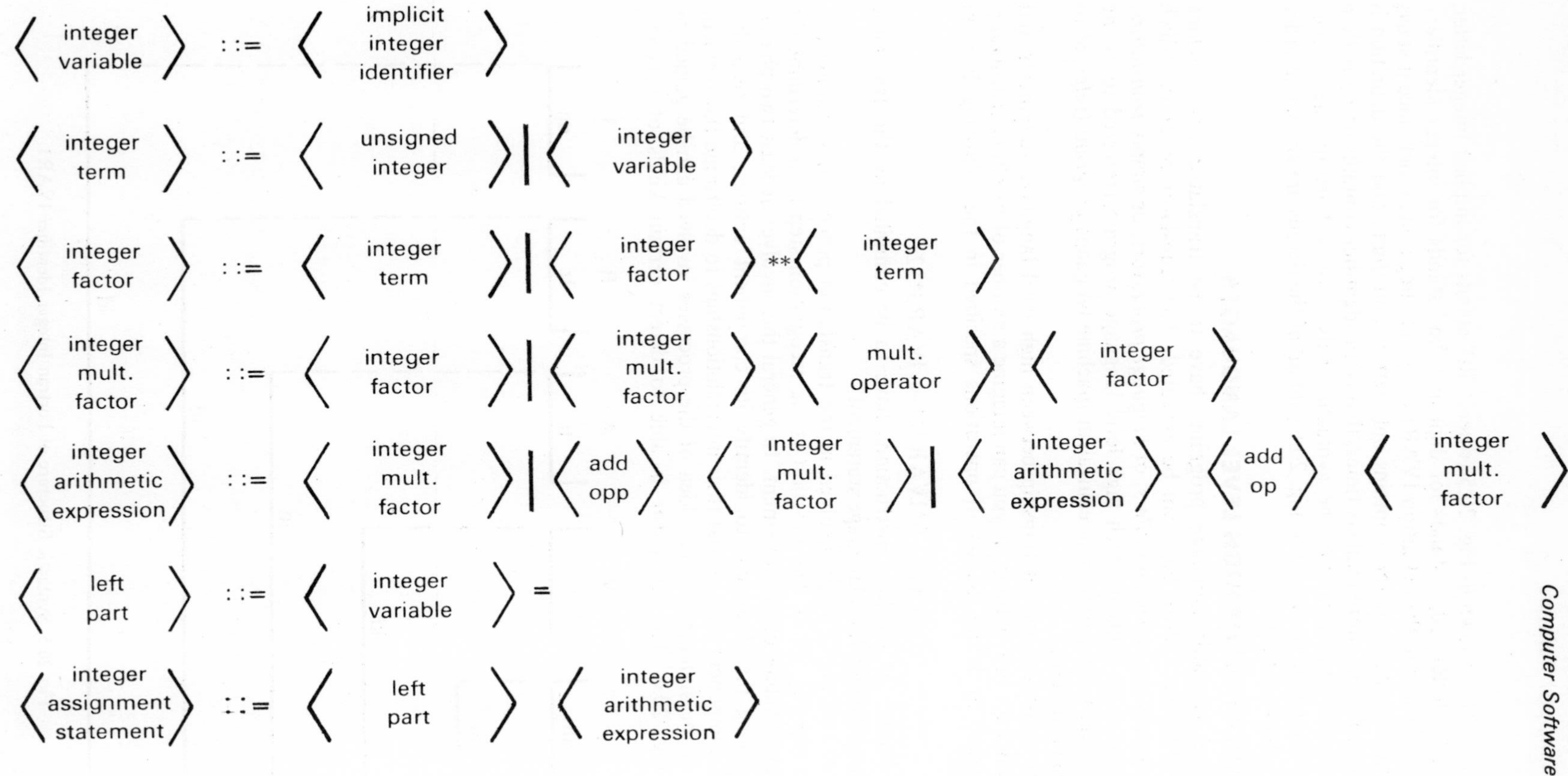

FIG. 20.4 BNF Definition of the Syntax of some Integer Elements of the Fortran Language.

identifier IVAR1 is shown in Fig. 20.5 (where "ilt" stands for implicit integer letter, "lt" stands for letter, "dg" stands for digit and "id" stands for integer identifier). Note that according to the definition IVAR1 is a valid integer identifier since it starts with an implicit integer letter, contains only letters and digits and the definition is recursively applied no more than six times, that is the identifier contains nc more than six symbols. Another example, the syntactic analysis of a Fortran integer type assignment statement, is shown in Fig. 20.6 (not all of the details are shown, in order to avoid excessive detail).

D. TRANSLATION OF HIGH LEVEL LANGUAGES

High level symbolic language, programs have to be translated into machine language programs before they can be executed. The translator programs which perform this task are called *compilers*, or *language processors*, or simply *translators*, their function is to translate each high level language program (referred to as the source language program) into an equivalent machine language program (referred to as the object program).

There is no one-to-one relationship between high level language statements and machine language instructions, as you can imagine a number of machine instructions will be needed to define the set of operations specified in the following Fortran statement:

IVAR1 = 3*IVAR2/4 + IVAR3**2.

Hence a number of machine instructions have to be compiled by the translator program for each high level language statement.

There are two major steps involved in the translation process—recognition and generation (see Fig. 20.7). The function of the recognition step is to determine the structure and content of the statement. In general this involves at least two phases; firstly, scanning the statement to identify its component entities; and secondly, analysis of component entities and their interrelationships to determine the syntactic structure of the statement. Some idea of the processes involved can be gained by studying Fig. 20.6. The translator is able to detect certain kinds of syntactic

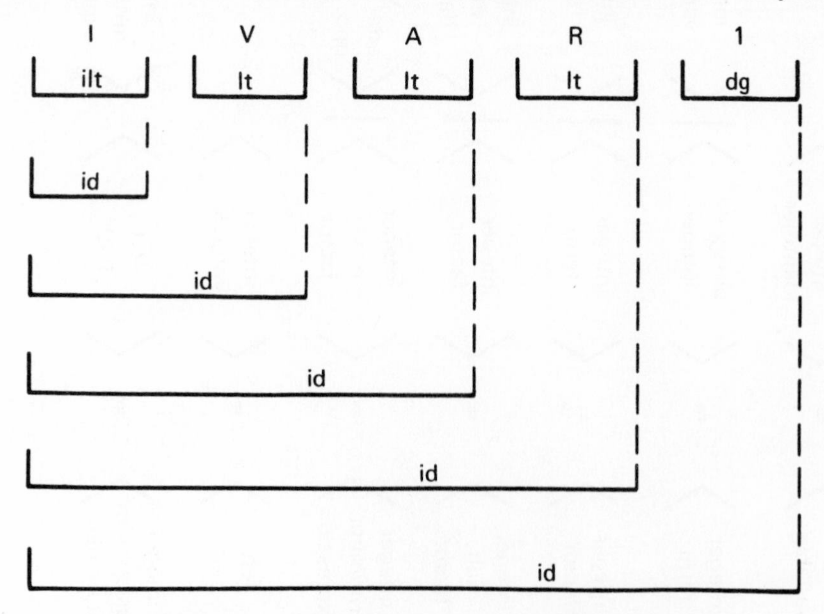

Fig. 20.5 Syntactic Structure of Fortran Integer Identifier IVAR1.

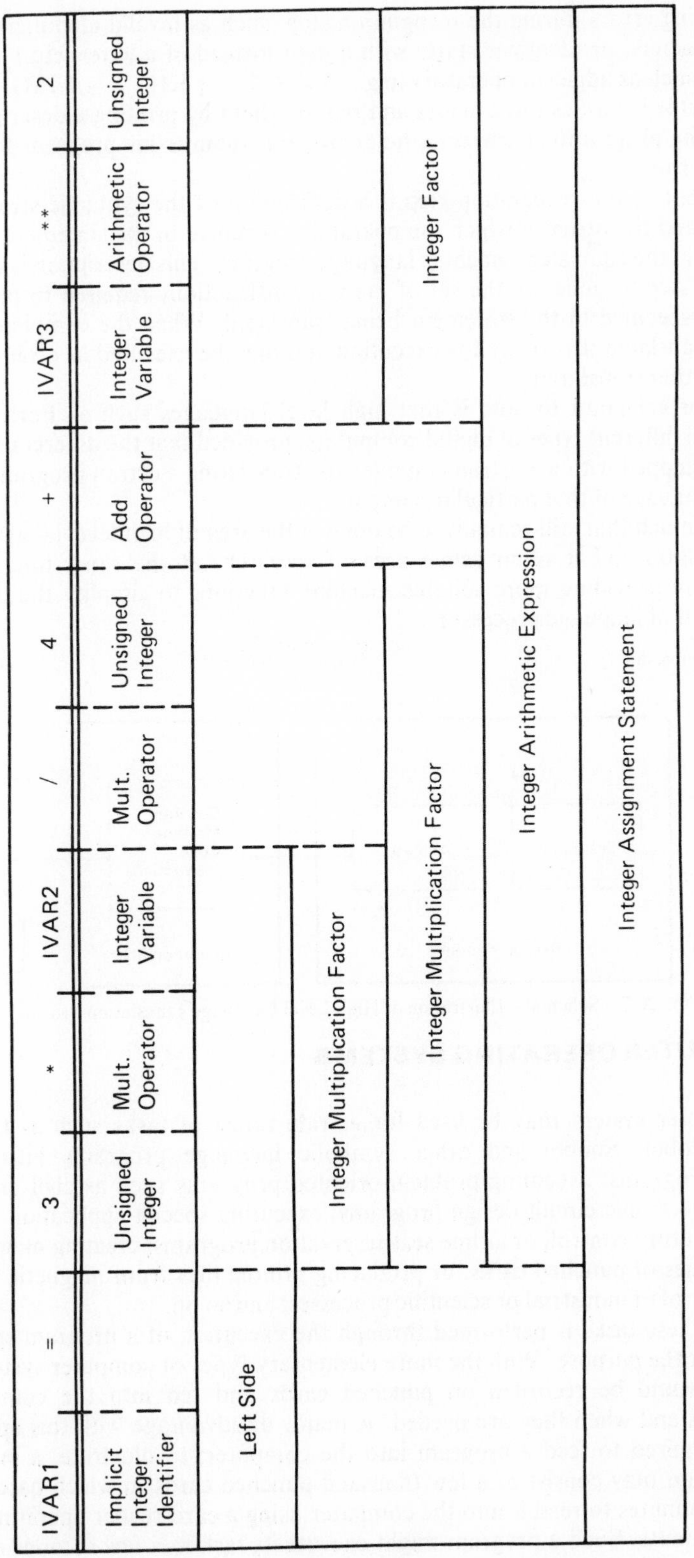

IVAR1	=	3	*	IVAR2	/	4	+	IVAR3	**	2
Implicit Integer Identifier		Unsigned Integer	Mult. Operator	Integer Variable	Mult. Operator	Unsigned Integer	Add Operator	Integer Variable	Arithmetic Operator	Unsigned Integer
Left Side										
		Integer Multiplication Factor								
		Integer Multiplication Factor						Integer Factor		
		Integer Arithmetic Expression								
Integer Assignment Statement										

FIG. 20.6 Syntactic Structure of the Fortran Integer Assignment Statement
IVAR = 3 * IVAR2/4 + IVAR3**2.

programming errors during the recognition step, such as invalid identifiers (e.g., too many characters, or identifier starts with a digit instead of a letter, etc.) and invalid structures such as adjacent operators (e.g., IVAR/*3, or KIN + − LIM), and so on. The translator identifies these errors and reports them by printing a descriptive error message, the programmer corrects the errors and submits his program for another translation run.

The output of the recognition step is a description of the syntactic structure of a statement and the order in which the operations specified in the statement should be performed in the equivalent machine language program. This description is used in the generation step to generate the set of machine instructions required to perform the operations specified in the statement being translated. When the complete program has been translated it is ready for execution and may be executed as often as needed without further translation.

An important point to note is that high level languages such as Fortran can be executed on different types of digital computers, provided that the different computers are each equipped with a Fortran compiler for translating Fortran programs into the machine language of that particular computer.

There is much that still remains to be done in the area of high level language design and translation and it is an active area of research, at the same time hardware designers are providing more suitable machine functions to simplify the design and development of language processors.

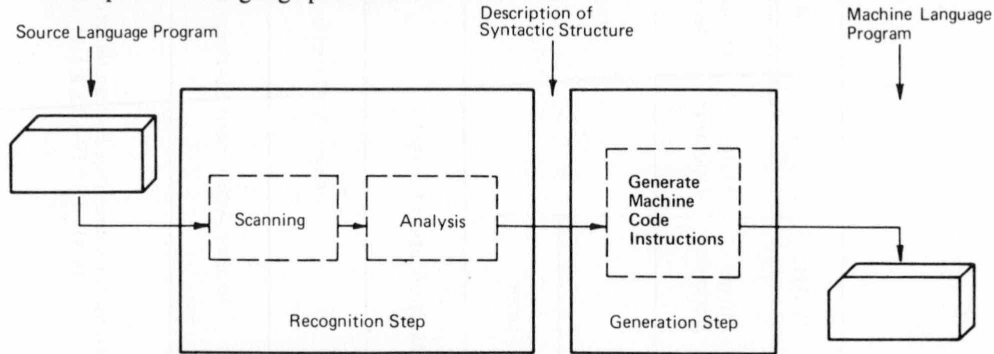

FIG. 20.7 Schematic Illustration of High Level Language Translation Process.

E. COMPUTER OPERATING SYSTEMS

1. General

A computer system may be used for a wide range of tasks such as translating Fortran, Cobol, Snobol and other symbolic language programs into machine language programs; executing problem-oriented programs such as civil engineering design or electronic circuit design programs; executing special application programs such as inventory control, or airline seat reservation programs; creating magnetic tape files from files of punched cards, or producing printed files from magnetic disk files; on-line control of industrial or scientific processes; and so on.

Each of these tasks is performed through the execution of a program specifically designed for the purpose. With the more elementary types of computer systems these programs would be recorded on punched cards and fed into the computer for execution as and when they are needed. A major disadvantage with this approach is the time required to feed a program into the computer; to illustrate, a moderately sized program may consist of a few thousand punched cards in which case it would take a few minutes to read it into the computer using a card reader operating at 1000 cards per minute. Such a program might execute its task in a few seconds or tens of seconds, the net result being that most of the computer's time is spent in non-productive input activities. Another disadvantage is that the operator has to handle large quantities of punched cards—taking them from their storage cabinets, loading them into the card reader and then returning them to their cabinets—all of this takes time, during which the computer probably stands idle. Further disadvantages arise from damaged or worn cards which give rise to reading errors, or because the operator may make a mistake and feed the cards in in the wrong order, and so on.

With more eleborate computer systems sets of programs are stored within the system, usually on a magnetic disk unit, to form an *on-line program library*. When the user wants to execute a particular program it can be fetched automatically from the on-line program library and loaded into main storage for execution. Since information can be transferred from magnetic disk storage at very high rates, it will take only a fraction of a second to read a program from a magnetic disk which would otherwise take a few minutes to read through a card reader. With this approach the amount of time spent reading programs into the computer is considerably reduced resulting in an increased amount of time spent executing programs, and hence more effective utilisation of the computer. Other advantages arise from the fact that the operator is not involved in the continuous loading of large decks of punched cards into the system, this further minimises computer idle time and reduces the possibility of errors arising from worn or out-of-sequence cards.

When operating with an on-line program library a computer system can be made to process automatically a sequence or *batch* of user jobs. Each job is preceded by a set of job control cards which specify the program to be executed, the computer system reads and analyses the information on these cards and then proceeds to load the specified program from the library into main storage for execution. Immediately that job is completed the system proceeds to read the control cards for the next job, and then to fetch and load the specified program and initiate its execution. In this way the system is automated to execute a sequence of jobs, thus simplifying the operator's task and eliminating non-productive computer time.

2. Operating System Components

(a) GENERAL. A computer operating system consists of two major component parts—an on-line library of processing programs and an integrated set of system control programs.

The processing programs are those which perform the various tasks required by the computer users—they perform the productive work done by the system. These programs include language translators, standardised problem-oriented and application programs and the general range of user-written programs. The set of processing programs included in the library can be changed with programs being added or deleted as required. Usually the library is arranged to contain only those programs that are in fairly constant use, in order to conserve storage space and minimise library overheads. Other infrequently used programs may be recorded on punched cards or perhaps magnetic tape and loaded into the system for execution only when required.

The system of control programs is designed to automate the operation of the computer system to achieve maximum throughput of a continuous stream of processing jobs. The control programs supervise the fetching of processing programs from the on-line library, or the input of processing programs from some specified input device, and the execution of these programs. The work to be done by the system consists of a job stack which constitutes a continuous stream of jobs to be executed under the control and supervision of the system of control programs.

The control system is directed by user prepared job control statement cards which specify such things as the processing program(s) to be executed for each job, which input/output units are required to be used for each job, and the data files to be processed by each job. The control system programs read and analyse the job specification statements recorded on these cards to determine the processing requirements for each job. A typical control system consists of three major components:

(i) an *input/output control system* which is designed to supervise and perform the input and output of information;
(ii) a *job control program* which controls the scheduling and execution of processing jobs; and
(iii) a *supervisor program* which controls and coordinates the overall operation of the system.

(b) INPUT/OUTPUT CONTROL SYSTEM. As indicated in Fig. 20.8 medium to large scale computer systems may have a variety of input/output and auxiliary storage devices. Information (programs and data) is transferred between these devices and the main storage unit as required (for execution or procession). The programming of input/output devices is quite a complex task and for this reason a library of standardised input/output programs are usually supplied to relieve the average user of this complex task—this standardised set is called the Input/Output Control System. When the user wants some specific input/output operation performed he simply calls upon the appropriate standardised program to perform the operation for him (such as the GET and PUT macro-routines described in Chapter 19).

The various types of input devices may include: devices which *sense* or *read* data recorded on various types of media such as punched cards, paper tape; magnetic tape or disk or drum; devices which read characters printed with magnetic ink on paper documents such as cheques; specially formed printed characters can be read by mechanised optical reader devices from ordinary paper documents or from photographic images on 35 mm microfilm. These devices read the input information and transmit it, under program control, to the main storage unit where it becomes available for processing or execution. Various types of output devices are used to record information (from main storage) on punched cards or paper tape, or on magnetic tape or disk or drum, or produce printed copies on paper or microfilm, or produce graphical (diagrammatic, pictorial) outputs on paper or cathode-ray-tube displays and so on.

Some devices are used for manual, on-line input of information, the two most important of which are: a *keyboard device* similar to a typewriter through which information is entered by a manual "typing" operation; and a graphical input device where a *light pen device* is used to enter information by drawing on a cathode-ray-tube display screen.

Some input/output devices may be situated in close proximity to the computer, in the same room for example, and connected by cable directly to the computer. Other devices may be in remote positions such as on different floors in the same building, or different buildings in a complex of buildings such as an industrial complex or university complex, or they may be hundreds or even thousands of miles away from the central computer location. Where the peripherals are not too far away they are usually connected by cable direct to the computer, where they are a long distance away they may be connected via the standard telephone network, including submarine telephone cables from one country to another, or even by radio and satellite communications systems.

Input/output control systems are designed to cope with all these forms of communications between a computer and its peripheral devices.

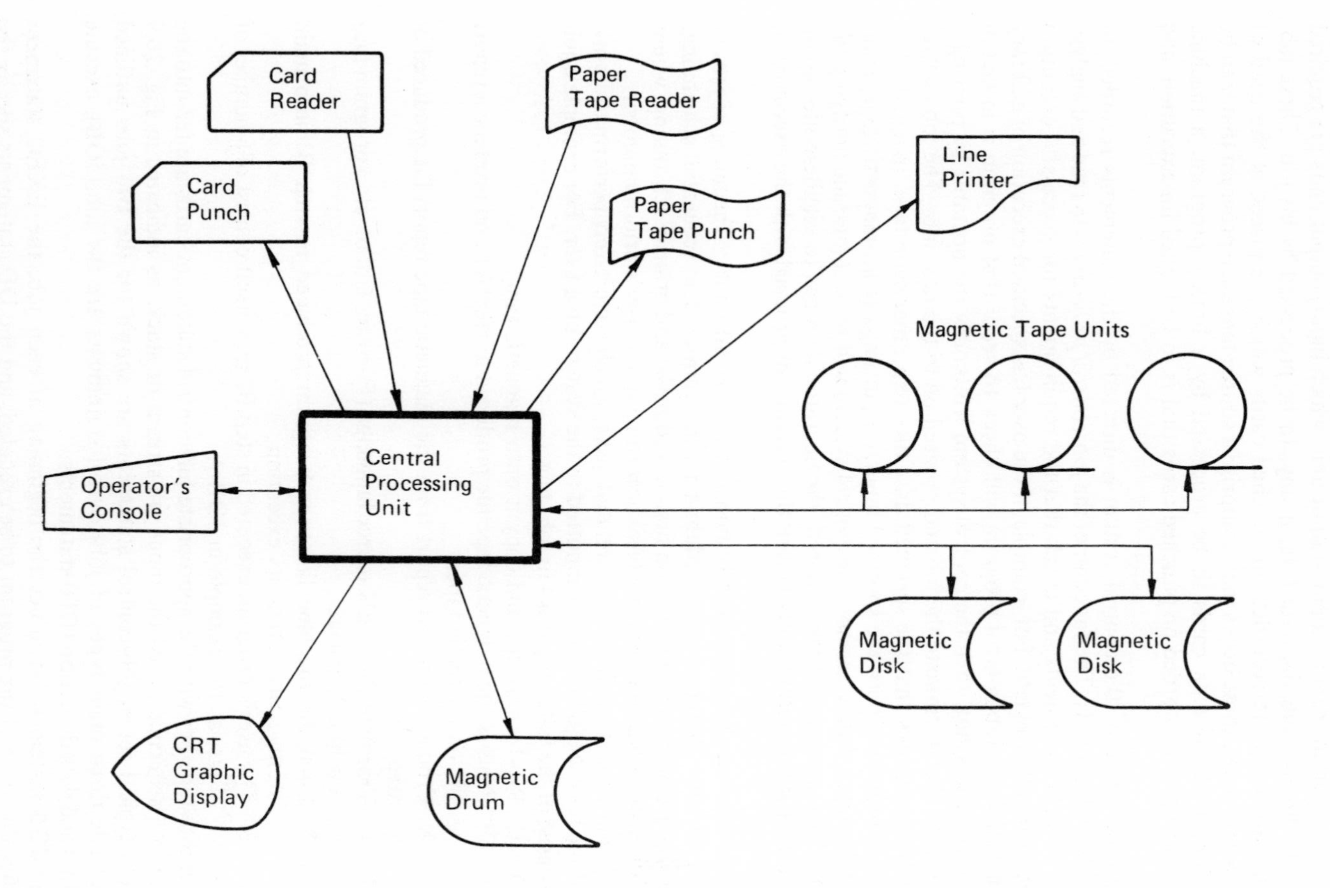

FIG. 20.8 Schematic Diagram of Computer System showing some of the various kinds of Input/Output Units and Auxiliary Storage Units.

(c) JOB-CONTROL PROGRAM. Each user specifies the requirements of his job, by using a job-control language. The user prepares job-control language statements, specifying the processing program to be executed, whether that program is in the on-line library or has to be read in from an input/output unit, which input/output units are required for use by his job and the data files, if any, to be processed by his job. These job-control statements are recorded on punched cards which are placed at the head of each job card deck, these decks may comprise a source language program that is to be translated, a file of data cards to be processed by a library program, a machine language program recorded on punched cards that is to be loaded for execution, and so on.

The functioning of the control system is directed by the statements recorded on these job-control cards. The function of the *job-control program* is to read and analyse the job-control statements and check that the requirements for successful execution of the job are all available. For example, if the user has specified execution of a library program then the job-control program will check to see if that program is in fact in the library, if it is not, then the control system will notify the operator by printing a message and then terminate the job and proceed on to the next job. The job-control program also checks that the specified data files (e.g., magnetic tape, magnetic disk files) have been set up on the specified input/output units (e.g., magnetic tape, magnetic disk units), if they are not, the control system will print a message notifying the operator that specific files are required, the operator then has to set these files up on the specified input/output units before the system can proceed with the execution of the job.

The job-control program is also responsible for the automatic scheduling of jobs in the input stream. Two types of scheduling are in common use: sequential scheduling, where jobs are processed on a first-in first-served basis; and priority scheduling, where the jobs are processed on a priority basis according to the user's priority number.

A job may consist of one or more related tasks, which collectively perform the work to be done for each job of work submitted to the system by a user. For example a job may consist of the following tasks or job steps:

JOB A
1. Read a card file and copy it onto magnetic tape.
2. Process the magnetic tape file produced in step A1 and produce a report file on magnetic tape.
3. Produce a printed report from the magnetic tape report file produced in step A2.

JOB B
1. Translate the user's source language (Fortran, Cobol, etc.) program into machine language.
2. Load the machine language program produced in step B1 into main storage and initiate its execution.
3. The user's program executed in step B2 may itself consist of a number of tasks, as for example in JOB A.

User jobs together with the appropriate job-control statement cards are fed into the system in a sequence called the input job stream or stack, as indicated in Fig. 20.9 where a typical set of job-control statements are shown for the two jobs outlined above. The three main types of job-control statements are the job (JOB), execute (EXEC) and data definition (DD) statements.

The JOB statement identifies the beginning of each job, the EXEC statement specifies the name of the program to be executed, and the DD statements specify the name of data files to be processed, the characteristics of these files (e.g., magnetic tape, disk, punched cards, record size, etc.) and the specific input/output units on which these files should be mounted. Where a job consists of a number of job steps a separate EXEC statement and DD statements are required for each step as shown.

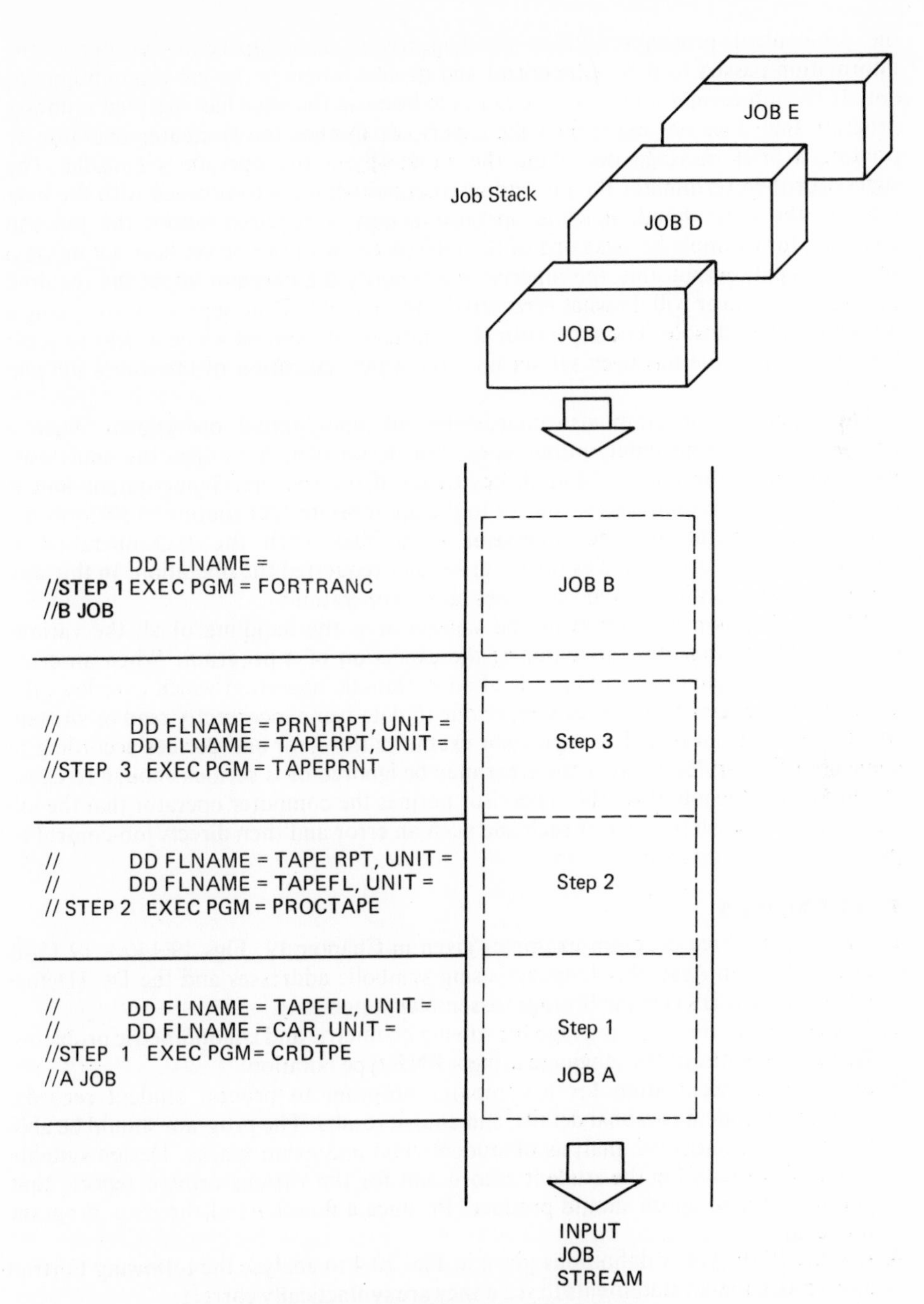

FIG. 20.9 Job Stack and Job-Control Cards.

(d) SUPERVISOR PROGRAM. The supervisor program is the control centre of the operating system, it coordinates and supervises the operation of the various control programs and the execution of processing programs. For example, when the execution of a job is completed the supervisor initiates the job-control program which proceeds to read and analyse the control statement cards for the next job, and then checks to see that all the requirements are available for successful execution of the job; if not, the job-control program notifies the supervisor. The supervisor examines the information passed to it by job-control and decides whether the job can continue or not. If the job cannot continue, for example because the user has specified a library program that does not exist, then the supervisor notifies the computer operator by printing a brief message describing the situation on the operator's console. The supervisor then terminates the job and instructs job-control to proceed with the next job. On the other hand, if some operator action is required before the job can continue, for example because one of the job's data files has not yet been set up on a specified input/output unit, the supervisor will notify the operator about the required action, the operator will do what is required and then notify the supervisor by typing a message on his console. The supervisor then notifies job-control which checks to see if the correct data file has been set set up after which execution of the user's job can proceed.

The supervisor program also coordinates all input/output operations. When a program wants some input/output operation performed, it notifies the supervisor which examines the request, and checks to see if the specified input/output unit is available for use, it then passes control to the appropriate I/O routine to perform the input/output operation. The supervisor is notified when the I/O operation is completed and in turn it notifies the program that requested the operation. In this way the supervisor coordinates and supervises all I/O operations.

Another function performed by the supervisor is the handling of all the various error conditions that may arise during the execution of a program. When an error occurs, for example due to the result of an arithmetic operation which overflows the capacity of a general purpose register, or due to data being incorrectly read or written, the supervisor is notified. The supervisor examines the error and decides according to some predefined rules whether the error may be ignored or is serious enough to cause termination of the job, if so, the supervisor notifies the computer operator that the job has been terminated because of such and such an error and then directs job-control to proceed with the next job.

F. EXERCISES

1. Rewrite the three program examples given in Chapter 19, Figs 19.14(a), 19.15(a) and 19.16(a) in Assembly language using symbolic addresses and the DC (Define Constant) and DS (Define Storage) assembler statements.
2. Design a programming language for solving geometric and trigonometric problems. Define the syntax of the language using a BNF type notation.
3. Draw up a specification for a computer program to process student records, containing student personal details, and course results. The program should be able to perform a statistical analysis of student class and exam marks. Design suitable standard formats for the student record and for the various printed reports that you think the program should produce. Produce a flowchart of the main program procedures.
4. Use the BNF syntax definitions given in Fig. 20.4 to analyse the following Fortran integer assignment statements to see if they are syntactically correct.
   (1) INUM1 = NUM3/2 + 5/NUM16
   (2) M = N + L + J/ – K
   (3) I1569 = K1**2/CONST + K3
   (4) IFACT6 = I6*I6 – 1*I6 – 2*I6 – 3*I6 – 2
5. Correct any syntax errors in the Fortran statements given in Exercise 4, then write a set of IBM/360 Assembly language statements to perform the operations specified in the Fortran statements.
6. Draw up an outline design (just a few pages) of a computer system including

input/output units and auxiliary storage units and a computer operating system suitable for a high school or a University. The system is to be used for processing administrative records such as payroll and student records and for teaching purposes, so include the sorts of programs and input/output units suitable for this purpose.

SUGGESTED READING MATERIAL

WEISS, E. A. (ed.) *Computer Usage: 360 Assembly Programming*. McGraw-Hill, New York, 1969.

This book is recommended to those who want to make a more detailed study of IBM/360 Assembly language programming. It begins by explaining the fundamentals and proceeds to more advanced topics of Assembly language programming.

A Programmers Introduction to the IBM System/360 Architecture, Instructions and Assembler Language. International Business Machines Corporation, New York, 1967. Publication No. C20-1646.

This is a very useful introduction to the general operating principles of the IBM/360 and its Assembly programming language.

BLATT, J. M. *Introduction to FORTRAN IV Programming: Using the Watfor Compiler*. Goodyear Publishing Co., California, 1968.

A comprehensive, readable text designed for student use which deals with basic and more advanced aspects of the Fortran programming systen for IBM/360 computers.

MCCRACKEN, D. D. and DORN, W. S. *Numerical Methods and Fortran Programming*. John Wiley and Sons, New York, 1964.

This text will be of interest to students interested in mathematics, science and engineering, it deals with numerical methods for solving mathematical problems and is well illustrated with examples and case studies.

MCCAMERON, F. A. *Cobol Logic and Programming*. Richard Irwin, Illinois, 1966.

An introductory text on the Cobol programming language suitable for the novice who wants to learn more about this language.

GRISWOLD, R. E., POAGE, J. F. and POLANSKY, I. P. *The Snobol 4 Programming Language*. Prentice-Hall, New Jersey, 1968.

This text on the Snobol 4 programming language is readily understood and well illustrated with examples.

General Information Manual, Introduction to IBM Data Processing Systems. 2nd Ed. International Business Machines Corp., New Jersey, 1968. Publication No. C20-1646.

This manual provides a very readable, well-illustrated, general introduction to computer hardware and software systems.

RICHARDS, R. K. *Electronic Digital Systems*. John Wiley and Sons, New York, 1966.

A very interesting book which deals with the history, theory, structure and operating principles of computer hardware and software systems.

Scientific American, September, 1966.

The special issue of *Scientific American* on computers and automatic information processing is recommended reading.

References

BACKUS, J. W. *et al*. "The Fortran Automatic Coding System" *Programming Systems and Languages*. S. Rosen (ed.) McGraw Hill, 1967.

NAUR, P. and ROSEN, S. (eds) "Revised Report on the Algorithmic Language Algol 60" *Programming Systems and Languages*. McGraw-Hill, 1967.

Part III

Actual Systems

Foreword

Part I of this book is concerned primarily with those concepts which are necessary for an understanding of the operation of systems. In Part II these concepts were dealt with in greater depth and various aspects of modern technology were examined, such as integrated circuits and computer organisation and programming, which exercise dominating influences on today's technology. In contrast, Part III is devoted to discussing a selection, of typical systems. This part of the book should not be regarded as a quantity of material which students must absorb in order to progress, or worst still to pass examinations—but rather as reading material in preparation for a discussion at a tutorial. Better still, students should be encouraged to visit the library and read a topic related to the subject, which is of interest to them. A tutorial can then be devoted to a discussion on topics introduced by the students.

In this way the particular biological system described in Chapter 21 is one example, chosen from a numerous group of biological systems. Instead of choosing to discuss power systems (Chapter 25) or communication systems (Chapter 23), the authors could have chosen a particular chemical plant, such as an oil refinery, or a traffic control system of a large city, etc. The particular choice is not so important, but a thorough discussion of the various interconnecting forces and their effects on the overall characteristics of the system is very important indeed. A period for tutorial work should be reserved, if practicable, for solving problems and discussing results. If students can pose their own questions so much the better.

For a reader who prefers to follow the material suggested in Part III some advice can be given: do not rush through the material, but rather take time to reason on the more obscure points with your colleagues or your tutor. More insight will be obtained by frequently referring back to the related introductory and basic technical work in Parts I and II. This action is suggested in a number of places in the text and, whereever possible, the reader should try to associate each statement in Part III with some appropriate section or statement in the earlier Parts of the book. Referring back in this way will help greatly when attempting the exercises at the end of each chapter and the separate groups of problems at the end of the book. Again, frequent cross-reference between related material in different sections of the book (i.e., jumping *forwards* as well as *backwards*) will be found to be of considerable help.

Some useful data are given in Appendixes 1 and 2 and the brief outline of matrix algebra in Appendix 3 is relevant to the work in Chapter 22. The index of key words will help in this process. Finally, and following the appendixes, is a group of problems on the material of the book as a whole. Again it will be found helpful to refer back to relevant sections when attempting these problems.

A.E.K. R.M.H.

CHAPTER 21

The Crab Eye

(An Ideal Sensor, freely moving, stabilised against tilt, stabilised against rotation, analyser for e-vector plane, wide-range motion-sensitive, for wavelengths λ = 300 to 600 nm.)

G. A. Horridge

"He had been Eight Years upon a Project for extracting Sun-Beams out of Cucumbers..."

J. Swift
A Voyage to Laputa

A. INTRODUCTION

Almost all of you will be familiar with the active, scavenging crabs of rocky seashores, and you may know that among their peculiar attributes they have remarkable eyes (Fig. 21.1). Some shore crabs can easily see a flying seagull at a distance of about 50 metres, but it is less well-known that crabs must see the direction of the movement of stars across the sky, as we see the moving lights of an aeroplane, and that they see the movements of the edges of shadows cast by the sun much as we see the approach of a black car. Moreover, for some reason unknown to us, their freely moving eyes are stabilised against tilt. Their eyes are also remarkable in that they are sensitive over a wide range of illumination; for example, some facets of the eye cannot help looking at the centre of the sun but remain undamaged, while the same facets, when adapted to the dark, are about as sensitive as our photomultiplier tubes, with a photon capture efficiency of several per cent.

The two eyes move separately by nine eye muscles (Fig. 21.2) in two dimensions through horizontal and vertical angles of about 15°. Each eye muscle has tonic or postural muscle fibres and also phasic or twitch fibres, which are separately supplied by motor nerve fibres. The eyes are able to compensate when the crab is tilted in pitch tilt (i.e., tilt in the fore and aft direction). Stabilisation against tilt in other plans also tends to keep the facet rows horizontal. These responses are defined in Fig. 21.3. When the crab turns itself voluntarily, the eyes move in such a way that the surroundings stay approximately stationary on the eye. This is true even in the dark. Normally the eyes are stabilised in the horizontal plane by any contrasting objects in the surroundings, and when these objects move all together (as when the crab is at the centre of a rotating striped drum) the eyes follow the movement. This is called the optomotor response. In this following motion they flick back at intervals and start a new traverse. This is called optokinetic nystagmus.

B. STABILISATION AGAINST TILT

In response to tilt of the crab, in any direction, the muscle tensions are changed so that the eyestalk remains close to its original position in space. Over a small range the compensation is 90 per cent of the imposed tilt. The sense organ consists of a pair of

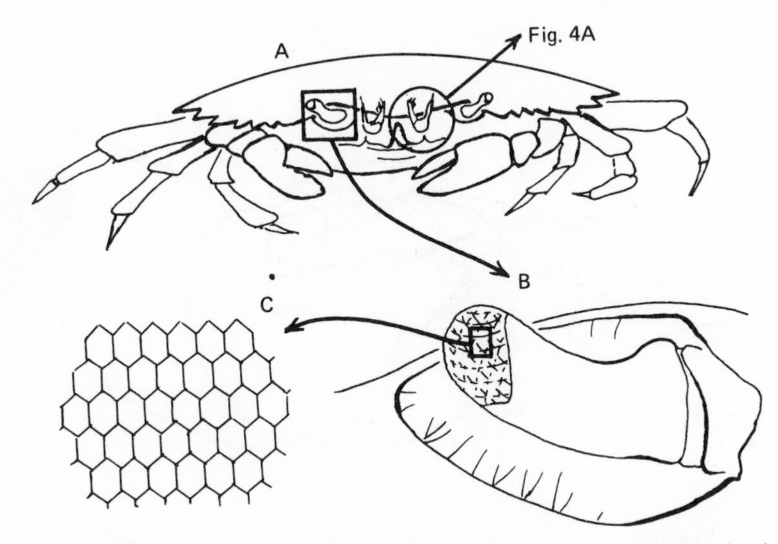

FIG. 21.1. A. Crab from the front. The statocyst at the base of the antennule, located in the circle, is illustrated in Fig. 21.4A. B. The right eye enlarged. The eye pivots in all directions about a universal joint which is a membrane distended by internal pressure. C. The lines of hexagonal facets are always held horizontal when the crab is near its normal posture.

statocysts which are rigidly mounted at the base of the crab's antennule, and vision is not necessary for the response. Each statocyst is a group of small heavy grains surrounded by sensory hairs so that the direction of tilt is coded according to which hair is touched by the tilted grains. The amount of tilt is also coded by the choice of hair because different hairs are excited over different parts of the range of tilt (Fig. 21.4). The interesting point is that the angle at which the eyestalk is held, is the result of a pattern of motor nerve impulses to the 9 muscles, that is determined in the brain of the crab. Part of this output is illustrated in Figs 21.5 and 21.6. There is an eye position for every body position but the pattern of impulses to these muscles is unchanged by clamping the eye in the wrong place or even by cutting it off, i.e., there is no feedback to the eye control system about the eye position.

C. EYE STRUCTURE

Behind each facet is a structure known as an *ommatidium* (Fig. 21.7), which is built in such a way that it is sensitive to a narrow pencil of light. The aperture of each ommatidium is about 30 μm. Each ommatidium is tilted about 2° to the axis of its neighbour, making about 6000 facets over a total field of view greater than a hemisphere. The rigid isotropic transparent *cone* acts as a spacer. The position of the principal focal plane within the cone is determined mainly by the curvature of the outer surface of the *cornea*. The amount of light reaching the receptor (which is called the *rhabdom*) is determined by its angle of incidence on the facet. As the angle of incidence changes, the intensity distribution moves across the receptor (Fig. 21.8). This effect, as seen by the receptor which simply catches all the light it can, is responsible for the angular sensitivity of the receptors (Fig. 21.8E). There is no question of a useful image being divided up between many different receptors as in our own eye, in which receptors lie in the plane of the real image.

The optimum dimensions of the ommatidium depend on the following conflicting interests.

(a) Increase of the aperture of the lens narrows the image of a point source and therefore narrows the pencil of light that is *possible* in the cone. But this is true only

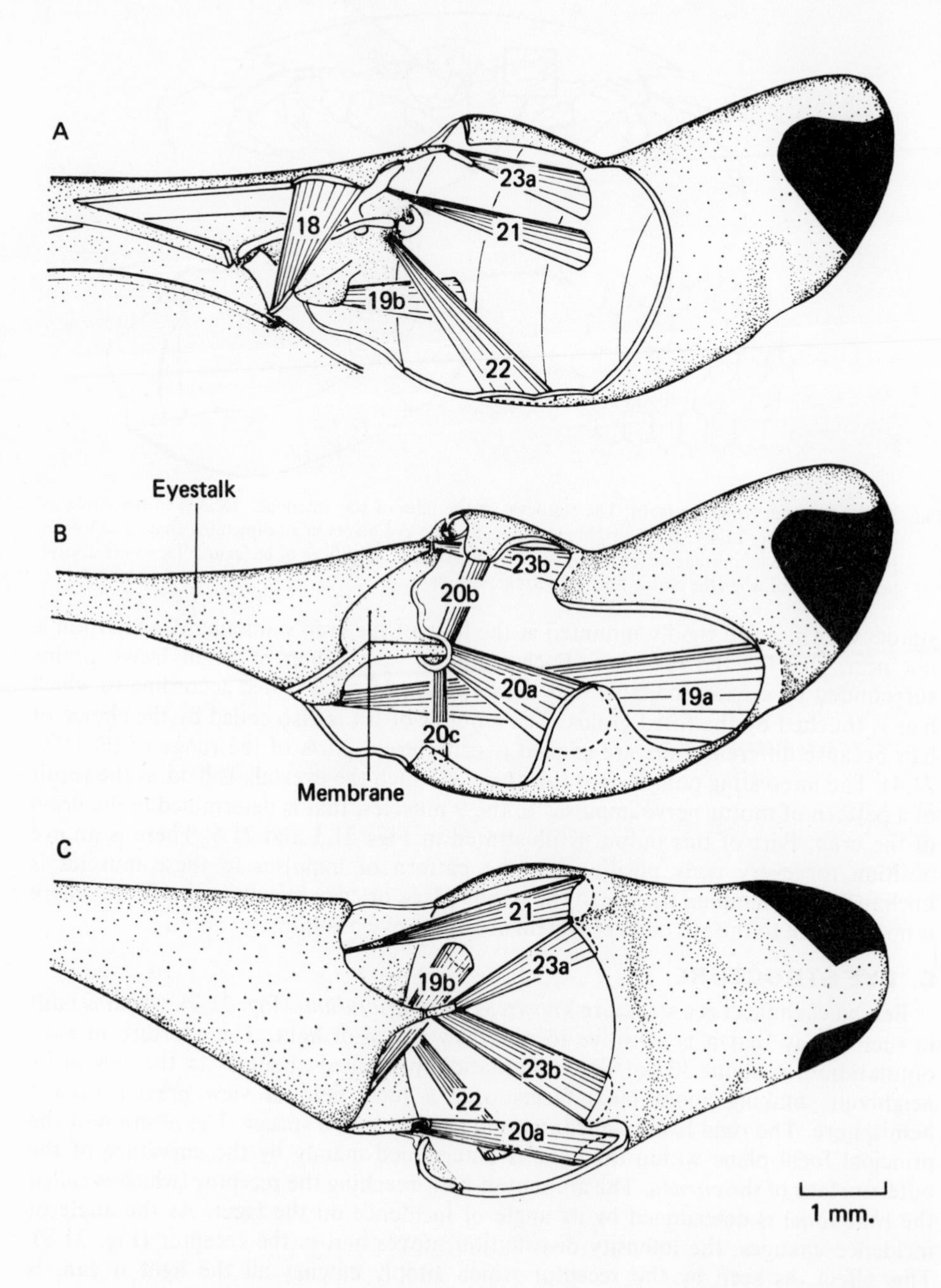

FIG. 21.2 Musculature of the eyecup and eyestalk in the crab *Carcinus*. The joint is surrounded by a continuation of the membrane which attaches the eyestalks to the main body skeleton. There is no hinge; the universal joint floats on a membrane and is kept rigid by internal blood pressure. All muscles exert a background pull all the time, and all participate to some extent in all movements (except 19a, which blinks the eye). A. Right eyecup dissected from the lateral side to show the medial muscles. B. Dissection from the same side to show the lateral muscles. C. Dorsal view with muscles 18, 19a, 20a and 20b omitted.

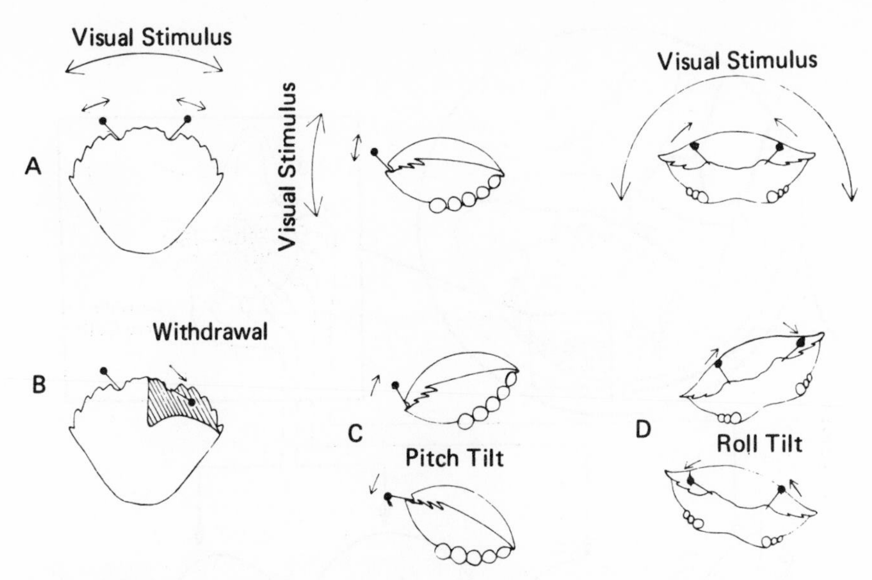

FIG. 21.3. Definitions of stimuli and responses. A. Movement of the surroundings causing optokinetic nystagmus in the horizontal, vertical pitching and vertical rolling planes. B. Protective withdrawal of the eye. when the shaded area is touched. C. Compensatory eye response to pitch tilt. D. Compensatory eye response for roll tilt.

for a perfect lens. Decrease of aperture means that more facets can be crowded onto the eye.

(b) Narrowing the pencil of effective light reduces the amount of light reaching the receptor, and so cuts down its sensitivity. Widening the pencil allows in more light but reduces the sharpness of angular sensitivity, i.e., makes motion perception more difficult.

(c) There is no point in having an admittance function that is narrower than the cross-section of the end of the rhabdom.

(d) The diameter of the end of the rhabdom is also a compromise between a narrow rhabdom which preserves acuity, and a wide one which allows in more light.

The rhabdom, or receptor structure, is a light guide which is composed of a long stack of plates (Fig. 21.7C). Each plate is composed of tubules, which are extensions of the membrane of the photoreceptor cells and contain the visual pigment. All of the many plates contributed by a single cell point in a constant direction, which is either horizontal or vertical with reference to the animal's normal posture. The visual pigment in the tubules absorbs light, and this process causes electrical depolarisation of the membrane of the receptor cell. The efficiency of capture of photons is about 10 per cent in some insects investigated. Absorbtion of light is about a half per cent per micron travel through the receptor, whereby most of the light is absorbed. This composite light guide has the interesting property of dividing the light between the two types of cells. If the orthogonal plates are equally balanced then the plane of polarisation wili not be rotated as it passes down the guide. Absorption of waves polarised in one plane will be as effective as in the plane at right angles. This makes possible the discrimination of the polarisation plane of the incident light by the differential response of the two types of cells, and *at the same time all the light can be absorbed.*

Light absorbed in the receptor, depolarises the standing potential of about 50 mV across the bounding membrane of the receptor cell. The potential change is approximately proportional to the log of the light intensity over a range of about 10^5

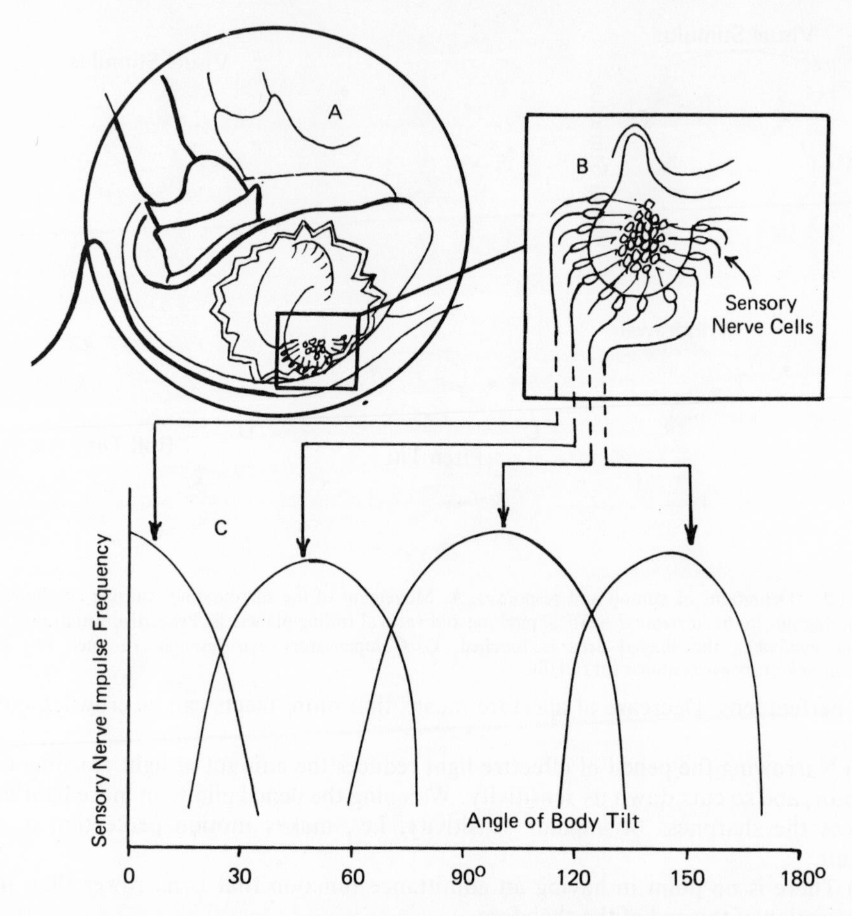

FIG. 21.4. The statocyst, organ of balance. A. The organ from the location shown by the circle in Fig. 21.1A. B. The organ consists of a group of sensory hairs encircling a few rock grains which move the hairs when the body is tilted. C. Impulse frequencies of the nerve cells of the sensory hairs showing their different ranges of action according to their position.

of intensity. The potential change is conducted down the axon processes of the receptor cells as a graded potential, not as nerve impulses. The equivalent circuit is that of a submarine cable and the law* governing transfer of potential from the photoreceptor to the next nerve cell, resembles that of the transfer of heat along a conducting rod. The receptor axons end in the optic ganglia beneath the eye, and almost nothing is known of the mechanisms beyond them.

D. ACUITY

The angular sensitivity-curve is the window function through which the sense organ sums all rays reaching it from the outside world, reducing their effectiveness at greater angles off the axis (Fig. 21.9). It is not quite like a photocell placed at the end of a straw, because the angular sensitivity curve is a smooth bell shape approximating $ae^{-b\theta^2}$ where θ is the horizontal axis in Fig. 21.8E, and a and b are constants.

Alternate equal black and white stripes moving across the field of view cause oscillation in the intensity of light entering the receptor. This is called the contrast transfer (Fig. 21.9). This light changes the potential across the membrane of the

* See Chapter 14 and references to diffusion equations.

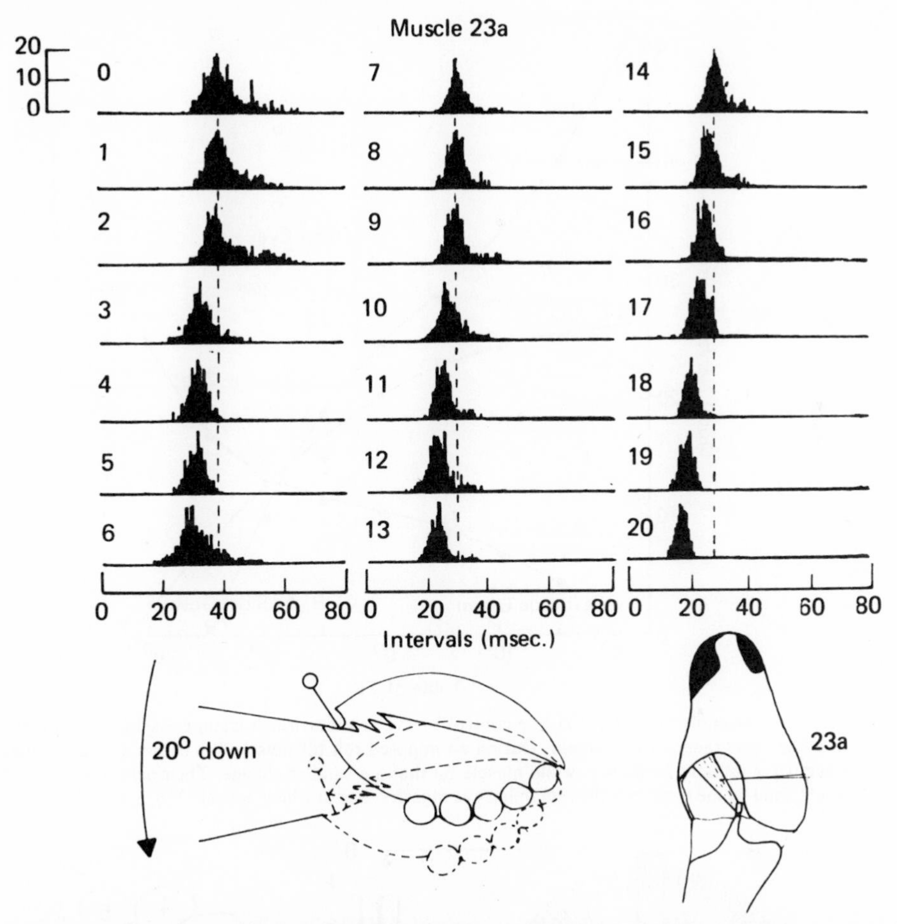

FIG. 21.5. The extreme sensitivity of the tonic motoneuron to muscle 23a is shown when histograms of intervals between successive nerve impulses are plotted for each 1° change in body position as the anterior of the body is pitched downwards. A dashed line is drawn through the centre of the first histogram in each column.

receptor. Therefore, wider stripes are more effective than narrower stripes in causing excitation, and the limit of vision of a striped pattern is found when the stripes are so narrow that moving them, in the field of view, causes an oscillation that is indistinguishable from the electrical noise level (Fig. 21.10). Clearly,

(1) there is no sharp threshold,

(2) this definition of acuity applies to the receptors but sets a limit upon which the animal as a whole cannot improve except by taking averages over many receptors and thereby reducing the noise level,

(3) movement of a regular striped pattern is an essential component of a test of acuity. From the above it appears to follow that vision is dependent on movement of the eye, and very extensive cross-correlations are necessary for any kind of form vision.

E. MOTION PERCEPTION

Because neighbouring facets on the eye have axes tilted at about 2° to each other, their angular sensitivity curves overlap. When a light moves across a pair of adjoining facets it will be illuminating one facet progressively more, just as it is passing away

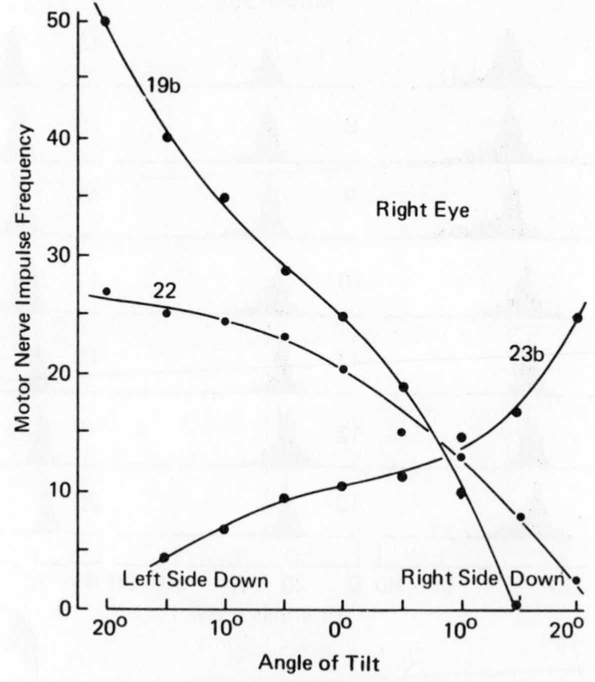

FIG. 21.6. An example of coarse and fine control by eye muscles which change tension simultaneously over the same range of movement. During an imposed roll tilt muscle 19b shows a large change of tonic nerve impulse frequency, while muscle 22 shows a smaller change. Their action is opposed by gravity and by the tension in the remaining eye muscles, one of which, muscle 23b, is shown here.

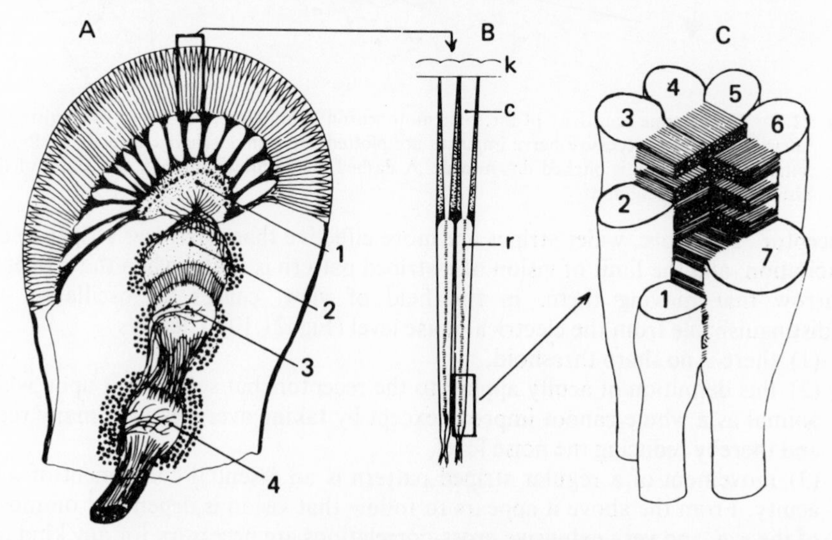

FIG. 21.7. Structure of crab's eye. A. Longitudinal section showing retina and four successive optic nervous centres 1-4 leading to the optic tract. B. Two ommatidia of the retina in longitudinal section, with cornea (k) crystalline cone (c) and retinula cells (r) drawn to scale. The light sensitive element is a solid dielectric cylindrical light guide, called the rhabdom, down the centre of the retinula cells, formed by a complicated system of tubules extending from the retinula cells. C. The organisation of the components of retinula cells 1 to 7 into the rhabdom. The layers alternate in tubule direction. Four retinula cells participate in the vertical direction, three in the horizontal, and the combined structure acts as a light guide. An eighth basal cell (not shown) is rudimentary.

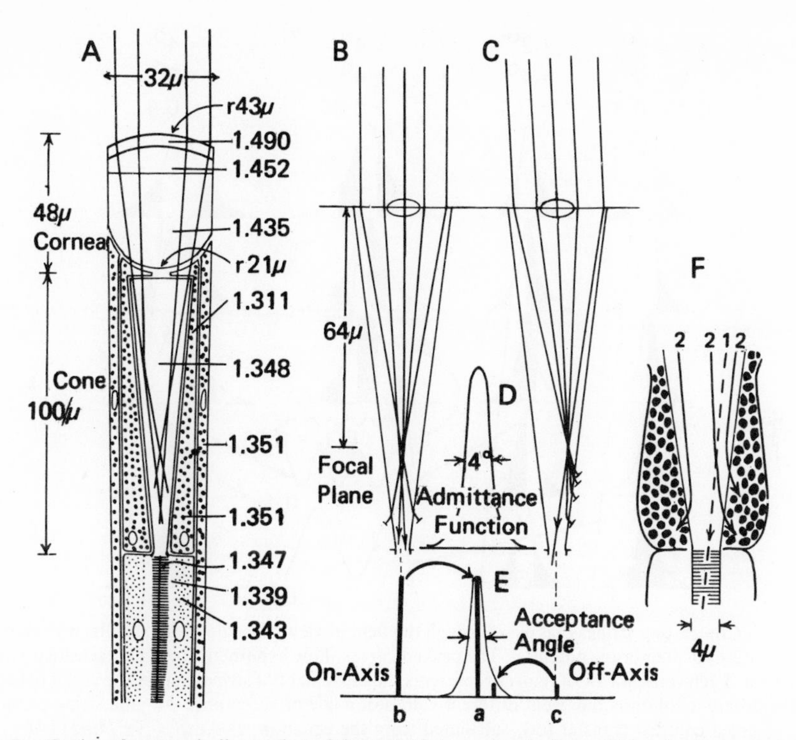

FIG. 21.8. Optics of an eye similar to that of the crab (data are for the bee, which is better known). A. The anatomical arrangement as in Fig. 21.7B, with refractive indices. B. Rays from a distant point source on axis produce an intensity pattern at the tip of the cone as shown in D. C. Movement of the source causes a sideways shift of the intensity distribution so that fewer rays pass the tip of the cone. D. The admittance function is the intensity distribution across the plane of the cone/rhabdom boundary. In D and E the horizontal axis θ is the angle of incidence of the light on the facet. E. The angular sensitivity of the receptor cells. This is the effective sensitivity to light from a distant point source that is moved at different angles to the axis. The curve in E is derived from the curve in D in the way shown by the thick curved arrows. The transfer function between cone tip (D) and rhabdom (E) is not yet known in any compound eye. F. A likely complication at the tip of the cone is the removal of rays (2) that approach the boundary between cone and pigment cell, by the higher refractive index of the latter. Data for the bee are from Varela and Wiitanen (1970).

from the other. A *spatial* pattern of illumination which moves in front of two facets causes two *temporal* patterns of excitation in the two sets of receptors. These two temporal patterns are out of phase because the axes of the two receptors are at an angle of 2° to each other, Fig. 21.11. *Movement perception is a correlation between slightly out-of-phase parallel channels.* It is not known how the optic ganglia abstract the direction and velocity of the seen motion from this data, although clearly an effective mechanism for doing so must exist.†

In the optic lobes one can record impulses from nerve fibres which are sensitive to movement in one direction or another. To stimulate each movement perception neuron there is an optimum speed, and a high and low cut-off speed. Different nerve cells are sensitive over different ranges of speed. This *range fractionation* is typical of many nervous mechanisms.

† Compare this statement with the idea expressed in Chapters 7, 11 and 23 that information, when received, still needs to be encoded in a form suitable for use.

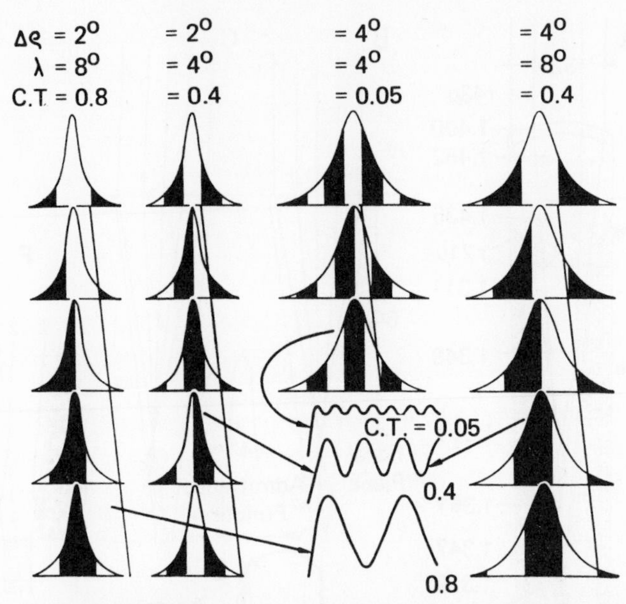

FIG. 21.9. Contrasting stripes that pass through the field of view of single retinula cells, with the resulting oscillations in membrane potential. The peaked curves show hypothetical angular sensitivity of the cell to light. Each vertical row of figures represents the movement of stripes across the visual field of a cell. The different columns represent different values of angle of acceptance $\Delta\zeta$ and stripe period λ. The theoretical contrast transfer (ct), calculated from the equation $ct = exp - (\pi^2/4ln2)\,(\Delta\zeta/\lambda)^2$, as in Fig. 21.10, is equal to the maximum contrast in illumination within the retinula cell. This can be found experimentally as $(I_{max} - I_{min})/(I_{max} + I_{min})$, where I_{max} and I_{min} are the illuminations that produce receptor potentials of P_{max} and P_{min}. The relation between P and I is found separately by varying the intensity of axial illumination. The correspondence between values calculated from the fields of view and those measured with electrodes in receptor cells seeing stripes is shown in Fig. 21.10.

F. EYE MOVEMENT

Impulses in the movement perception neurons arrive at the brain and cause a movement of the eyes. The eyes follow the direction of the movement of a rotating drum in the centre of which the crab is placed. When they have traversed in the slow forward phase about 15° across the eye socket, the eyes *swing quickly back in a fast phase and start the cycle over again.* Faster movement of the drum causes faster following by the eye, and more frequent flick backs over about the same angle. The movement of the eye on its stalk is subtracted from the drum movement. Fig. 21.12. The forward gain *G* is high and the velocity range of response is wide, Figs 21.13, 21.14.

The astonishing fact about this system is that the control system receives no indication of the actual movement of the eye, apart from the visual feedback. An eye that is held rigid (or even cut off) receives exactly the pattern of motor impulses to its muscles that would cause the appropriate eye movement for the stimulus to the crab. The horizontal movement superficially resembles the output of a relaxation oscillator which is powered by a variable potential. But the eye position is determined by nine muscles; its position is controlled in two dimensions and its position at any one time is the result of the complicated stream of nerve impulses in about 36 motor axons to these muscles. There is, therefore, a centrally determined pattern of impulses for every postural and visual stimulus situation: the eye movement is the result of an appropriate centrally controlled change in this pattern. If the eye is painted over and driven by the vision in the other eye, this pattern is unchanged by forcibly waggling the

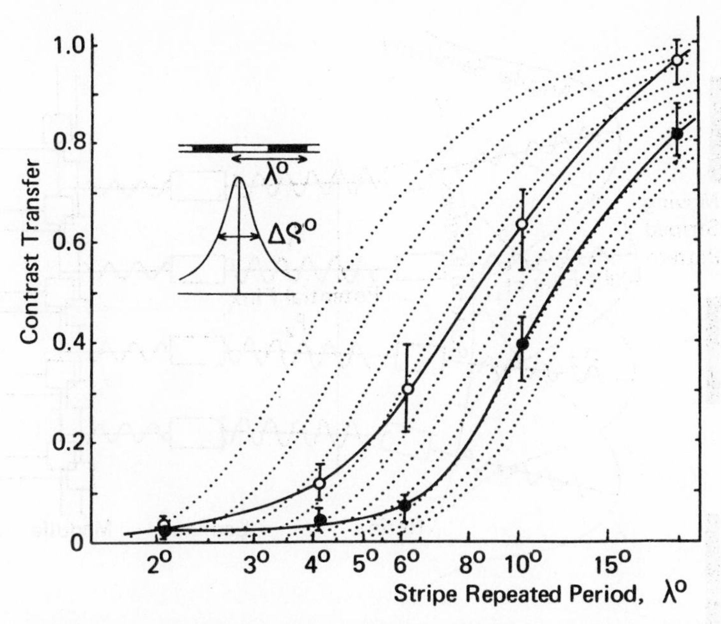

FIG. 21.10. Theoretical (dotted line) and experimental (solid line) contrasts between peaks and troughs when stripes of various repeat periods (λ°) are moved in front of a window with a Gaussian angular sensitivity curve, as in the inset at upper left. The contrast transfer is defined as $ct = (I_{max} - I_{min})/(I_{max} + I_{min})$, where I_{max} and I_{min} correspond to the light intensities that would produce the maximum and minimum retinula cell membrane depolarisations. These curves show how the contrast of illumination produced by the stripes increases with increasing λ°, and that retinula cells with narrower angular sensitivity give, with finer patterns, the same contrasts as cells with wider fields do with wider patterns. Solid circles show experimental points for dark-adapted eye; open circles, for the same eye when light-adapted. The position of the experimental curves agrees reasonably well with the angles of acceptance of the retinula cells, as measured directly with a movable point source. The small disagreement between theory and experiment (shown by crossing of dotted and solid curves) is presumed to originate in the non-Gaussian shape of the experimental angular sensitivity curve. This data is for the locust.

eyestalk at random, while the impulse pattern is recorded. The brain acts as if it generates a complex tape-recording for every situation without reference to the actual eye position. How the complex integration of excitation upon the motoneurons, illustrated in Fig. 21.15, produces an appropriate output, is quite unknown. The variety of optokinetic responses in the horizontal plane alone is shown in Fig. 21.16.

This system stabilises the eye visually against movement of the whole background. But a surprising feature is that when the crab turns itself voluntarily, it is not necessary for the eye to see for it to be stabilised relative to the surrounding world. A blind crab moves its eyestalks in exactly the appropriate way, during a voluntary turn, so that the eye would be stabilised upon the surrounding world. No reason is known why it should work like this unless visual stabilisation is important at night.

The low cut-off speed limit of movement perception is also astonishing. A crab placed in a rotating drum, which moves slower and slower, goes on responding down to an angular velocity of about 1 revolution of the drum in 3 days, with optimum gain in the system at about 1 revolution per day, i.e., earth speed. The eye of a crab in a dark room follows the motion of a single light, when that is all it has to stabilise on, Fig. 21.17. By reducing the brightness of the light it is easy to show a response to the motion of 1 candle at 100 metres. Therefore, crabs respond to motion best when it is at earth rotation speed, and they can see bright stars. Therefore, they can pick up the direction of the earth's axis from the night sky and can easily see the movement of

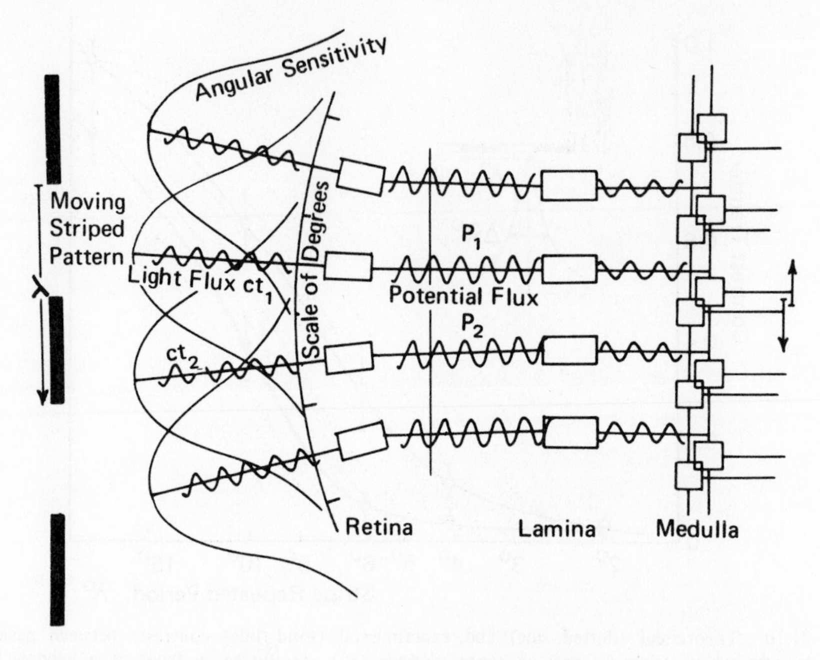

FIG. 21.11. The moving striped pattern on the left causes a contrast transfer of light flux *ct* in the two receptors A and B which have angular sensitivities as shown. The phase lag in the two stimulus patterns persists in the two response patterns. By some nervous correlation system as yet unidentified the two response patterns are converted into a nervous signal of the direction and velocity of movement.

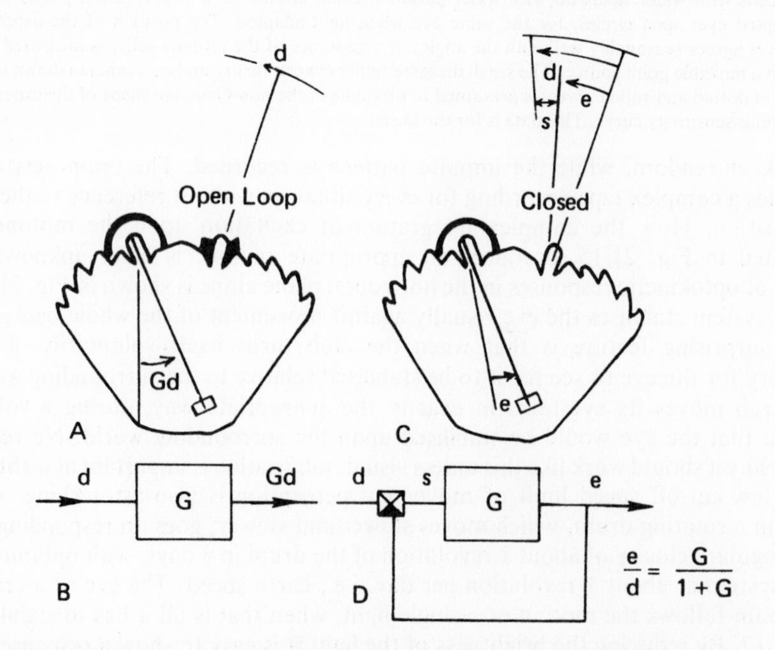

FIG. 21.12. A, B. Movement of the stimulus d and the response of the eye Gd under open loop conditions, i.e., with the seeing eye fixed. C. The same under closed loop conditions, with eye speed *e*. D. Explanation of symbols in C, and the formula relating the forward gain G to the ratio response/stimulus.

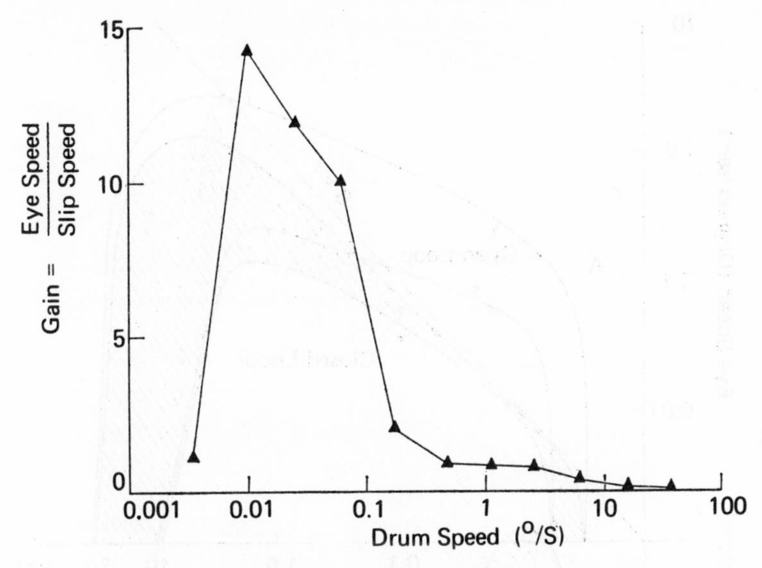

FIG. 21.13. The ratio of the eye speed to the slip speed, expressing the gain of the forward nervous control system, whether the visual feedback loop is open or closed. At slow drum speeds the gain increases but then decreases suddenly at the lowest drum speed. Fixation of the eye on the moving pattern would occur only at infinite gain, which is never reached.

shadows caused by the sun or moon. Again, why they should be built in this way is a mystery.

As for the human eye, the crab's eye is continually jittering with a frequency of about 3 Hz over an angle of about 0.02-0.05° (Figs 21.18, 21.19). This has the effect of moving contrasting edges in the visual field on and off the receptors. As in ourselves the jitter has the effect of sharpening up edges but, unlike our eyes, the jitter is only in the horizontal plane (Fig. 21.19). A special muscle with quick response to an irregular impulse pattern causes the jitter, therefore, crabs are for some reason more interested in vertical than in horizontal edges.

G. CONCLUSION

This account provides a definite example of an animal control system, with two major types of input from statocyst and eyes, a complex central processing of nervous excitation and an easily recorded output in the form of eye movement. As stressed in Chapter 8, the main lessons to be learned are:

(a) the system is of unknown complexity,

(b) the discovery of the components and their interactions will for a long time be the main goal of analysis,

(c) as stressed in Fig. 21.21, a working model or exact mathematical description of the mechanisms is not possible because it is not known how the information flow at each stage is divided out among many pathways in parallel, and

(d) an equation which fits the input/output relation is of no assistance in analysis of the actual components.

H. EXERCISES

1. The crab eye (and human too) jitters continually in order to improve vision, and gives the brain no information about which direction it is moving at any instant. Design a mechanism without proprioceptive feedback from the eye or eye muscles, so that it can tell which side of an edge is black and which side is white. Clue: the

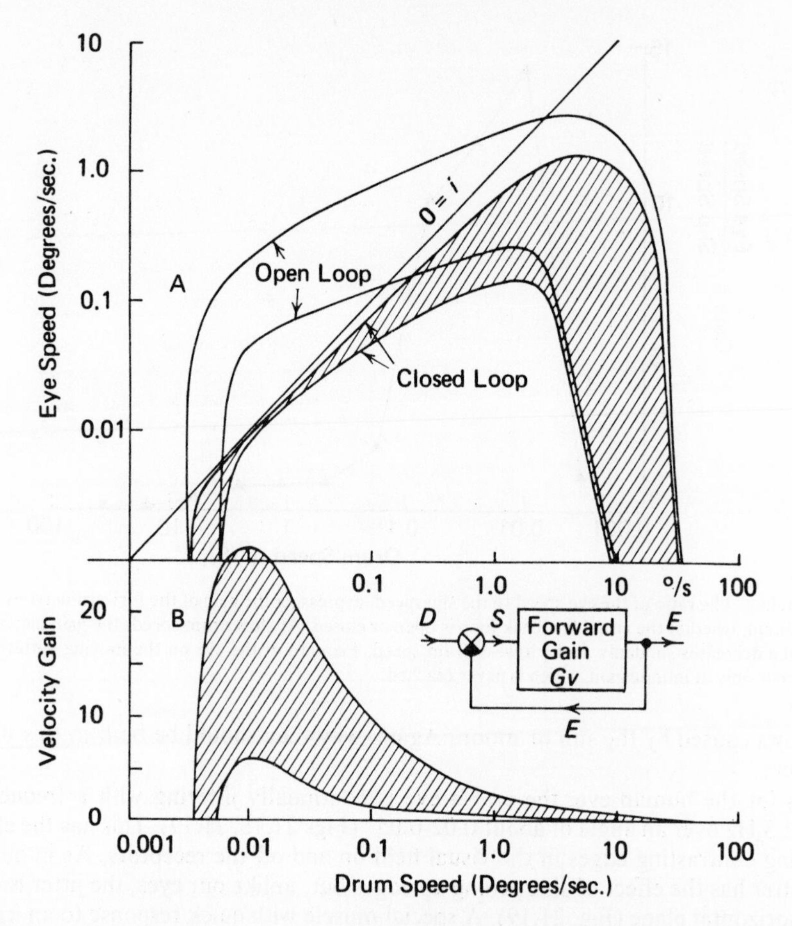

FIG. 21.14. A. Eye speeds over a range of five orders of magnitude of drum speeds. The straight line (o = i) shows when the eye speed equals drum speed. The upper and lower curves show the approximate range of variation. The open and closed loop curves show typical responses for the arrangements of Figs 21.12A and 21.12C. B. The spread of values of velocity gain over the same range of drum speeds, as in Fig. 21.13. The shaded area shows the limits within which the responses fall; note how the range of variation shown by the response is small under closed loop conditions at low velocities when the velocity gain is high.

optic pathways to the brain can be coded in the time dimension and of several kinds.

2. Design an ideal interaction between the movements of two eyes; remember that they move in opposite directions relative to the mid-line and consider the muscle locations.
3. When you voluntarily move your eyes you act as if the world stays stationary. When you forcibly push your eye your nervous system reports movement of what the eye sees. It is the same for the crab. Design a system which does this, and which also allows separate voluntary movement of each eye although the eyes are stabilised on the surroundings, see Figs 21.5 and 21.20, which are both inadequate as they stand. Design for (a) 2 eyes fixed on a movable head, (b) 2 independently movable eyes. (c) Suggest why no animal has more than 2 independent eyes.
4. Predict the movement of the crab eye if it is in a drum that is moved to a new position, during a brief dark period but which is otherwise illuminated and left stationary. What about visual feedback while the eye moves?

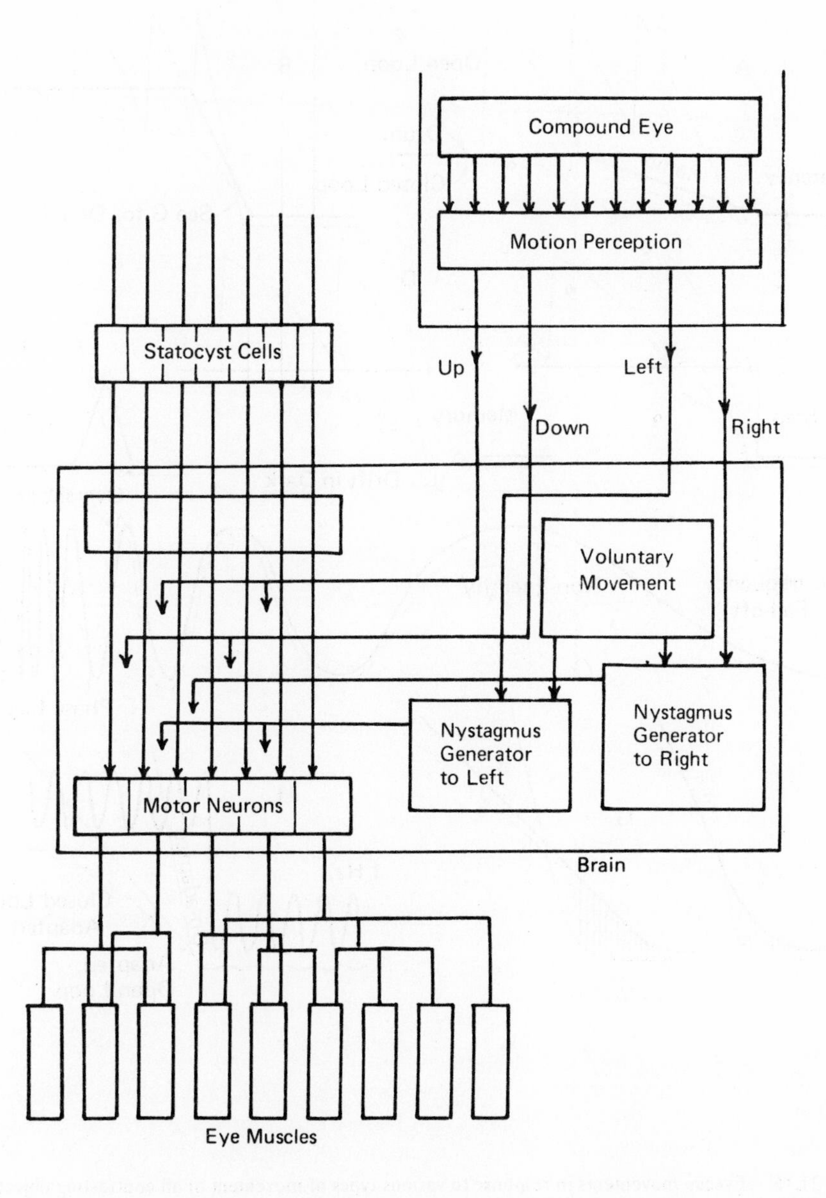

FIG. 21.15. Simplified flow diagram of excitation in the control of crab eye movement by statocyst and visual inputs. There are numerous simultaneously active parallel lines at every stage, and it is not known how many stages of interaction lie in the brain. The box labelled "voluntary movement" is added to represent two interactions (a) voluntary turns while the eyes see, (b) the movement of the eyes which keeps them pointing at a fixed target while the crab turns voluntarily *even in the dark*.

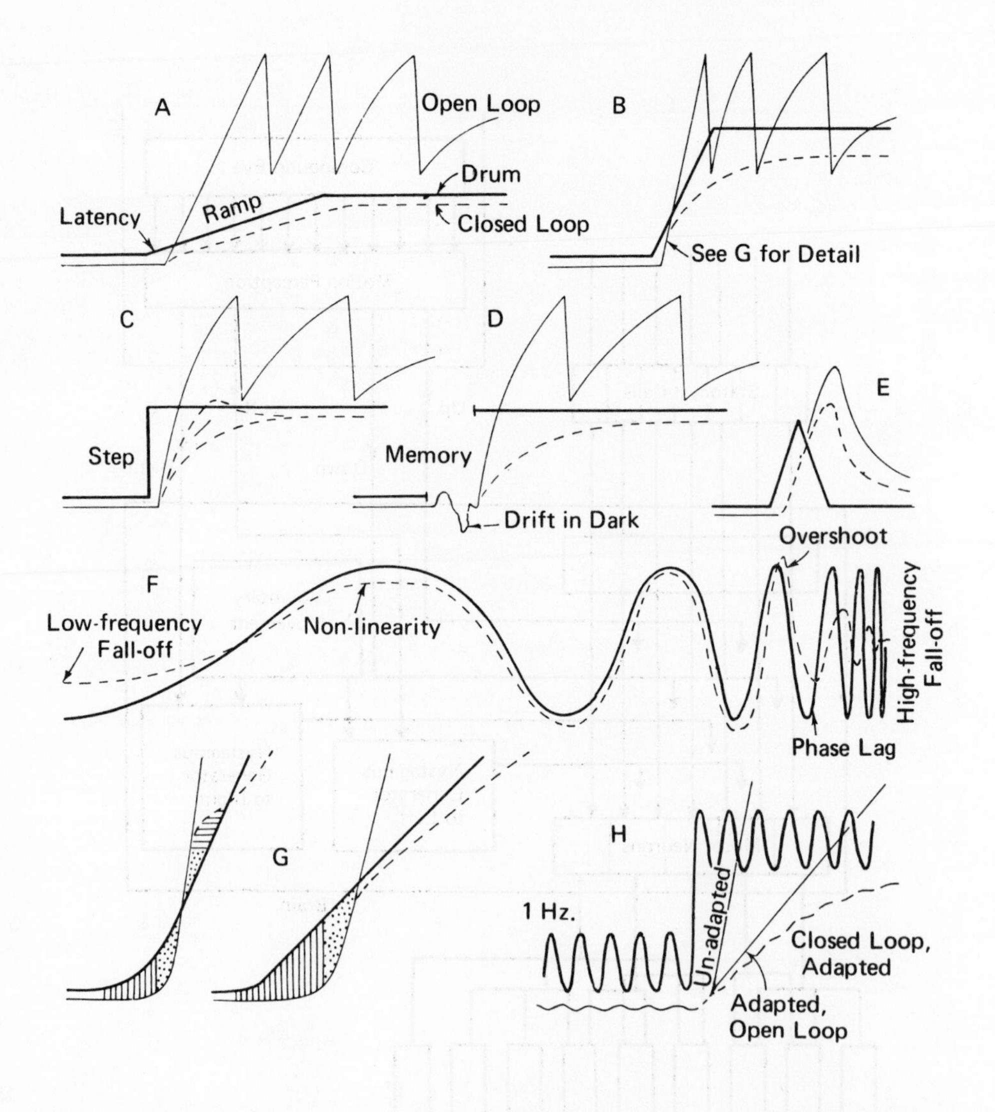

FIG. 21.16. Eyecup movements in response to various types of movement of all contrasting objects upon which the eye is stabilised. Closed loop responses, dashed line; open loop, thin line; drum movement, thick line. Note in A-D the optokinetic nystagmus (oscillatory movement of the eye) caused under open loop conditions by the high gain and small lag in the motion perception system. In D. there is a period of up to 10 minutes when the light is off. F. illustrates results that would yield a Bode plot. G. (detail of B.) illustrates latency (vertical hatch), high velocity gain (dotted area) and velocity overshoot (horizontal hatch). H. shows that the fast component of the response is adapted out by previous oscillations of the stimulus.

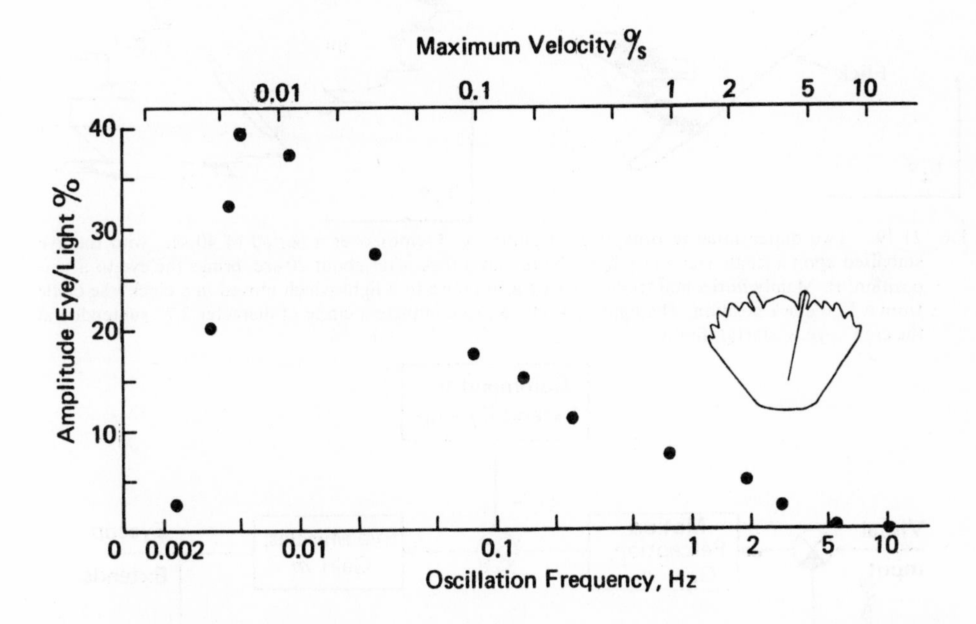

FIG. 21.17. The percentage amplitude of the steady-state response of the eye to a horizontal sine-wave oscillation of a single small light of 0.05 lux moved through a constant amplitude of 0.4°. The corresponding maximum velocities ($\pi \times$ angular amplitude $\times$ frequency) of the lamp as it passes through the centre of its path are plotted along the top.

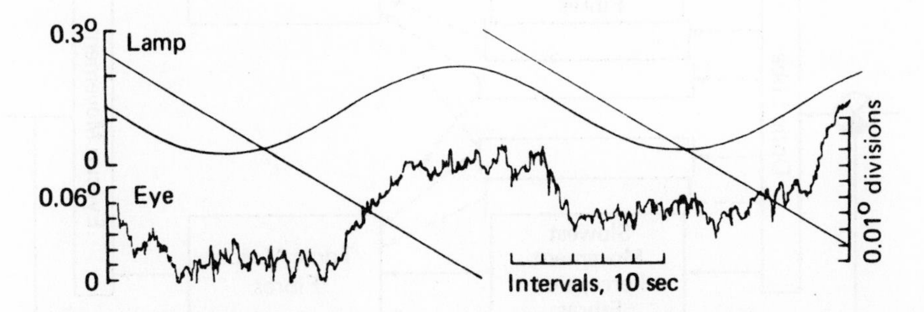

FIG. 21.18. The lower limits in frequency and amplitude in the response of the eye to a lamp moving sinusoidally back and forth along a straight horizontal line. The smooth sinusoidal curve shows the movement of a sub-miniature pin-light placed 57 cm from the eye. The noisy line shows the response of the eye (closed loop). Note how the response flattens out during the periods of low velocity. The low velocity limit is much higher than for larger or brighter lights or for striped drums. The typical movement of the sun (15°/hr) plotted on the same scale is shown by the straight sloping lines.

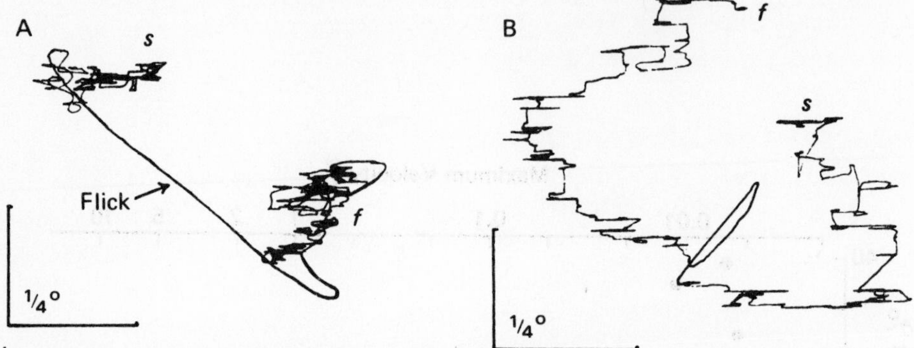

FIG. 21.19. Two dimensional records of eye tremor. A. Tremor over a period of 40 sec. with the eye stabilised upon a single stationary light. Note that a flick after about 20 sec. brings the eye to a new position. B. Mainly horizontal tremor during a response to a light which moved in a clockwise circle from a 12 o'clock position. The light took 48 sec. to complete a circle of diameter 2.7° subtended at the crab's eye. *s*, start; *f*, finish.

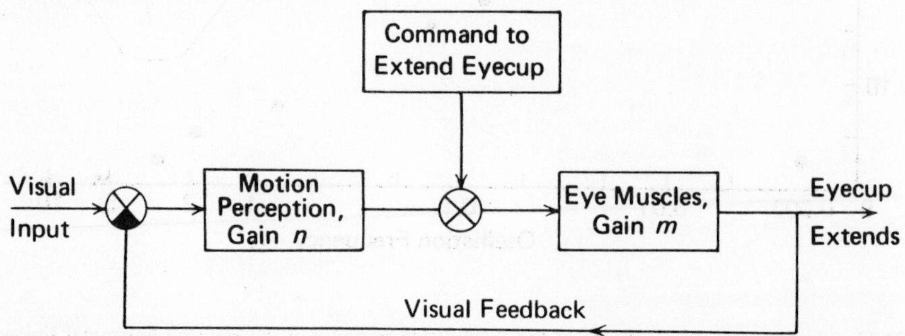

FIG. 21.20. The relation between the optomotor mechanism and the central command to extend the eye so that a single movable eye (or head with fixed eyes) can move voluntarily while vision is in no way hindered. The visual feedback loop reduces the effect of the command but cannot prevent it. This system would respond to an oscillation during the course of a voluntary eyecup movement, but is not adequate for separate voluntary control of two eyecups while the optomotor mechanism is linked between them.

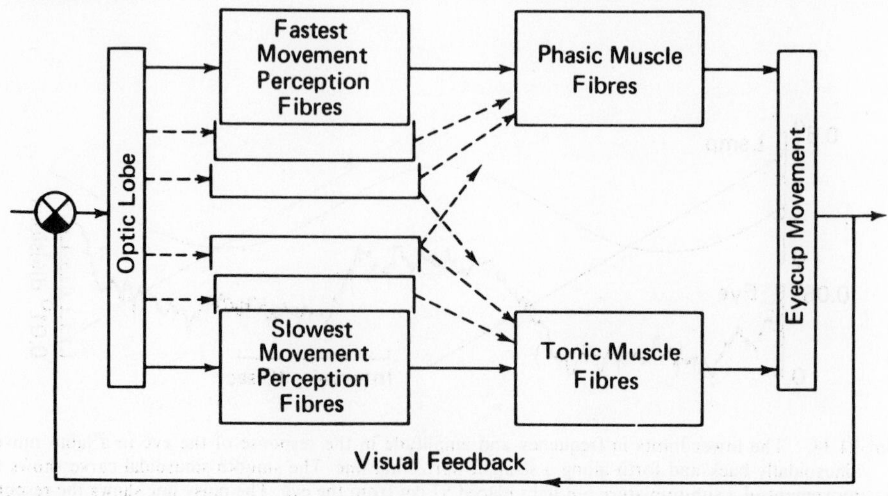

FIG. 21.21. Parallel pathways in the optokinetic system. The fast movement-perception system influences mainly the phasic (twitch) muscle fibres while the slowest movement-perception system controls the tonic (postural) muscle fibre tension. As indicated, there is a spectrum of time-constants on the movement-perception side and also interaction between fast and slow systems.

5. Design an experiment which observes only the external behaviour of the crab, to show that in all parts of the eye control system there are at least two communication lines in parallel with differing time constants, see Figs 21.16H and 21.21.

SUGGESTED READING MATERIAL

HORRIDGE, G. A. *et al. Journal of Experimental Biology.* **44,** pp. 263-295 (1966); **49,** pp. 223-324 (1968); **50,** pp. 651-682 (1969).

VARELA, F. G. and WIITANEN, W. *Journal General Physiology.* **55,** 336-358, (1970); on the optics of the bee's eye.

Glossary

Bode plot: a graph showing log (gain) against log (frequency) for sinusoidal signals.

Cone: a clear glassy conical structure behind the facet of an arthropod eye. It acts as a spacer so that the corneal lens can produce a first image in front of the receptor and so is analogous to the fluid filling the human eye.

Correlation, cross-correlation: a mathematical process in statistics used to discover a regular relation within one or between two, time series or functions of the same variable.

Electrical depolarisation: the electrical sign of the response of a nerve cell or receptor cell. The inside of the cell briefly becomes less negative than its normal resting potential.

Nystagmus: the movement by which a freely moving eye follows a movement of the whole visual field, with an intermittent fast recovery stroke whenever the limit of eye movement is reached (as with ourselves in a train).

Ommatidium: the group of light receptor cells and screening pigment cells that lie behind a single facet.of the compound eye.

Optic ganglion (pl. ganglia): the group of visual processing neurons behind the eye.

Orthogonal: in geometry at right angles, referring here to patterns of nerve fibres.

Proprioceptive: related to sensing the animal's own position or movement.

Rhabdom: the specialized portion of the visual receptor cell which contains visual pigment and absorbs light to cause the response.

Statocysts: herein, sensors for orientation relative to the direction of gravity forces.

CHAPTER 22

Control of Large Systems

C. B. Speedy

"The stage is now set for the creation of automata and automated systems with their brains (computers), arms (controllers), connections (feedback loops), sensors and actuators, and working modes. The master sorcerer, having accomplished the feat of creating Wiener's Golem (Wiener, 1964), is about to depart, leaving the stage of history to his apprentice, who will press the switch to activate the cybernated systems and extend man's dominion over nature."

John Rose, in the epilogue to Automation: its anatomy and physiology, *1967.*

A. INTRODUCTION

For present purposes a *system* is regarded as being a particular portion of our total environment. Thus, as in Fig. 22.1, we may think of our environment as having two parts: that which lies within the system and that which lies outside of it. The total environment is considered to comprise both *material* and *energy*, some of which is able to move across the boundary between *the system* and *its environment*. For instance, in an electrical power generating station considered as a system, we observe that fuel possessing both material and energy is transferred to the system from its environment, and electrical energy is transferred from the system to its environment. In the study of systems we are generally concerned with understanding the interaction across the boundary between the system and its environment.

For convenience, the flow of material and/or energy into the system is referred to as the *input* to the system and the reverse flow as the *output* of the system. The input may have several components, some of which are amenable to human control (the *control inputs*) and others which are not (the *disturbances*). In a similar way some of the outputs are able to be measured (the *measured outputs*) whereas others are not (*non-measured outputs*). For instance, if we consider coal to possess both mass and energy, then in an electric power station we may regard the mass flow rate of the coal as being a *controlled input* whereas, because of the variability of its calorific value and the practical difficulty of measuring this quantity continuously, as the coal flows to the furnace, the energy flow into the boiler is a *non-controlled input*. Likewise, the electrical energy which leaves the power station is an accurately *measured output* whereas the waste heat energy dissipated from the power station into its environment is inconvenient to measure and often, because of practical difficulties, is a *non-measured output*.

Systems generally possess significant inertia, characterised by energy storage and time-lags, so that the effect upon the output from the system of a sudden change in input takes place over a significant interval of time. Such systems are termed *dynamical systems*. The dynamic output/input relationship is called the system response, and is defined with respect to a specified input; many of the major system

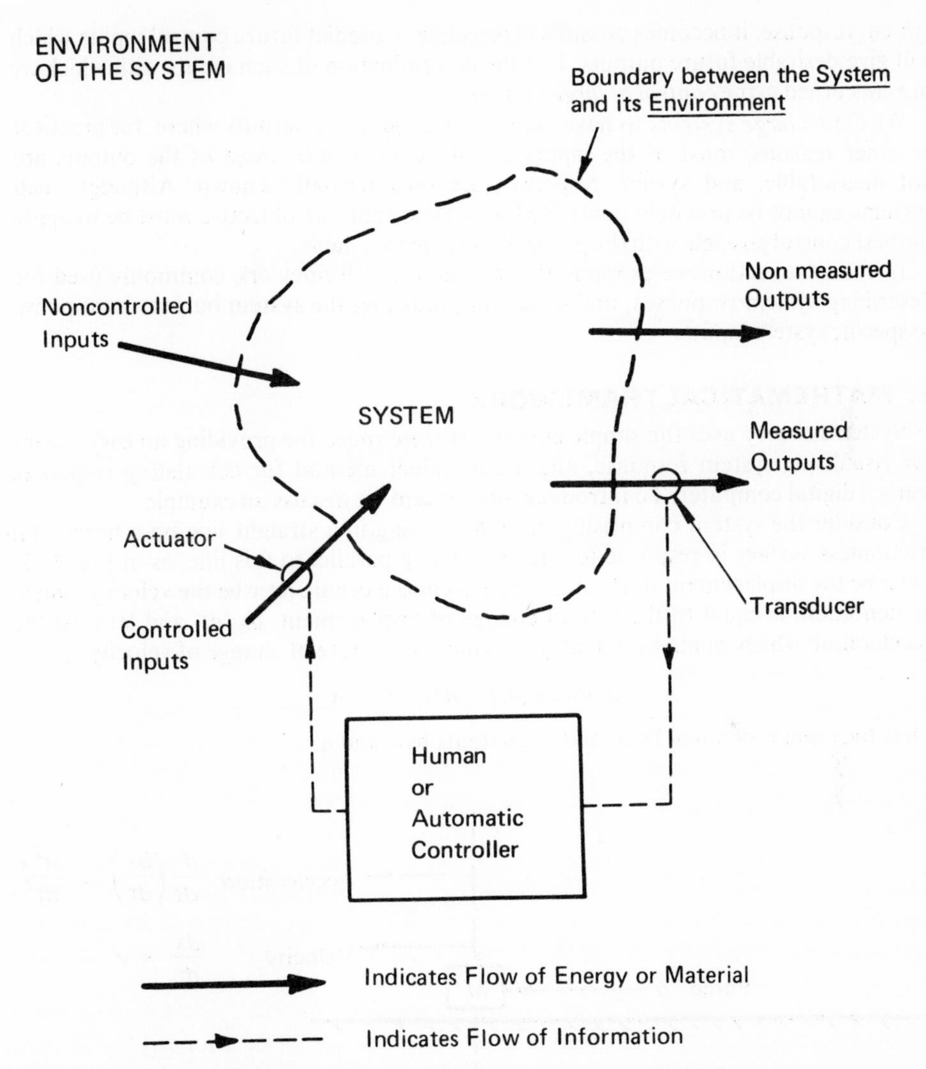

FIG. 22.1 The Flow of Energy and Material across the Boundary between a System and its Environment.

characteristics may be described in terms of responses to well-defined test inputs such as step functions and sinewaves.

So far we have described systems qualitatively in terms of inputs, outputs and responses. For engineering design purposes we need a quantitative description in which numerical values are assigned to the variables concerned, and where the dynamical relationships between the variables are described in mathematical terms. To achieve this we require *transducers* (measuring instruments) which give electrical signals proportional in strength to the amplitude of the physical quantity being measured. A typical transducer may give an electrical signal proportional to steam pressure in a boiler. For controlling inputs to a system, *actuators* (*motors, hydraulics,* etc.) are needed to supply control forces proportional to input electrical signals. It is usually assumed that the *electrical signals* possess negligible energy, but perform the vital task of conveying *information* from the transducers at the output of the system, and to the actuators at the input to the system. With quantitative information continuously available from the output of the system, and with a knowledge of the system response, it becomes possible in principle to predict future control inputs which will give desirable future outputs. It is the determination of such inputs with which we are concerned in the *control of large systems*.

We define *large systems* to have many inputs and many outputs where, for practical or other reasons, most of the inputs are not controllable, most of the outputs are not measurable, and system responses are only partially known. Although such systems cannot be precisely controlled, it is clear that our objective must be to apply the best control possible with the partial knowledge available.

In the next section we examine the mathematical framework commonly used for describing system responses, and hence for predicting the system outputs in response to specific system inputs.

B. MATHEMATICAL FRAMEWORK

Systems theory uses the simple concept of *state space* for providing an easy means for visualising system response, and a convenient method for calculating responses using a digital computer. To introduce this concept we discuss an example.

Consider the system comprising mass M moving in a straight line on a horizontal frictionless surface in response to a force u acting parallel to this line, as in Fig. 22.2. Let x be the displacement of the mass from a chosen origin, let v be the velocity which, by definition, is equal to the rate of change of displacement, dx/dt, and let α be the acceleration which, again by definition, is equal to the rate of change of velocity.

$$d/dt(dx/dt) = d^2x/dt^2 = \alpha$$

Thus the input is denoted by u, and the outputs by v and α.

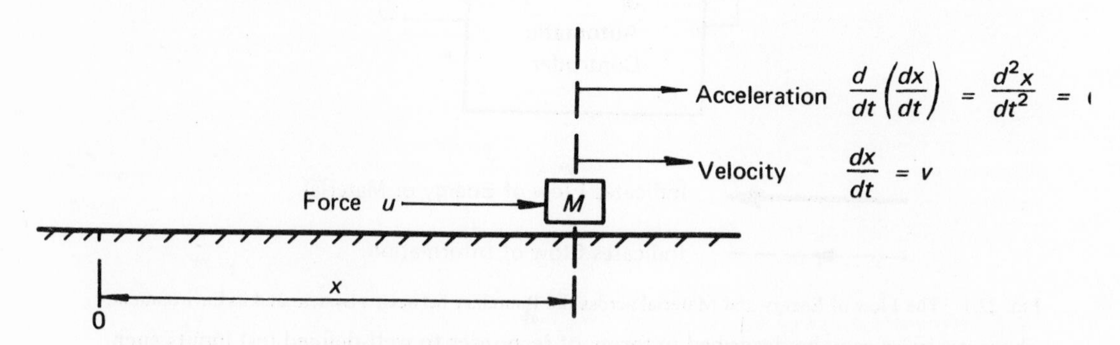

FIG. 22.2 A Mass moving on a Horizontal Smooth Surface.

To carry out an experiment to determine system response, it would be quite practical to apply a control force of varying magnitude over a period of interest, and to observe with the aid of transducers both the displacement and velocity. The techniques for doing this are fully established. However, the dynamical properties of this simple system are completely understood so that we may determine the response by calculation. Therefore, to obtain the responses necessary to describe the concept of state space in this example, we use calculation rather than experimentation.

Let us choose as the system input $u(t)$ over a time interval $t = 0$ to $t = 3$ seconds, the waveforms sketched in Fig. 22.3(a), namely the input

$$\left.\begin{aligned} u(t) &= 1\ ,\ 0 \le t < 1 \\ u(t) &= 0\ ,\ 1 \le t < 2 \\ u(t) &= -1\ ,\ 2 \le t < 3 \end{aligned}\right\}. \tag{22.1}$$

Let us assume also that at time $t = 0$ the displacement and velocity of the mass are zero. That is

$$\begin{aligned} x(0) &= 0 \\ v(0) &= 0 \end{aligned} \tag{22.2}$$

From Newton's Law of Motion, we know that

$$\text{force} = \text{mass} \times \text{acceleration}$$

If we assume for simplicity in this example that the mass is unity ($M = 1$), it follows that

$$\alpha(t) = u(t) \tag{22.3}$$

Since $u(t)$ is defined in this example by (22.1), and since also by definition $\alpha(t) = d/dt\ v(t)$, we may calculate the velocity by integration. Hence,

$$v(t) = v(0) + \int_0^t \alpha(\tau)\,d\tau \tag{22.4}$$

from which

$$\left.\begin{aligned} v(t) &= t &, \quad 0 \le t < 1 \\ v(t) &= 1 &, \quad 1 \le t < 2 \\ v(t) &= 3 - t &, \quad 2 \le t < 3 \end{aligned}\right\} \tag{22.5}$$

By similar reasoning the displacement may be obtained from

$$x(t) = x(0) + \int_0^t v(\tau)d\tau \tag{22.6}$$

from which

$$\left.\begin{aligned} x(t) &= \tfrac{1}{2}t^2 &, \quad 0 \le t < 1 \\ x(t) &= -\tfrac{1}{2} + t &, \quad 1 \le t < 2 \\ x(t) &= -2\tfrac{1}{2} + 3t - \tfrac{1}{2}t^2 &, \quad 2 \le t < 3 \end{aligned}\right\} \tag{22.7}$$

Figure 22.3(a) shows the control input $u(t)$ given by Equation (22.1). Figure 22.3(b) shows the calculated velocity response given by Equation (22.5) and the displacement response given by Equation (22.7). In each of these diagrams, we note that time t is the independent variable.

Figure 22.3(c), the *state space diagram*, is quite different in that the locus of points is a trajectory generated by the variation of the displacement and velocity of the mass. For instance when $t = 1\frac{1}{2}$ seconds, we observe from Fig. 22.3(b) that $x = 1$, $v = 1$, which are coordinates of a point on the trajectory.

The coordinates of the state space diagram are called the state variables, in this case, the displacement x and the velocity v. In state space it is convenient to refer to the *motion of the system* as corresponding to the movement of a point along the trajectory. Thus, in Figure 22.3(c), we observe the motion of the system from time $t = 0$, when displacement $x = 0$ and velocity $v = 0$, to time $t = 3$, when displacement $x = 2$ and velocity $v = 0$. Note that if the control were to be changed at any point in time along the trajectory specified in Equation (22.1), the trajectory would change accordingly from that time onwards. Thus we may think of the control input as providing a means for "steering" a *representative point* through the state space. At each point in time, the direction of motion of the representative point is determined within limits governed by the momentum of the system and by the allowable sudden change in control input which may be applied.

So far we have considered a system with two variables, referred to as a second-order system. With higher-order systems the same concepts of state space, state variables, trajectories, representative points, and motions of a system may be applied.

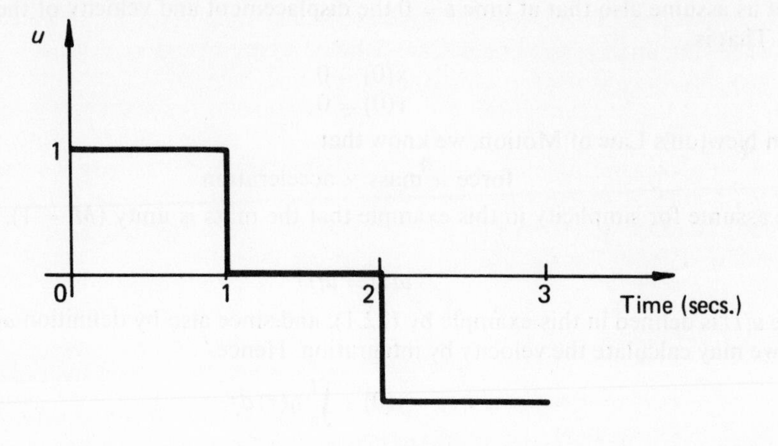

(a) Control Input.

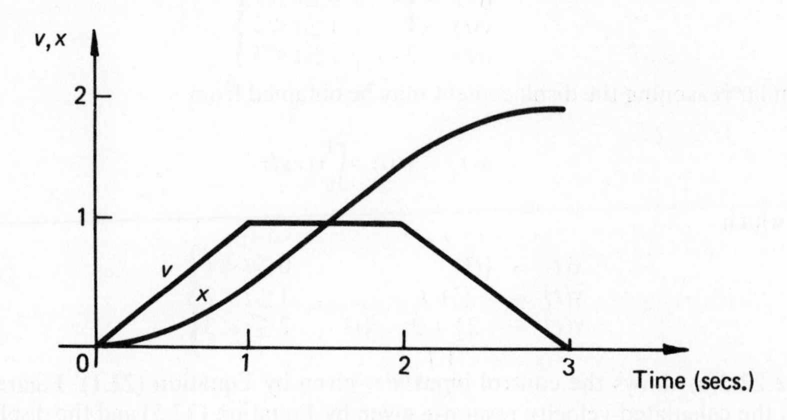

(b) System Output.

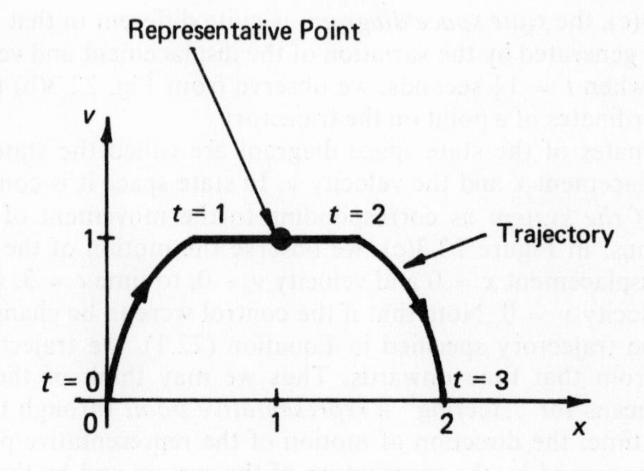

(c) State Space Diagram.

Fig. 22.3

A third-order system requires a three-dimensional state-space as indicated in Fig. 22.4, where the three state variables are denoted by x_1, x_2 and x_3. ABCD is a typical trajectory resulting from a particular control input applied over the interval from $t = t_1$ to $t = t_4$. At point B the *direction of motion* of the system is given by the tangent BE. The "steering" effect of the control may be visualised by supposing that at time $t = t_2$, the control input was changed slightly from that causing the direction of motion BE to that of BF. Because of the change the point is steered away in the direction of the new trajectory BG.

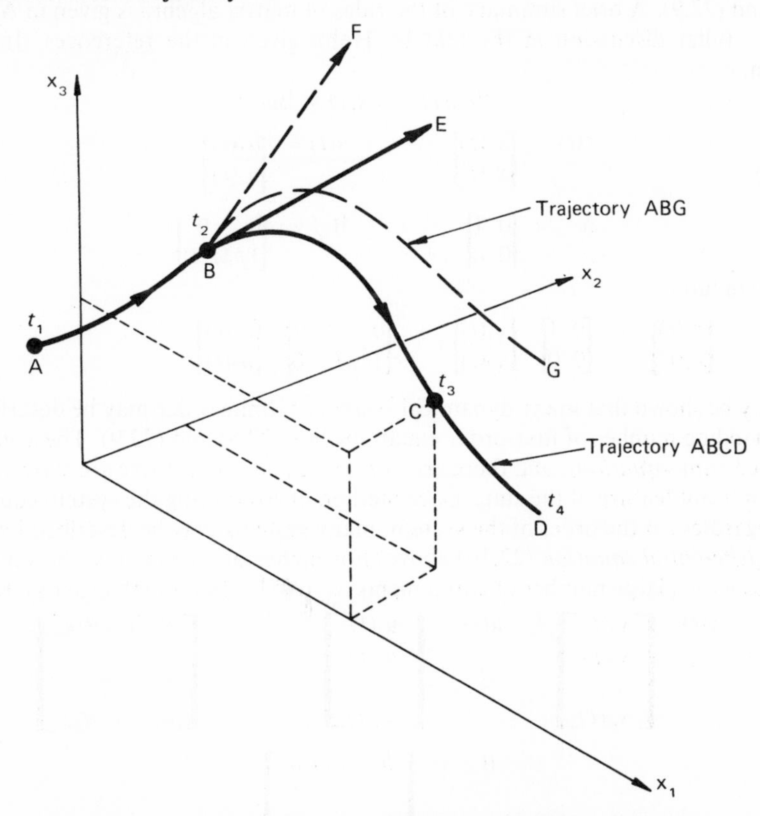

FIG. 22.4 A 3-dimensional State Space Diagram showing Two Trajectories.

Since our physical world has three spatial dimensions, we are unable to visualise fourth and higher-order spaces. Despite this difficulty we may continue with the concept of state space and trajectories within it. In fact, the mathematical framework employed is the same for all systems of whatever dimension, and this is the great virtue of the state space method. We will now take the second-order example and show how it may be formulated in a way which will enable an easy extension to systems of any finite order. Since it is our intention now to generalise, we will introduce the two state variables x_1 and x_2 defined such that

$$x_1 = x = \text{displacement (output)}$$
$$x_2 = v = \text{velocity (output)}.$$

Further, we allow the possibility of up to two controls u_1 and u_2 defined as

$$u_1 = u = \text{force (input)}$$
$$u_2 = 0$$

since in this particular example there is no second input.

For reasons which will become apparent we may write the equation governing the motion of the system as follows. From definition of velocity,

$$d/dt\, x_1(t) = x_2(t), \tag{22.8}$$

and from Newton's Law of Motion (acceleration = force/mass),

$$d/dt\, x_2(t) = 1/M u_1(t) \tag{22.9}$$

At this point we invoke matrix algebra, which simplifies the writing of Equations (22.8) and (22.9). A brief summary of the rules of matrix algebra is given in Appendix 3, and a fuller discussion in the text by Hohn given in the references. In matrix notation,

$$d/dt\, x(t) = \mathrm{A}x(t) + \mathrm{B}u(t) \tag{22.10}$$

where

$$x(t) = \begin{bmatrix} x_1(t) \\ x_2(t) \end{bmatrix}, \quad u(t) = \begin{bmatrix} u_1(t) \\ u_2(t) \end{bmatrix}$$

$$\mathrm{A} = \begin{bmatrix} 0 & 1 \\ 0 & 0 \end{bmatrix}, \quad \mathrm{B} = \begin{bmatrix} 0 & 0 \\ 1/M & 0 \end{bmatrix}$$

Written in full

$$\begin{bmatrix} x_1(t) \\ x_2(t) \end{bmatrix} = \begin{bmatrix} 0 & 1 \\ 0 & 0 \end{bmatrix} \begin{bmatrix} x_1(t) \\ x_2(t) \end{bmatrix} + \begin{bmatrix} 0 & 0 \\ 1/M & 0 \end{bmatrix} \begin{bmatrix} u_1(t) \\ u_2(t) \end{bmatrix} \tag{22.11}$$

It may be shown that many dynamical systems of finite order may be described by a corresponding number of first-order equations like (22.8) and (22.9). These equations are called *state equations*, and there are as many equations as there are state variables. The significant feature of the state space method of structuring the system equations is that, regardless of the order of the system, many systems may be described by a single *vector differential equation* (22.10) above. For higher order systems, the vectors and matrices have a large number of components as follows. For an nth order system,

$$x(t) = \begin{bmatrix} x_1(t) \\ x_2(t) \\ \vdots \\ x_n(t) \end{bmatrix}, \quad u(t) = \begin{bmatrix} u_1(t) \\ u_2(t) \\ \vdots \\ u_n(t) \end{bmatrix}, \quad \mathrm{A} = \begin{bmatrix} a_{11} \dots a_{1n} \\ \vdots \quad\quad \vdots \\ a_{n1} \dots a_{nn} \end{bmatrix}$$

$$\mathrm{B} = \begin{bmatrix} b_{11} \dots b_{1n} \\ \vdots \quad\quad \vdots \\ b_{n1} \dots b_{nn} \end{bmatrix},$$

For computation, general-purpose programs are available for solving Equation (22.10), given the nth order control vector $u(t)$ defined over the control interval from an initial time, say $t = t_0$, to a final time, say $t = t_1$, and given the values of the elements of the A and B matrices, and the initial conditions $x(t_0)$.

Thus we see that the state space approach to the quantitative description of large systems, provides a simple conceptual basis for visualising the motion of a system in response to a control input, and also leads to a form of vector differential equation ideally suited to a digital computer solution.

C. THE CONTROL PROBLEM

We have seen in the previous section how it is possible in many cases to set up a vector differential equation describing the motion of a given physical system. Provided the elements of the matrices A and B are known, digital computers may be used to calculate solutions to these equations and thus obtain the system responses to chosen control inputs.

Of course, as in all engineering problems, the practical situation is seldom as simple as the previous paragraph might suggest. Firstly, with large systems it is very difficult to determine a set of state variables which fully describes the system. Secondly, even if all the state variables are known, it is usual to select only the more important ones in order to reduce the amount of computational work. Thirdly, in large systems, it is difficult to obtain good *mathematical models*. Fourthly, there are generally disturbance inputs which by definition are unknown, but which may be of sufficient amplitude to have a marked effect upon the output of the system. Despite all the approximations, uncertainties and unknown quantities, the control engineer's task is to seek out the best control inputs to apply to the system, commensurate with the partial knowledge of the system.

It will be apparent from the above remarks that, whereas the responses of simple, precisely-known systems may be calculated exactly, those of large systems can only be approximately computed. Needless to, say, the great challenge when working on large systems is to find ways and means of obtaining simple, yet sufficiently accurate, *mathematical models* from which system responses can be calculated in order to determine the best controls to apply.

Before going on to consider a series of examples of large systems, it remains to draw attention to the need for a *performance index.* If it is supposed that an adequate mathematical model of a given system is available, then it follows that, for any arbitrarily chosen control input function (defined over the future interval of interest), the corresponding output can be calculated. However, the problem is to find that control which gives the best output during the period of interest. Because of this, it is necessary to set up a criterion by which to judge the quality of a chosen control input. This criterion is called the performance index. For instance, there are many ways of controlling a vehicle using accelerator and brake to drive it from one point to another, and the choice of the control depends upon the objective involved. If, for instance, the objective were to move the vehicle between the two points in, say, 100 seconds with the least fuel consumption, we would set up a performance index J of the form

$$\text{total fuel consumption,} \quad J = \int_{t_0}^{t_0+100} u(t)dt \tag{22.12}$$

where

t_0 = initial or starting time

$t_0 + 100$ = final time

$u(t)$ = fuel flow rate.

In this example the control functions are accelerator position (fuel flow rate) $u(\tau)$, $t_0 < \tau < t_0 + 100$, and brake position. Time waveforms for the controls would be chosen on some arbitrary basis, tested out on the mathematical model, and the total fuel consumption J computed. This would then be repeated with other controls which caused the vehicle to move between the chosen points in 100 seconds, and eventually the control which gave the least value of the performance index J, would be chosen as the best control input. Such an input control function is called the *optimal control.*

To show some of the challenging problems arising in the control of large systems, we consider below a number of examples. We do so with particular reference to the mathematical model of the system, the performance criteria involved, and the nature of the optimal control input.

D. SATELLITE ATTITUDE POSITIONING SYSTEM

In order that the antennae aboard satellites should have the correct orientation with respect to earth, it is necessary to be able to align the satellite axis with a particular reference axis. This may be achieved by small gas jets located as indicated in Fig. 22.5.

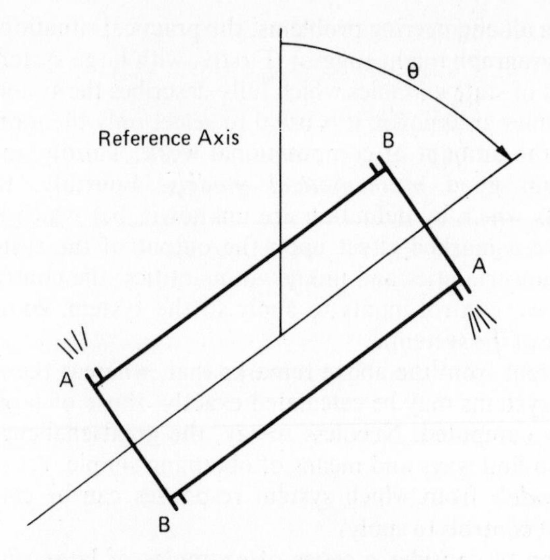

FIG. 22.5 Satellite Attitude Positioning System.

Because of the limited gas supply, it is clearly essential to execute the manoeuvre with the minimum usage of gas.

This is a system which is precisely known, and for which the mathematical model may be established exactly. As such it does not fall into the category of large systems, but will be examined to reinforce the concepts of state space, state variables, state space trajectories, performance index and optimal control inputs already defined.

Being a frictionless inertial system, the *state equations* are

$$d/dt\, x_1(t) = x_2(t) \tag{22.13}$$

$$d/dt\, x_2(t) = u(t) \tag{22.14}$$

where it is assumed that the moment of inertia is unity, and where

$x_1 = \theta =$ angular displacement
$x_2 = d/dt\, \theta =$ angular velocity, and
$u =$ torque.

Since the gas flow rate is limited, the torque which can be exerted is also limited between positive and negative limits. Suppose for example that the greatest torque is unity (positive or negative) then we can write

$$-1 \leq u \leq +1 \tag{22.15}$$

If $t = 0$ denotes the beginning of the control interval and $t = T$ denotes the end of the interval, and if the initial error in displacement is unity say, then the end conditions are

$$\left.\begin{aligned} x_1(0) &= 1 \\ x_2(0) &= 0 \end{aligned}\right\} \text{ initial conditions} \tag{22.16}$$

$$\left.\begin{aligned} x_1(T) &= 0 \\ x_2(T) &= 0 \end{aligned}\right\} \text{ final conditions.} \tag{22.17}$$

The control interval T is chosen to be of acceptable duration, noting that the longer the interval the less gas need be used.

Since the object of the control is to transfer the state of the system from the initial conditions (22.16) to the final conditions (22.17) with the least fuel consumption, a suitable performance index is

$$J = \int_0^T |u(t)|\, dt \tag{22.18}$$

where $|u(t)|$ is the modulus of $u(t)$, which is defined as

$$|u(t)| = u(t) \text{ when } u(t) > 0$$
$$|u(t)| = -u(t) \text{ when } u(t) < 0 .$$

This is necessary since the quantity of gas stored can only decrease, not increase, regardless of whether the torque is positive or negative.

It may be shown by calculation that if, for example, the control interval is chosen such that $T = 4\frac{1}{4}$ seconds, there is an infinite set of controls which will move the system from the initial state (Equation 22.16)) to the final state (Equation (22.17)) in $4\frac{1}{4}$ seconds.

Two of these controls $u_a(t)$ and $u_b(t)$ are shown plotted as functions of time in Fig. 22.6. These are

$$u_a(t) = \left.\begin{array}{ll} -1, & 0<t<0.25 \\ 0, & 0.25<t<4.00 \\ +1, & 4.00<t<4.25 \end{array}\right\} \quad (22.19)$$

and

$$u_b(t) = \begin{cases} -0.222, & 0<t<2.125 \\ +0.222, & 2.125<t<4.25 \end{cases} \quad (22.20)$$

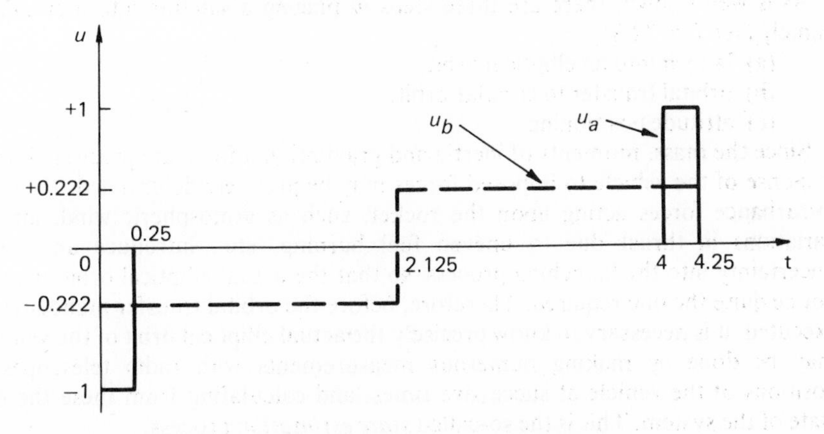

FIG. 22.6 Two possible Control Decision Sequences $u_a(t)$ from Equation 22.19 and $u_b(t)$ from Equation 22.20.

The corresponding *trajectories* are shown in Fig. 22.7, where the trajectory (a) is in the response to control input $u_a(t)$, and trajectory (b) to $u_b(t)$. In both trajectories the system is moved from its initial state of unit displacement and zero velocity to its final state of zero displacement and zero velocity in $4\frac{1}{4}$ seconds. In so doing, the maximum allowable control amplitude of plus or minus unity is not exceeded.

To determine which of these two controls is the better, the performance index (Equation (22.18)) needs to be evaluated. When this is done we find that,

$$\text{for trajectory (a), } J_a = 0.50$$
$$\text{and, for trajectory (b), } J_b = 0.94.$$

This clearly shows that trajectory (a), corresponding to control input $u_a(t)$, is the better control trajectory. We note, however, that we have considered only two possible controls from the infinite set of admissible controls, and the reader might reasonably ask whether there are better controls than $u_a(t)$. In answer to this question it may be stated that there is a theoretical proof that $u_a(t)$ is the best possible control (or the *optimal control*) in this case. The proof is quite difficult, but necessarily of great importance in modern system control theory.

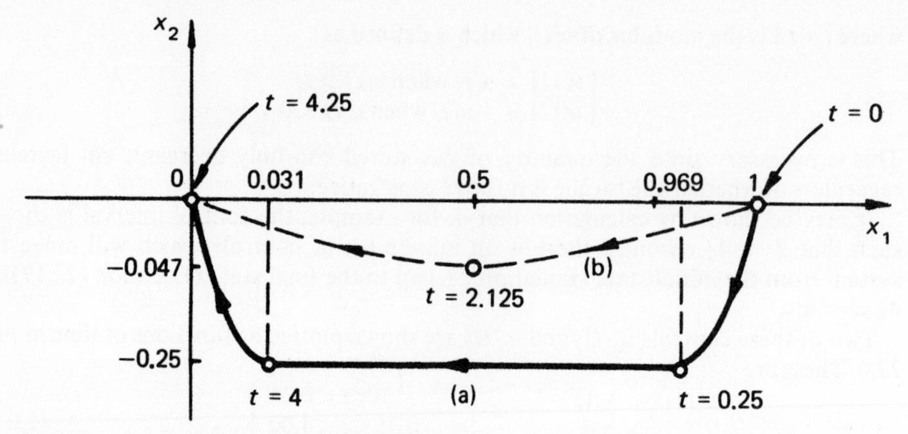

Fig. 22.7 Two Trajectories in State Space corresponding to u_a and u_b from Equations 22.19 and 22.20.

E. SATELLITE LAUNCH SYSTEM

As is well known, there are three steps in placing a satellite into a circular orbit, namely (see Fig. 22.8),

(a) launch into an elliptical orbit,
(b) orbital transfer to circular orbit,
(c) attitude positioning.

Since the mass, moments of inertia and gravitational force are precisely known, the response of the vehicle to imposed forces may be precisely determined. However, the disturbance forces acting upon the rocket, such as atmospheric wind, air friction, variations in thrust due to uneven fuel burning, etc., introduce an element of uncertainty into the launching process, so that the actual elliptical orbit acquired will not be quite the one required. Therefore, before the orbital transfer manoeuvre can be executed, it is necessary to know precisely the actual elliptical orbit of the vehicle. This may be done by making numerous measurements with radio telescopes of the positions of the vehicle at successive times, and calculating from these the orbit, or state of the system. This is the so-called *state estimation* process.

Knowing the actual elliptical orbit and the desired circular orbit, the control forces of the rocket engines and the precise times at which they are to be applied may be calculated, and at the appropriate times these forces may be impressed upon the system to implement the orbital transfer. Due to unpredictable forces on the satellite, its orientation will be displaced from the desired one. Again the error is measured in some way, and the appropriate forces applied to correct the attitude with minimum fuel consumption.

In this example, we have seen that, although the dynamical properties of the vehicle are precisely known, and hence its mathematical model is precisely known, there are unknown disturbance inputs which prevent us from predicting exactly the system outputs. This, in turn, gives rise to the need for a state estimation procedure to find the precise state of the system after the known control inputs have been applied.

F. ELECTRIC POWER SYSTEM

Unlike the satellite launching system, the electric power system (see Fig. 22.9), comprising generators, transformers, transmission lines and a wide variety of electrical load devices consuming electrical energy, has numerous complex components and is thus not amenable to precise description. Thus, any mathematical model of the system must be of a very approximate nature.

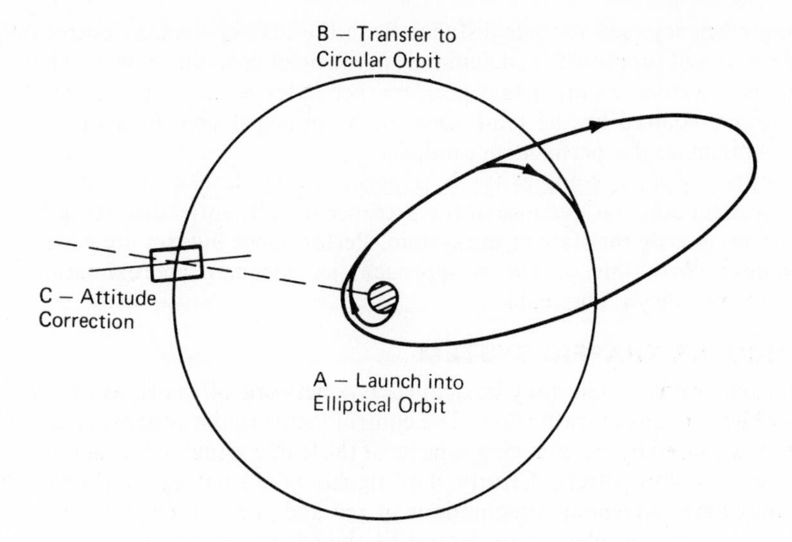

FIG. 22.8 Successive Steps in Satellite Launch into Circular orbit.

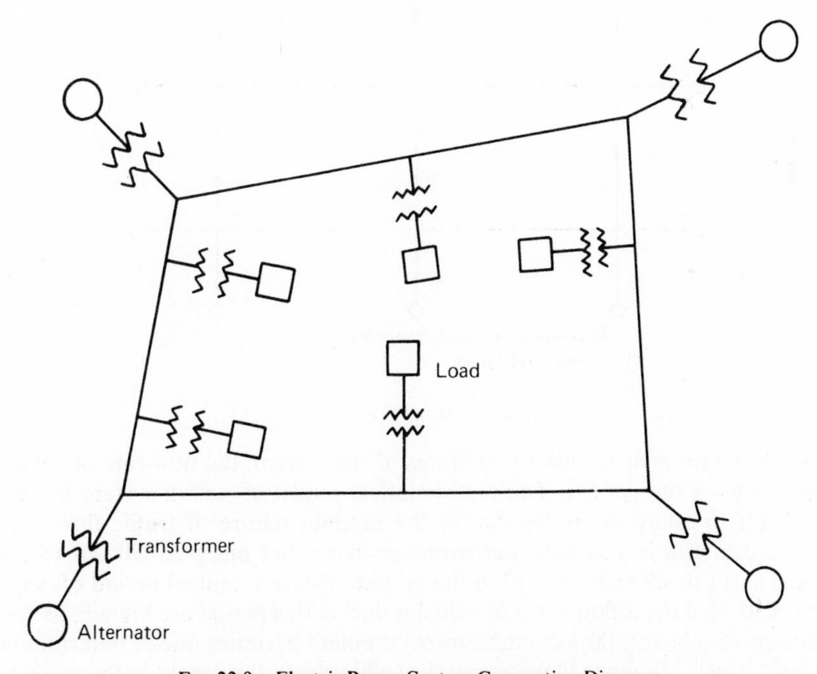

FIG. 22.9 Electric Power System Connection Diagram.

Furthermore, since the load upon the system depends upon the users' requirements at any particular time, the disturbance inputs to the system are of major proportions. Upon the basis of previous records, the future demand upon the system may be anticipated in a approximate way, and a control input policy established. At the same time, the policy must allow for the unexpected demands upon the system.

The principal inputs to the system are the fuel supplies to the steam/electric generators in the power stations, and the principal outputs are the electric power components supplied to the main load centres. Since power stations differ in their generating efficiency, and in their distance from the load centres, a control policy is required which will supply the load demands in the most economical way. Thus, in the electric power system, an important performance index is the cost of generating the electric power required by the load. One of the principal objectives of the control policy is to minimise this performance index.

In a system of such complexity, it is necessary to develop and use simplified mathematical models and, because of the presence of substantial disturbing forces, to measure continuously the state of the system. Performance indexes are generally of a simple nature. With this simplified approach, an adequate approximation to the optimal control policy is obtainable.

G. VEHICULAR TRAFFIC SYSTEM

A vehicular traffic system may be defined as a network of roads, as in Fig. 22.10, through which streams of traffic flow. The control inputs to the system are the control of traffic flow caused by the changing aspects of the traffic signal lights, and the output is the consequent flow pattern. Clearly, if all signals have a red aspect, then the flow is zero. If they have sequential combinations of red and green, then traffic flow is permitted. The control problem is to determine the sequence of signals in the system which will minimise (or maximise) some system performance index.

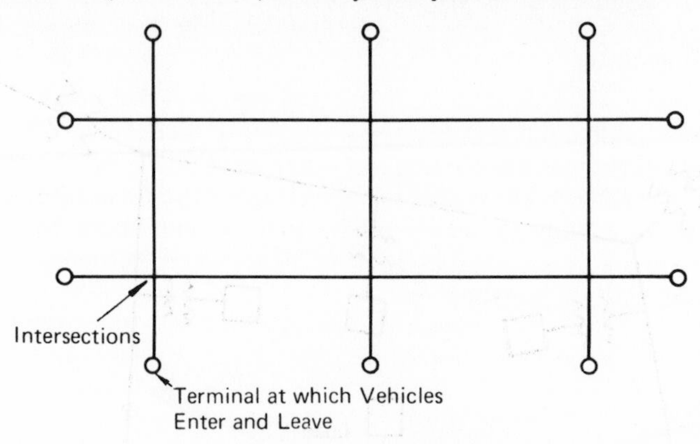

FIG. 22.10 Road Traffic Network.

It may be convenient to choose, as states of the system, the flow rate of vehicles at selected points in the system. The mathematical model of such a system has proved very difficult to establish, partly due to the random nature of traffic flow. Equally difficult to establish is a suitable performance index, but one possibility is to use the total time spent by all vehicles within the system during a control period of, say, half an hour. Provided the inflow rates of vehicles during this period are known, as also the destinations of vehicles, the systems control problem becomes one of determining the signal-light aspect sequence to minimise the total vehicle-time spent in the system.

This has proved to be a very difficult control problem. The dynamics of the system are very complex, and hence the mathematical model is difficult, if not impossible, to establish; the traffic flow into the system contains a large unpredictable component; the performance index is difficult to establish; and the flow rate of traffic is expensive to measure. The flow rates have large random fluctuations, and the flows themselves consist of discrete events rather than a continuous stream.

Despite these somewhat formidable difficulties, traffic flows are controlled with fair efficiency, using empirical procedures based upon human experience in dealing with the problem.

H. HYDRAULIC SYSTEM

The control of the movement of sand in a river estuary, as in Fig. 22.11, often involves the placement of a breakwater to deflect the water currents and cause the shipping lane to be clear. This differs from other large-system control problems in that the solution involves a single control decision, namely the size and location of the breakwater wall.

Such systems are so complex that analytical mathematics cannot be employed. Mathematical models would be of impossible complexity, and the states would be astronomical in number. Furthermore, the performance index would be extremely difficult to establish in mathematical terms. So the problem is solved using an analogue model rather than a mathematical model. A small-scale physical model of the estuary is built with all the components scaled in a very special way. Water flow is regulated to simulate the river and ocean currents and sand is introduced to accord with the real system.

In such models, events occur a great deal faster than in the real system, so that the determination of the control decision may proceed on a systematic trial and error basis. Breakwater walls of various lengths and positions are introduced, and an understanding of the influence of the wall placement upon sand flow is obtained within a few hours. On completion of the experimental study, the best position for the breakwater is established and, after its construction in the real estuary, the systems control problem is solved.

This is an example which defies a mathematical approach, but which is particularly amenable to the experimental attack using an analogue model. The effect of control inputs, also the disturbances due to irregular water input flows, may be predicted with good accuracy. The performance index takes on a subjective form with account being taken of many considerations which might well be impossible to formulate in mathematical terms. This is a good example of a case where an analogue approach is superior to an abstract mathematical one.

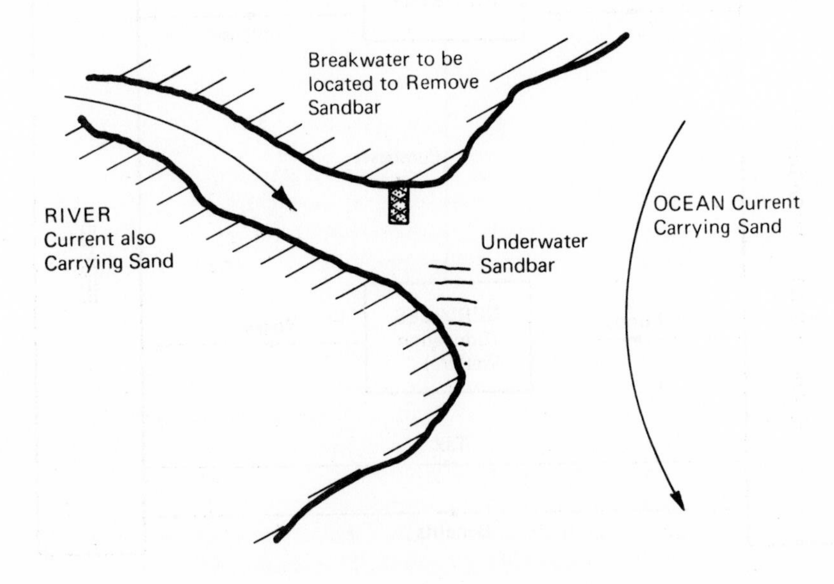

Fig. 22.11 Harbour Silting System.

I. THE ECONOMIC SYSTEM

The economic system is the mechanism by which goods and services flow between people in our community (see Fig. 22.12). It is in many ways self-regulatory, but comes under the strong influence of government, which attempts to control the major money flows with the object of producing a steady growth in the wealth of the nation without excessive oscillations in economic activity.

The dynamics of this system are being intensively studied and increasingly understood. The development of mathematical models is hampered by the intrinsic presence of the human element in the system, which introduces unpredictable and often emotional factors. The control inputs, namely tax and credit controls, are well-defined, as are the outputs in terms of gross wealth. The performance index is seldom spelt out in precise mathematical terms, but the subjective measure of steady growth is the principal criterion.

The system is too complex for description with sufficient accuracy by either abstract mathematical models or analogue models. Digital-computer simulation may provide a partial answer.

So, rather than to experiment with models, it remains to experiment with the system itself. Over the years, a body of knowledge on cause and effect has been built up, so that it becomes increasingly possible to predict with some reasonable degree of certainty the effect of a particular control action. Such control systems suffer from the fact that they cannot predict the effect of some new disturbance not previously encountered.

Clearly, the smooth running of the economic system is so fundamental to the health of our community that we can expect to see increasing development of digital-computer simulation models, enabling better predictions and hence better control strategies to be devised.

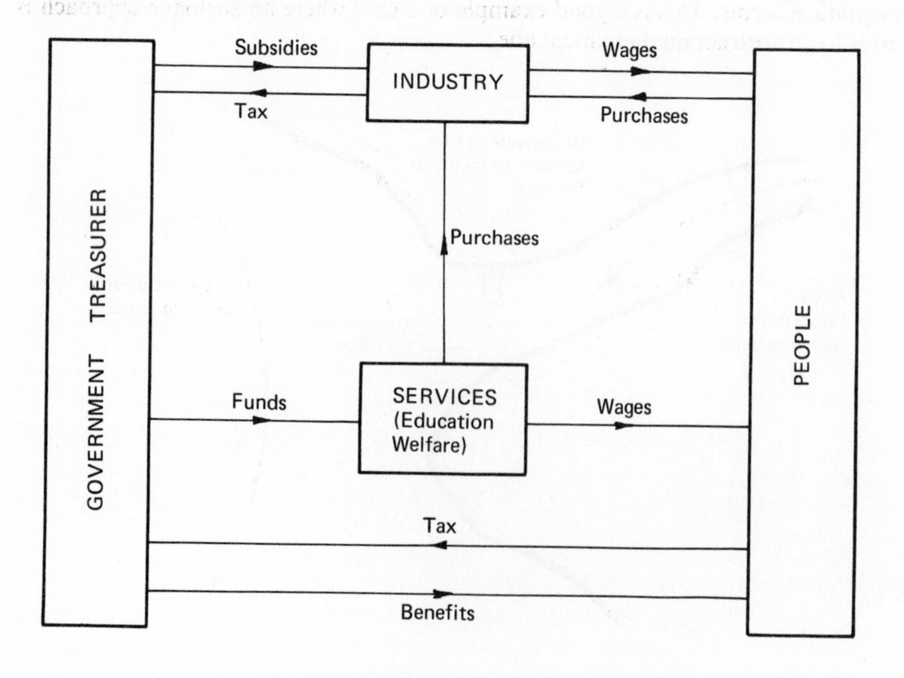

FIG. 22.12 Money Flow in the Economic System.

J. CONCLUSIONS

The science of the control of large systems is still in a rudimentary stage of development in many areas. However, as an aid to a logical and analytical approach to particular systems, it is useful to select a set of principal variables to describe the behaviour of the system. Some of these variables are input variables, of which some are amenable to deliberate adjustment and are control inputs, whereas the others are beyond such adjustment and are disturbance inputs. Of the remaining variables, some are outputs in the sense that they can be measured, whereas the remainder are inaccessible outputs.

Some of the less complex systems can be described by abstract mathematical models, in which the state equation, comprising a set of first-order differential equations, provides a standard, convenient method of description of the system dynamics. Otherwise analogue models, digital-computer simulation models, or the system itself must be used as the vehicle for predicting the effect of proposed control inputs.

Perhaps the single most important event in the science of the control of large systems is the advent of the powerful digital computer which opens the doorway to large scale system simulation, and hence to the development of more comprehensive control strategies for large systems.

K. EXERCISES

1. Show that the state equations of a system comprising a mass M moving horizontally in response to a control force u, against a viscous frictional force (proportional to velocity), are

$$d/dt\, x_1(t) = x_2(t)$$
$$d/dt\, x_2(t) = -(1/M).f.x_2(t) + (1/M)u(t)$$

where f is the friction coefficient
x_1 denotes the displacement
x_2 denotes the velocity.

If, at the initial time $t = 0$, the displacement and velocity are both zero ($x_1(0) = 0$, $x_2(0) = 0$), calculate the response to a unit control force applied for a period of 2 seconds. Assume $f = 1$. To perform the integration use increments of time $\Delta t = 0.2$ seconds and calculate

$$\Delta x_1\,(k\Delta t) = x_2\,(k\Delta t)\,.\,\Delta t$$

and hence $$x_1(\overline{k+1}\,\Delta t) = x_1(k\Delta t) + x_2(k\Delta t).\Delta t$$

and, similarly, $$x_2(\overline{k+1}\,\Delta t) = x_2(k\Delta t) + [-(1/M)x_2(k\Delta t) + (1/M)u(k\Delta t)]\,\Delta t$$

where $$k = 0,1,2\ldots 10.$$

Plot the state space trajectory of the system during the 2 second period.

2. Considering a joinery factory to be a large system, draw a block diagram showing the state variables, control inputs and system outputs. Is it possible to construct a mathematical model of the system? What performance criterion might one use to assess the quality of proposed control strategies?

SUGGESTED READING MATERIAL

SHEARER, J. L. MURPHY, A. T. and RICHARDSON, H. H. *Introduction to System Dynamics.* Addison-Wesley Publishing Co., Mass., 1967. For a further study of the dynamics of a wide range of physical systems.

CHAPTER 23

Communication Systems

A. E. Karbowiak

"Over the mountains
And over the waves..."
Anon.

A. TYPES OF SYSTEMS

Communication systems can be conveniently classified into two broad classes:

Class 1: Includes all those systems which are designed primarily to carry information from one point in space to another. Figure 23.1 illustrates the principle. The action performed by such systems is *transportation* of information.

Class 2: Includes all systems whose primary objective is to extract (and process) the information about the environment and finally, in the case of the more sophisticated systems, to take action on the basis of the received information, in accordance with a built-in plan. Figure 23.2 illustrates the principle of such a system with reference to a radar defence system. A system belonging to this class is principally concerned with *extraction* of information.

The sole objective of a communication system of Class 1 is to convey information from one point of space to another using various transmission media and devices. A telephone network is a typical example of a simple system belonging to this class. A radio-telephone is another example, although here the medium of communication is different. Whereas with a common telephone the medium (channel) of communication might be a pair of wires, with a radio-telephone the information is conveyed by electromagnetic waves carried in the space around us. The medium of communication profoundly affects the characteristics of the system as a whole. The choice of the medium is frequently dictated by geographic or economic factors and the objective of the communication engineer is to design and engineer the system to meet the assigned objectives: the resulting system is usually a compromise which emerges from among a number of conflicting requirements, such as minimum installation cost, minimum maintenance cost, reliability, life of the system, etc.

With most communication systems various ancillary apparatus is needed (Fig 23.1) to perform a variety of functions such as encoding of the input data into a form suitable for transmissions along the channel. At the receiving end the signal is processed by means of a suitable decoder into a form required at the destination.

Radar is a communication system in a somewhat different sense. The source is a high power directive transmitter, emitting signals of known form. The receiver, usually placed alongside of the transmitter, detects the signals reflected by distant objects (such as an aircraft) and the signal is invariably accompanied by noise as well as being distorted in form, due to numerous contributing factors. With such a system

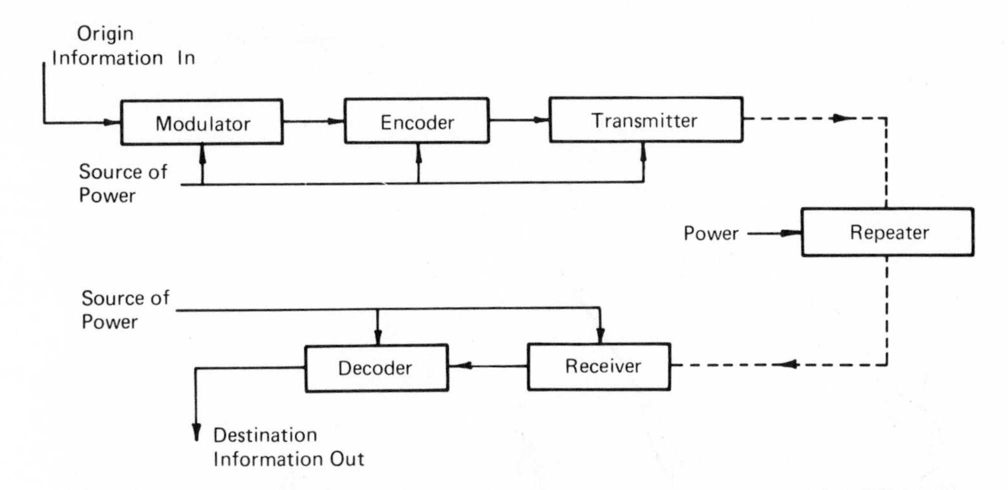

FIG. 23.1 A Class 1 Communication System.

the information received is really not "information transmitted", since the transmitted signal (the transmitted data) is a periodic series of pulses and, as such contains no information (see Chapter 11). The receiver is required to trace a different kind of information; information on the position of the target in terms of distance, bearing and elevation. This information is obtained in an indirect manner by correlating the received pattern of pulses with that of the transmitted ones, and there exists a whole range of devices and instruments which can achieve this.

In effect a radar system consists of two sub-systems (transmitter and receiver) interconnected by a data link (radio path) in such a way as to determine the information about the position of the target. It is an example of a relatively simple system.

One objective of radar measurements is to determine the position of an enemy target so that subsequently, through human actions, defensive means such as anti-aircraft guns can be brought into operation to intercept the enemy target. The complete operation consists of a radar system linked to a defence system using human intervention, with all the attendant disadvantages. The next step in improving the overall performance of such a defence system would be to link the radar system directly to anti-aircraft guns or missiles by means of a data link, thus permitting a direct control of the anti-aircraft guns in accordance with the data received by the radar systems, under an overall control of a suitable computer. Figure 23.2 shows the system schematically with provision being made for direct human control if required.

There are, of course, many non-military applications of radar but the above example is particularly illustrative of the concepts involved.

An automatic pilot system is another example of a system of Class 2. Here, electronic devices sense the environment and process the data collected, which is applied to the electronic devices for direct and wholly automatic control of the aircraft flight. There is provision for intervention by the pilot, but under normal flying conditions the machine would operate entirely without instructions from a human being.

Another, though less obvious example of a communication system is that of a scientist and his instruments engaged in investigation. Here, the scientist is transmitting various signals of his choice into the physical environment, whether it is a biological cell or the nucleus of an atom. He then observes the response of the system and by interpreting (correlating, extrapolating, etc.) the patterns received he *gathers*

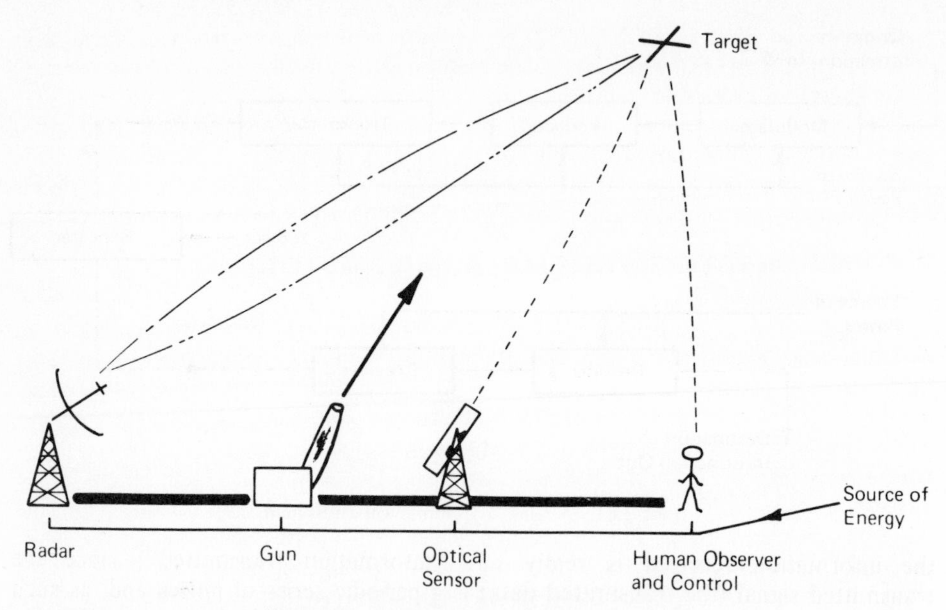

FIG. 23.2 A Class 2 Communication System.

information on the physical environment. Clearly such a system can be looked upon as an embodiment of a communication system of Class 2 with the data or measurements forming the output.

It is at this juncture that we are likely to have a conflict between the two notions of information: the mathematical and the semantic. As far as communication theory is concerned we can analyse the system in minute detail. We can also determine the limits of significance of the data (whether it is in digital or analogue form), we can determine the maximum permissible flow of data and even suggest means for improving the system so that it increases the accuracy of the data. But we cannot say anything about the value, meaning or significance of the data as far as the scientific observer is concerned for this would be semantic information. The data could be a meaningless thermal vibration waveform or the key signature of the DNA chain, this however, would be of no interest to the communication theory, it is outside its province.

To put this important aspect in yet a different way. The output data of an elaborate scientific equipment will wander aimlessly, even with the help of the communication theory and the most sophisticated computer, but when processed by the skilful thoughts of a scientist it acquires meaning and beauty in the form of the patterns so created.

B. CHARACTERISTICS OF SYSTEMS

Communication systems can exist in a variety of forms, but intrinsic to them all are three fundamental characteristics. These are:

1. Usually the source of information and the destination are physically separated. Therefore an instantaneous transmission of the information is impossible because of the finite velocity with which signals travel (the highest velocity being the velocity of light). This imposes fundamental limits due to delays in transmission.
2. Because of imperfections in the system, signals will be mutilated by the channel as well as by the ancillary apparatus. There are in fact definite limits on what can be achieved.
3. All apparatus and processes of communication are subjected to external inter-

ferences and noise (see Chapter 2), on account of which a truly unequivocal communication is an ideal which cannot be realised in practice. Errors in communication are therefore inevitable.

The aim of the communication engineer is to design a system so as to transmit a given class of information-bearing patterns with minimum equivocation while optimising a required number of key parameters, such as cost. Naturally, the design is a compromise and communication theory can help to strike a better bargain.

C. COMMUNICATION SYSTEMS: A BRIEF OUTLINE OF DEVELOPMENTS

When discussing the history of telecommunications one needs to be careful in defining what "electrical communication" means. We know from historical accounts that electrical communication systems, as we know them today, were preceded by various other communication systems. For example, the electrical telegraph system was anticipated by such systems as the semaphore system. Indeed, the two systems have a great deal in common, both use a method of encoding individual letters of the message to be transmitted into a system of symbols suitable for transmission from one point of space to another. The only difference seems to be that in one case one could use radio waves as the carriers of information whereas in the other case the symbols would be generated by mechanical means and observed by optical means from a distant point. However, in both cases the transmission of symbols is taking place by means of electromagnetic waves (with radio systems it is the radio waves which carry the information, whereas with the semaphore system it is electromagnetic waves in the visible part of the spectrum, see Fig. 23.3). While discussing the merits of communication systems against the background of history it may be instructive to

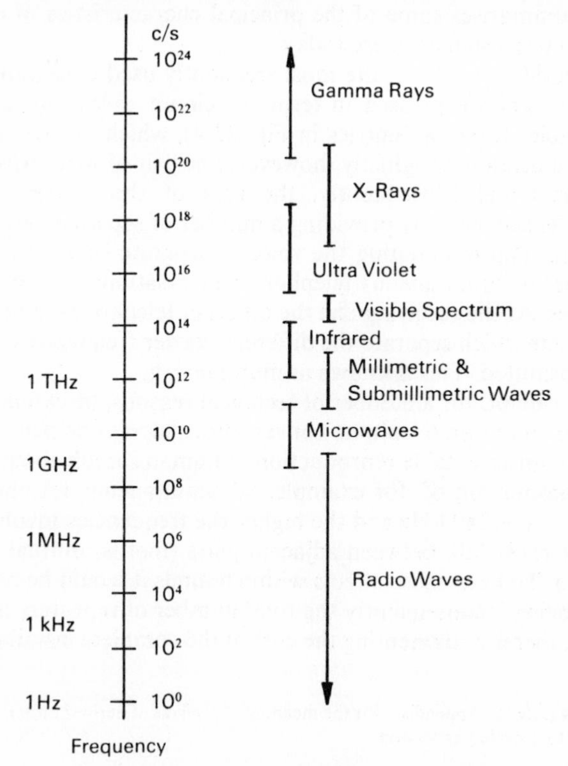

FIG. 23.3 The Electromagnetic Spectrum.

observe that the early heliograph system had, in principle, a great deal in common with some simple communication systems of today, using laser sources.

The principal shortcoming of early communication systems such as semaphore or heliograph was their slow rate of communication, as well as their unreliability. The system was only capable of being used under good weather conditions and would totally fail when the visibility, due to fog or heavy rainfall, was restricted.

Electrical communication systems can be said to have started with the invention of the electrical telegraph system, which seems to have taken place simultaneously on both sides of the Atlantic: in U.S.A. by Morse and in England by Wheatstone. The essence of the idea here was to transmit electrical signals in the form of pulses of defined duration (see Chapter 3). By encoding individual letters of the English alphabet into a sequence of dots and dashes it was possible to send messages along a pair of wires (or even one wire and the earth). As such, the system was an early example of encoding from one finite set of symbols into another; essentially a digital system.

Telephony is an example of an analogue communication system. The system came into being through the invention of two devices; a transducer for transforming the soundwave into electrical impulses, a microphone, and a transducer for converting electrical signals into soundwaves—an earphone.

Communication systems of today relate to one or the other of the above described communication systems. The differences such as there are, relate to sophistication in equipment and different methods of transmitting the signals. However, there are only two methods of transmitting signals: (i) by radio waves, i.e., by free space propagation; (ii) by various types of waveguides such as transmission lines and coaxial cables, i.e., by guided propagation.

Figure 23.4 summarises some of the principal characteristics of different methods of conveying electrical signals in use today.

Taking the world as a whole, the most frequently used communication media for telephone traffic (when expressed in terms of circuit miles) are the open-wire and balanced-pair cables (first two entries in Fig. 23.4), which incidentally are the oldest means of communication. Originally, however, one pair of wires was needed for every telephone channel and consequently, the cost of the system was substantially proportional to its capacity. By providing a number of separate carrier frequencies on each pair of wires and modulating the voice of separate subscribers onto a different carrier, a number of simultaneous telephone conversations can be transmitted along each pair of wires. At the receiving end the different telephone channels are unravelled by the use of filters which separate the different carrier frequencies and with them the intelligence transmitted. This is known as multiplexing.

Multiplexing cannot, for a number of technical reasons, be extended indefinitely by cramming more and more telephone conversations along one pair of wires. The chief reason is that for an acceptable reproduction of human speech about 4 kHz bandwidth is needed.* Transmission of, for example, 60 simultaneous telephone conversations necessitates $60 \times 4 = 240$ kHz and the higher the frequencies involved, the larger the attenuation, the cross-talk between adjacent pairs (that is, mutual interference), and signal distortion. To keep these effects within bounds it would be necessary to reduce the repeater spacing;† consequently the total number of repeaters needed would have to be increased, thereby augmenting the cost of the complete installation.

* 1 kHz = 1000 Hz (refer to Appendix 1 for the meaning of prefixes of derived units). 1 Hz is an oscillation of frequency equal to 1 cycle per second.

† A repeater is a device inserted along a transmission line to amplify and correct the signals carried.

Type		Operating frequency	Channel capacity	Repeater spacing (attenuation)
Open wire	One pair	36 to 84 kHz (go) 92 to 140 kHz (return)	12 + 3 + (1) = 15 + (1)	60 to 70 miles (0.4 db/mile)
	Maximum 16 pairs	As above	240 + (16)	
Balanced pair	One pair	12 to 252 kHz	60	12 to 30 miles (4 db/mile)
	Maximum 24 pairs	As above	1440	
Coaxial G.B.		Up to 4 MHz	960 (One super channel)	6 miles
		Up to 12 MHz	3 super channels	3 miles
Coaxial U.S.		Up to about 3 MHz	720	8 miles
		L3 system	1 TV channel + 600 speech channels	4 miles
Single wire transmission line		About 100 to 1,000 MHz	Probably a few TV channels	About 2 to 10 miles
Microwave links		Below 500 MHz	Maximum of 60	40 to 50 miles
		500 to 1,000 MHz	Up to 120	40 to 50 miles
		2,000 MHz	240 × 6 or 1 TV × 6	30 to 40 miles
		4,000 MHz	600 × 6 or 1 TV × 6	25 to 30 miles
		6,000 to 8,000 MHz	600 × 6 or 1 TV × 6 Maximum 2 TV × 6	25 to 30 miles
		11,000 MHz	Less than 600	Less than 20 miles
Long haul waveguide (H_{01})		30,000 MHz up to about 100,000 MHz	1,000 super channels = several hundreds of TV channels = several hundred thousands of speech channels	20 to 40 miles (attenuation 2 to 4 db/mile)

FIG. 23.4 Communication Systems and their Capacity.

The design of a communication system is a careful balance between numerous, often conflicting factors to minimise the total cost for a given performance. With modern equipment, up to about 15 simultaneous telephone conversations can be transmitted along an open pair of wires and 60 along each pair of wires of a balanced-pair cable system. The permissible number of pairs of wires in any one system is also limited by various factors, principally the maximum permissible cross-talk between the pairs. For example, the present practice with balanced-pair cables is to provide 24 pairs each having a capacity of 60 telephone channels giving a total capacity of 1,440 telephone channels and a repeater spacing of the order of 20 miles.

This is just one illustration of the aims of engineering design: given the objective, to use existing skills and techniques so as to minimise the cost of achieving the objectives. In the above example of the balanced pair, the 1,440 telephone channels could, in principle, be provided by simple voice channels on 1,440 pairs of wires. Instead, using multiplexing techniques the same capacity can be provided on only 24 pairs of wires. Clearly, in effect, the cost of line is traded for the cost of somewhat more complex terminal equipment and repeaters—and on balance the transaction is well worthwhile. As a general rule, the heavier the communication traffic on a given route, the greater the likelihood that a system of great technological complexity will prove to be economically attractive. This is clearly supported by long years of experience. Figure 23.5 shows the general trend of cost of communication (per channel) as a function of density of traffic for different types of communication systems.

There are many other factors entering into the design of a communication system (such as accessibility to repeater stations, climatic conditions, established practices,

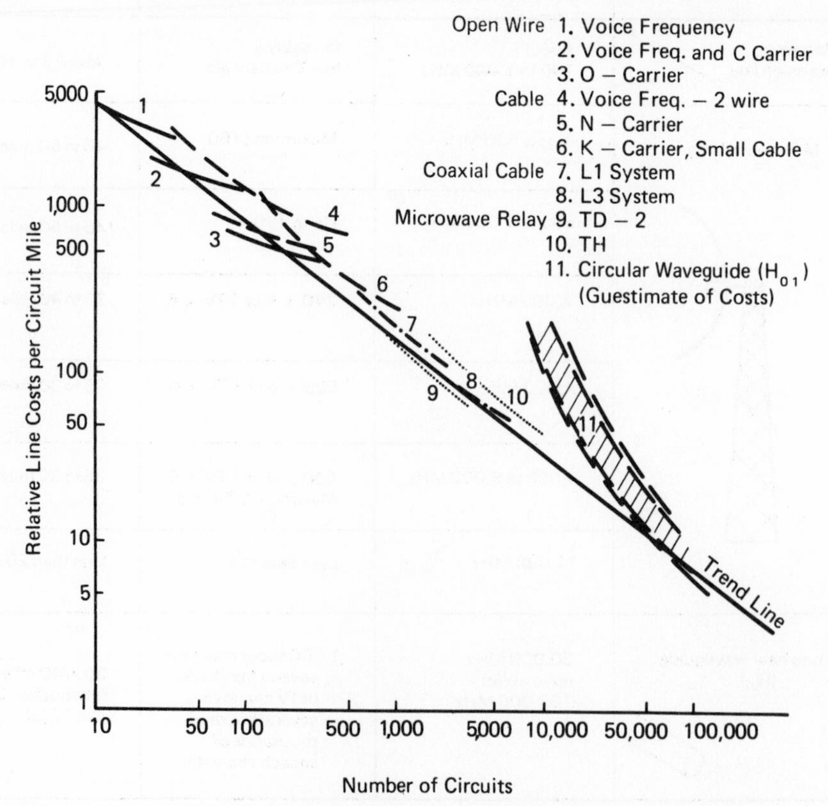

FIG. 23.5 Trend Line of Costs of a Communication Line Installation.

human prejudice, etc.) which need not be discussed here, but one additional factor must be considered: requirements for other than telephone transmission.

Since the Second World War there has been a great expansion in television. Whereas before 1939 there were only a few countries where television transmission had been attempted, today in many countries several independent and simultaneous television programmes are in operation and are served by a wide distribution network (e.g., Eurovision). Each television channel requires for transmission as much bandwidth as 1,000 or more telephone channels, and this bandwidth must be provided in a single communication medium. Clearly, open-wire or balanced-pair cables are unsuitable, because a single pair of wires cannot transmit efficiently such a wide bandwidth.

However, a modern coaxial cable or a microwave link (see Fig. 23.4) can easily accommodate several television channels in addition to the telephone traffic. These are the wide-band communication systems of today. For wide-band applications it is usually microwave links that are economically attractive. But, unlike coaxial cables, these use (for the carrier frequency) microwaves in the frequency range of thousands of megahertz (see Fig. 23.3). Furthermore, a microwave link is an example of an "open" communication system in that, like radio, it utilises free space as the vehicle for information transmission, and to avoid mutual interference careful planning is necessary as regards the frequency allocation. This frequency allocation is governed by international agreements, and in many highly populated areas free space is at a premium. In the next decade or two it is very likely that we shall witness a saturation of free space, particularly as this is more urgently needed for communication with moving objects, such as ships and aeroplanes.

Coaxial cables are at present the backbone of wide-band, large capacity, communication systems. With growing demands on communication channels we must look forward to new means of communication to meet the needs of the future.

D. SOME FUTURE POSSIBILITIES

In the last few decades we have witnessed an unprecedented rise in the demands on communication channels. There is every indication that the increase in demands of communication channels will continue at a rapid rate. These demands are so real that a phrase "explosion of communications" has been coined to describe the situation. Nowadays, many thousands of speech channels are required to provide for communication between large cities. In addition television requirements have risen enormously, and these demand channel capacities many times in excess of those needed for speech communication (one television channel uses as much of frequency bandwidth as 1,000 telephone channels).

We can foresee new demands for communication channels for various reasons which are already apparent. There are increased needs for facsimile transmission and data transmission for various business purposes such as in commerce or banking, in addition to the growing requirements for communication with computers. For everyday use video-telephone, that is a telephone which combines transmission of pictures as well as sound waveforms, has already been developed and could be put into operation provided that the greatly increased demands on communication channels could be satisfied. We foresee a great increase in the use of picture as well as speech transmission for purposes such as conferences between managers of different branches of the same company, or for direct communication of technical information within the structure of a company, and also for transmission of technical and other information from central computer memory store to various customers.

Indeed, in the years to come, there should be no need for people to spend hours in patent or technical libraries searching for information needed for their work. Such functions can be performed much more economically and efficiently by direct communication with a computer. The computer, in response to a request from a customer, would automatically search the stored information and then display the information requested on a television screen at the customer's home or office. Indeed, there seems to be no end to the varieties of services which could be provided in the years to come by direct communication with computers. At the moment, however, such possibilities must wait their implementation until the problem of providing large numbers of communication channels is solved.

When assessing the future possibilities for new communication channels, it is useful to examine the possible limitations arising from the fundamentals of physics. First of all, it is necessary to accept that in the future it will not be possible to make more extensive use of radio waves for point to point communication, because they will be needed for communication with moving objects such as aeroplanes and ships to an ever increasing extent. There are, however, distinct possibilities in the use of satellite communications using UHF frequencies in that, a properly planned satellite communication system would offer a more economical use of the radio frequency spectrum. There are however, definite limitations and it would appear, on the evidence before us, that satellite communication systems cannot offer a long term solution to the problems ahead of us. The reason being, that while a satellite can be designed to handle simultaneously 10 or 20 TV channels or the equivalent in other facilities, there is a limit to the number of satellites which can be accommodated in the space around the earth, without undue mutual interference. Let us then assume that satellite communication systems prove economically attractive, then each and every nation would wish to have a satellite or a share in one and if this were the case then only a small fraction of the customers could at all be satisfied. The outstanding problem of providing alternative communication means .than those available today would therefore remain.

What are the alternatives? Possibilities would seem to lie in greater exploitation of higher frequencies than UHF. Here we have in mind (see Fig. 23.3) the utilisation of the frequencies well beyond the microwave part of the spectrum, such as the millimetric spectrum (frequencies corresponding to the band 30-300 GHz or the sub-millimetric part of the spectrum 300-3000 GHz). It is also possible that frequencies even higher than those could be used for communication. In particular, with the advent of lasers, the exploitation of the infrared frequencies and the frequencies corresponding to the visible part of the spectrum* (the whole of these two spectra is known under the name "optical spectrum") has now become a possibility.

We can now generate optical frequencies almost as efficiently as we can produce radio waves. Thus, a point to point optical communication is one possibility, but there are even more exciting possibilities lying ahead of us in systems utilising guided optical waves in suitable waveguides not unlike radio waves which can be guided by coaxial cables (see last entry in Fig. 23.4).

Millimetric as well as sub-millimetric waves offer exciting possibilities. Here we have in mind guiding such waves inside metallic tubes of suitable construction (see Karbowiak, 1965). With such systems, it is estimated that it should be possible to transmit several hundreds of video channels in addition to many thousands of speech channels.

The recent advances which have been made in materials science and the technology of microminiature circuits (see Chapters 4, 16 and 17) have opened up new possibilities in all-digital integrated microwave communication systems. There is no doubt at all that in the next two decades we shall witness great progress in such communication systems.

* The frequency corresponding to red-orange visible light of wavelength 600 nm, is 500 THz.

Looking further ahead, there seems to be no reason at all why communication on a wider scale should not be made possible in the years to come: to provide hearing for the deaf and vision for the blind are distinct possibilities in the present day technology. But while these possibilities are under investigation, new possibilities emerge in communication by extending the capacity of the remaining human senses. Here we have in mind remote sensing by touch, smell or even taste. The basic issues here are the invention of suitable transducers, in the same way as telephone communication was made possible through the invention of the microphone, and the earphone.

We can also speculate that through the development in miniature computers new possibilities will open up. The personal computer of the future will be no bigger than the portable transistor radio receiver today. Yet such a computer will have sizable power and will be backed by a selection of miniature magnetic tape stores in addition to a sizable immediate access store. Among other things such a portable computer would represent a sophisticated "notebook", an immediate access library, and be a general mental aid for its owner. In addition the personal computer would be capable of being plugged into a communication network (accessible through a wideband outlet on subscriber premises, whether it is in his home or office) of a vast national giant computer capable of rendering a wide variety of services ranging from library or patent searches, education, entertainment, to advice on an unlimited variety of topics, including pictorial explanations available on consoles mounted on subscriber premises.

On another front, work is currently proceeding to try to expand the communication theory into the domain of semantics. This would seem to be a field full of difficulties ahead but nonetheless, once we learn to understand how to communicate semantics we shall have opened new possibilities in communication. Some philosophers even go so far as to foresee possibilities of direct communication of meaning as well as human feelings. It would seem that even the basic ideas behind human thoughts will be explored. While it is true to say that such possibilities are not, at the moment, within the scope of present day technology, they cannot be excluded from possible exploitation in the future..But, before such systems become a reality, we should do well to examine the likely effect which they might have on human societies.

E. EXAMPLES OF INFORMATION THEORETIC PROBLEMS ENCOUNTERED WITH COMMUNICATION SYSTEMS

1. Communication System of Class 1.

The transmitted sequence of symbols cannot be predicted (otherwise it would carry no information, see Part II) but must be random to some extent. With a binary system the transmitted sequence might be like

$$\ldots A\,B\,B\,A\,B\,A\,B\,A\,A\,A\,B\,B\,A\,B \ldots \qquad (23.1)$$

For example a sequence of this nature could be a binary encoded version of the message "Arrive tomorrow". On being transmitted through the channel and the receiving apparatus, the message could, on account of noise, turn out to be

$$\ldots A\,B\,B\,A\,A\,A\,B\,A\,A\,A\,B\,B\,A\,B \ldots \qquad (23.2)$$

where the fifth letter, "B" in the original sequence has been received as "A".

The receiver on decoding the message (23.2) must now make a decision: either

(i) that the message received is the transmitted message, or

(ii) that the message received is in error and request a transmission.

In this sense the receiver contains decision-making devices. The central problem is to design a system so as to minimise the number of erroneous decisions.

The effect of noise is to modify the sequence transmitted and make unequivocal decision difficult, or impossible to accomplish.

2. Communication System of Class 2.

Here, as it would be with a radar system, the transmitted sequence might be

$$\ldots \text{A B A B A B A B A B} \ldots \tag{23.3}$$

and would be known to the receiver *a priori*. Clearly, there is no question of information transmission.

On encountering a target the transmitted message becomes modified and could be received as

$$\ldots \text{A B B B A B A B A B A B} \ldots \tag{23.4}$$

Here the received sequence (23.4) differs from the transmitted one (23.3) in that the third symbol "A" has been changed to a "B". The receiver correlates the received message with the transmitted sequence (stored in the receiver) and learns data about the target (information). In a way the target betrays itself by unwittingly modifying a given sequence of symbols. It is in this way that the information is extracted from the environment.

SUGGESTED READING MATERIAL

SINGH, J. *Great Ideas in Information Theory, Language and Cybernetics*. Dover, 1966.

KARBOWIAK, A. E. *Theory of Communication*. Oliver & Boyd, 1969.

BECK, A. H. W. *Words and Waves*. World University Library, 1967.

KARBOWIAK, A. E. *Trunk Waveguide Communication*. Chapman and Hall, 1965.

KARBOWIAK, A. E. "Optical Waveguides". In *Advances in Microwaves*, Vol. **1**, Academic Press,London, 1966.

CHAPTER 24

Computer Systems

A. A. Thompson

"For a moment Vashti felt lonely. Then she generated the light, and the sight of her room, flooded with radiance studded with electric buttons, revived her. There were buttons and switches everywhere—buttons to call for food, for music, for clothing. There was the hot-bath button, by pressure of which a basin of (imitation) marble rose out of the floor, filled to the brim with a warm deodorised liquid. There was the cold-bath button. There was the button that produced literature. And there were of course the buttons by which she communicated with her friends. The room, though it contained nothing, was in touch with all that she cared for in the world. By her side, on the little reading-desk, was a survival from the ages of litter—one book. This was the Book of the Machine. In it were instructions against every possible contingency. If she was hot or cold or dyspeptic or at loss for a word, she went to the book, and it told her which button to press. The Central Committee published it. In accordance with a growing habit, it was richly bound."

E. M. Forster
When the Machine Stops, *1909.*

A. INTRODUCTION

Since it was first proposed, in detail, by Eckert, Mauchley, Goldstine, Braiwerd and Von Neumann, at the University of Pennsylvania in 1946 the stored program digital computer has proliferated into widespread use throughout the community of industrialised society, until today the majority of us have some contact or other with computers or the results of their operations. The effect and influence of the computer, on modern society and the lives and daily activities of its people, is increasing at what appears to be an accelerating rate. There is no doubt that this influence will bring many benefits and advantages that would be otherwise unattainable, there is also no doubt that it could bring disadvantages and dangers if used with bad intent—probably the computer's greatest potential for mischievous use would be as an information and reporting system on the life history, religion, associations, political interest, financial status, etc., of each and every member of the community so as to police their activities and freedom. So like most of mankind's tools—nuclear energy, motorised vehicles, drugs, aircraft, printing, radio and television, etc.,—the computer is a two-edged sword, its use for good or bad rests with us.

This chapter describes in broad detail a few of the uses to which computers are put in an attempt to illustrate some of the interesting and exciting things that are being done with them, their universal nature and the fundamental importance that they are assuming in modern society. When one considers that most, if not all, of their uses are still in their infancy it leads us to wonder and marvel at what the future holds for us.

The first stored program computer was made to work at Cambridge University

(UK) in 1949 by M. V. Wilkes and his associates. This device was followed in rapid succession by others at the University of Manchester, the University of Pennsylvania, and the Commonwealth Scientific and Industrial Research Organisation in Australia. These machines were put to work on scientific and engineering computations, and needless to say their use and the variety of functions that they perform in these fields has spread very rapidly up to the present time.

The first computer manufactured for sale, the UNIVAC I, was produced by the Eckert Mauchley Computer Corporation in 1951. The application of computers to business and administrative data processing moved quite slowly at first and it wasn't until the late 1950s to early 1960s, when reasonably priced, more highly reliable machines became available, that computers took on in this area. It was thought at that time that computers could only be applied to routine record processing and statistical tasks, and businessmen and administrators saw little use for computers except for these routine mechanical types of tasks. Today, however, a little over ten years later there are tens of thousands of computers in use by business and government organisations throughout the world. Much of their work is routine, such as processing payroll details, recording customer accounts, preparing and printing invoice details, stock control and analysis, and so on. However, they are also being applied to a wide variety of jobs in the business world, such as economic forecasting, project planning and control, modelling and simulation of the business organisation and its environment and management information reporting and control systems, to mention just a few.

The two most significant developments in recent times are the multi-access, time-shared, interactive computer system, and the combination of telecommunication networks of various kinds with computers to establish networks of computer input/output terminals connected to a central computer complex consisting of either one or a number of interconnected, multi-access time-shared computer systems. These two developments offer to make various forms of computer based facilities available to individual members of the community.

B. TIME-SHARED, MULTI-ACCESS, INTERACTIVE COMPUTER SYSTEMS

1. General

In the earlier days of computers when their use was not so highly organised, the computer programmer could take over the use of the machine for an hour or so to test and debug his programs. He could run his program, correct any errors discovered in that run, then run it again, correct any new errors that may arise, then run it again and so on. In this way, with full use of the machine, the programmer could test and prove his programs in a relatively short period of time.

With the advent of computer operating and control systems and consequently a highly organised production line approach to computer processing, jobs are fed into the system and stored in a queue. The computer control system then takes these jobs one at a time, one after the other, starting with the first in the queue and arranges their execution. Each job has to wait its turn regardless of whether it requires only a few seconds or a few hours of computer time.

Now, while this system resulted in more efficient computer utilisation, the user suffered because he might have to wait a number of hours for his job (which might require only a few seconds of computer time) to be processed and returned to him. This procedure became particularly time consuming when the user was involved in debugging and testing programs—with a job turn-around time of several hours it might take him a week or more to complete a program that he could finish in a few hours if he had direct access to the machine.

To overcome this problem C. Strachey, in 1959, proposed the time-sharing of computers to provide programmers with direct access to the computer once more. Strachey suggested that by taking advantage of the computer's operating speed and by devoting only a small fraction of computer time to each user in turn, on a round robin basis, the computer could in effect be made to service concurrently a number of users.

2. Time-Shared Computer Systems

A typical time-shared computer consists of a number of teletypewriter units connected to a central computer, the computer will also have the usual range of peripheral devices connected to it, such as card reader, line printer, magnetic tape and magnetic disk units. The users communicate with the computer through the use of a teletypewriter unit—they enter information into the system by typing on the teletype keyboard and the computer in turn communicates with the user by printing through the teletypewriter. The computer services each terminal user at regular periods for a fixed interval of time. For example if each terminal user received 20 milliseconds of computer time every second then allowing for computer overheads such as switching between users, the computer could concurrently service about 40 or more terminal users. Such a computer system might have a few hundred terminals connected to it, of which about 40 could be active at any one time; if more than 40 users wanted to use the computer concurrently the computer will ask the newcomers to wait, until some of the active users have finished.

A time-shared system can also be made to time-share its processing between a stream of batched processing jobs and terminal users; in which case it could service only a reduced number of terminals—if for example, it was arranged to spend a maximum of 50 per cent of its time servicing time-shared terminals and the remainder of its time processing the batched jobs, then on the figures quoted above it could only service about 20 terminals at any one time. The combination of batched job processing and terminal processing ensures that the computer is fully utilised at all times, since the computer can spend any spare time that may result from a low level of terminal activity, processing the batched input jobs.

Multi-terminal, time-shared computer systems are given the name *multi-access* systems because they can be simultaneously accessed by a number of terminal users. When active, a terminal and its user are said to be *on-line* with the computer, that is a "communications" link is established between user and computer whereby the user can request information and services and the computer can request information from the user. A very important aspect of this two way communication is that it enables the user and computer to interact with each other, in the solving of problems, or in the search for information, and so on—this has given rise to the term *interactive-terminals*. It should of course be noted and borne in mind that time-shared, multi-access, on-line, interactive-operations are made to operate by means of specially designed programs—the computer does nothing of its own accord; it simply follows programmed instructions.

With time-shared systems the terminal user sits at his teletype and enters information one line of type at a time—as he finishes each line he has to press the teletype "carriage-return" key to start a new line, a carriage-return signal notifies the computer that a new line of input is ready for transmission. When the computer next services that particular terminal it will accept the new line of input and process it. Whereas it will usually process each line of input in a few milliseconds or so, the user will usually take several seconds to type each line (and much longer if he has to scratch his head and think about what he is typing), during this time the computer will be busy servicing other terminals or processing batched jobs.

A common use, and the original objective, of time-shared computers is for program development and testing. To serve this purpose time-shared systems incorporate on-line interactive translator programs. The programmer sits at his terminal and types his program—usually one program statement per line of type. The translator program is designed to take each completed statement and analyse it to check if it is correct. If it is, the programmer simply continues on with the next statement of his program. If the program statement is incorrect the interactive translator immediately notifies the programmer by typing details of the error on the next line on the teletypewriter. The programmer then types a corrected version of the statement, which is accepted by the translator, after which he proceeds to input the next program statement. This system has the very significant advantage that the syntax of the user's program is being checked and corrected statement by statement, which leads to more rapidly developed error free programs.

One can imagine the overall benefits obtained from such a system with 40 or more programmers each using a time-shared on-line terminal to prepare his program. When each programmer has completed the input of his program (or some self-contained program section) he can have it executed by typing an appropriate command to the computer. His program (section) will be executed and the results printed on his teletype, in this way the programmer can perform repeated tests on his program and prove its correctness in a relatively short period of time. Before leaving this topic it should be pointed out that time-shared computers require specially designed translator programs which will accept programmed statements one at a time from terminal devices and which will check these statements for correct syntax, notify the programmer about any errors and subsequently accept the corrected statements.

Apart from facilitating programming activities time-shared, multi-access, interactive computer systems offer a number of other important advantages. For example, modern day computer systems have the capacity to store very large volumes of information. These may include such things as files of insurance company customer accounts and policies, or files of bank accounts, files of student or personnel records, or files of airline seat reservations, files of experimental data and results and so on. All of this information can be made available to accredited persons through a time-shared terminal. This means that a large number of people (e.g., bank officers, seat reservation personnel), each with access to a terminal device, can have simultaneous access to particular files of information (and can, therefore, provide customers, and others, with on the spot up-to-date information). Furthermore these accredited persons can add information to particular files from their terminals and thus for example update a customer's bank account on the spot, or enter a new airline seat reservation, and so on.

Another important feature of time-shared systems is that they are provided with libraries of programs which are available to all users. More importantly perhaps users can contribute programs to the library through their terminals, and thus, over a period of time the library is made more comprehensive, individual programs are made more reliable and generally applicable as a result of their constant use and improvement by a variety of users.

3. The MIT Time-Sharing System

The Computation Centre at the Massachusetts Institute of Technology quickly took up Strachey's time-sharing proposals and by November, 1961 the centre had implemented a prototype system; since that time MIT has continued to research and develop time-sharing systems (Fano and Corbato, 1966). Their present system can service about 30 simultaneous users through teletypewriter devices which are installed in various offices and laboratories on the MIT campus and in the homes of some of the staff members. About 160 of these terminals are connected directly to the computer system. In addition to this the MIT system is connected to the teletype communication networks of the Bell System and Western Union, making the computer available to thousands of terminals both in the U.S.A. and overseas countries. This enables the MIT time-shared computer system to be used by a wide-spread community with up to 30 users being serviced concurrently.

The MIT system contains a large store of information, equal to about 100 million computer words, comprising supervisory, utility and processing programs, programming language translators, input/output programs, libraries of sub-routines, files of data, and so on. The basic component of the system is a set of about 100 programs, each of which may be activated by a specific command issued by a user through a teletypewriter. These programs control such things as interactive communications, control and scheduling of computer processes, language translators, library supervisor programs and so on. In addition to libraries of generally available data files, programs and routines, the system also contains private user program libraries and data files, access to which is restricted to one particular user or a small group of accredited users. At the heart of the system is a complex of programs called "the supervisor" which coordinates and supervises the operation of the whole system including operation of the many peripheral units, terminal communications, allocation of processing time and system services, scheduling of users' requests, transferring control of the central processor from one user to another for present periods of time, moving programs and data in and out of the main storage unit, and managing public and private program libraries and data files.

When a terminal user wants to use the system he begins by typing the command LOGIN followed by his name and project identification. The machine answers by printing the time of day on the terminal and the user then has to type a password to enable him to gain access to the programs and data stored under the project identification. If the given password does not check with that stored against the persons' name and project number, or if he has exhausted his monthly allowance of time on the computer, then the machine prints a message stating that access is not available and why. This also happens if the computer is being used to full capacity, in which case the user simply tries again after a short interval of time.

If the user does gain access he may then proceed with further commands, for example, he may request an account of the amount of time and storage space that he has used up to date. The user may then proceed with his work in which he may write a new program, or modify an existing program to perform say a mathematical analysis on some data that he enters through his terminal, or a solution to some kind of engineering problem, perhaps in an interactive way.

The user can also command the system to print out the list of programs and data files listed in his file directory. He may also give a command authorising the system to allow other specified users access to one or more of his programs or data files, and conversely may gain access to other public or private files that he is authorised to use. Another convenient feature allows one user, or the system manager, to send messages to another user; on logging into the computer a user may be informed by the machine that there is a message in his "mail box" and the computer will print the message at the user's terminal when the user requests it.

Time-shared, multi-access systems are being applied in all sorts of fields ranging through science, business, engineering, education and medicine. They form the basis of computer networks and utilities which offer the potential of making computers widely available to all members of the community.

C. COMPUTER NETWORKS AND UTILITIES

1. General

Our modern community is serviced by a complex of telecommunication systems consisting of telephone, telex, radio, television, microwave, etc. networks which enable us to communicate information, in various forms, at the local, national and international level. This telecommunication complex is being put to a new use—the communication of information to and from computers.

Teleprocessing—the combination of telecommunications and computers—is a very powerful facility which enables a community of users, separated geographically by long or short distances, to use a central computer for their processing needs and to gain rapid access to files of information stored within a central computer system. The telecommunication links which carry information to and from computer systems are referred to as *data-links*. The central computers are usually multi-access, time-shared computer systems.

Teleprocessing systems first came into widespread use in the seat reservation systems developed by airline companies in the early to middle 1960s. These systems consisted of a large centralised computer connected by telecommunication links to offices in those major cities, both local and overseas, serviced by each particular airline company. The central computer stores information about the reservation and availability of seats on all aircraft flights scheduled by the company. This information could be interrogated from any office linked to the computer and accurate details of seat availability could be given on the spot to prospective passengers. The passenger could make a reservation and have it confirmed immediately, details of the reservation would then be communicated to the central computer which would enter those details into its files. Because all information about flight schedules and bookings was held centrally it could be kept accurate and up-to-date, thus avoiding the problem of duplicate bookings and enabling flights to be more fully booked. The gain in revenue from the resulting more intensive utilisation of the aircraft easily covered the cost of the world-wide telecommunication networks linking the central computer to the field offices.

Many other organisations, such as banks, warehouses, public utilities, transportation, and other large companies with many offices and establishments spread over a wide area, followed the airline companies in utilising the advantages of teleprocessing systems.

2. Data Links

When the requirement for computer data transmission first arose the telephone network was adapted for the purpose and is still the most widely used medium for data transmission (Pierce, 1966). However, it suffers two disadvantages. Firstly, it is designed for the transmission of analogue speech signals and not digital signals; secondly, the telephone network has only a narrow, low-frequency bandwidth which is adequate enough for speech transmission but inadequate for the high speed digital data transmission usually required for computer communications. Nevertheless, the telephone network provides a very valuable, economic, digital-data transmission network.

To enable transmission over telephone speech circuits, digital data signals are converted into analogue form for transmission and reconverted into digital form after transmission by electronic devices known as modems (modulators/demodulators). The most elementary of these modems enables data transmission rates up to 300 bits per second (i.e. about 40 eight-bit characters per second). More elaborate modulation/demodulation techniques enable transmission rates up to 2400 bits per second and current research and development indicates the possibility of transmitting perhaps up to 9000 bits per second over a single telephone circuit or line.

These transmission rates are not high enough for many applications—to overcome this problem techniques have been developed which enable higher transmission rates to be achieved using groups of telephone lines in parallel. In the U.S.A. these techniques enable transmission rates up to 19,200 bits per second over six parallel telephone lines, up to 50,000 bits per second over 12 lines, and up to 230,000 bits per second over 60 lines. For data transmission rates higher than this it is possible to use a standard television transmission trunk line for rates up to several million bits per second, and research and development is under way to provide even higher transmission rates.

3. Computer Networks

Computer networks take on a variety of forms of which three common forms are as follows. Firstly, the time-shared multi-access computer network described in the previous section, in which the user terminals are typically low speed teletypewriter devices which are connected to the central computer through the public telephone network. When a user wants to be connected to the central computer he dials the appropriate telephone number, when the computer responds to his call the user throws a switch which connects his teletypewriter, through a modem unit, on-line to the computer.

Another form of the time-shared, multi-access computer network is one in common use by banks, airline companies, wholesale distributors, and such organisations, where the system is dedicated to some specific task such as reservations, stock control, or account queries and updating. In this system a single high speed data transmission line serves several groups of terminals situated at different locations (e.g., bank branches, reservations offices). The terminals are connected to a concentrator unit which is connected, in turn, to the high speed line through a modem unit. The concentrator units collect the data being sent from local groups of terminals into "packages" and coordinates the transmission of these packages, with packages from other concentrators, to the main computer. The concentrators also handle packages of information from the central computer and deliver the contents to the appropriate terminals. Usually small to medium sized computers perform the task of concentrators, in addition these computers commonly perform routine data processing tasks such as maintaining local accounts or stock inventories, in which case they may only transmit summary information and/or the larger processing tasks back to the larger central computer for processing. Computer systems of this form are also to be found in scientific research laboratories or industrial situations, where a number of small computers, each controlling and/or analysing the results from some experimental or data logging apparatus (such as a gas chromatograph or a wind tunnel, or process heat and flow measuring instruments) are connected to a large central computer. In this sort of situation the small computer may transmit only summary information to the central computer for analysis and filing, or perhaps when the small computer wants some major processing task performed it will send the appropriate data and processing request to the central computer which will perform the required task and transmit the results back to the small computer.

A a third type of network is one in which a number of large, usually time-shared computers are interconnected by very high speed data transmission links. The computers may be situated in different parts of the state, or country or even in different countries. With this type of network when one computer is overloaded with work it can transmit jobs to one of the other computers which is not operating at full capacity, the jobs will be processed and the results sent back to the source computer which then distributes the results to the appropriate user. Also with this type of system the central computers can transmit files of information to one another on request, and hence if a job or request is entered into one computer, which does not have the relevant information, that computer can interrogate the other computers and obtain the required information if it exists anywhere in the system. An example of this type of system is the ARPA (Advanced Research Projects Agency) network currently being put into service in the U.S.A. This network will interconnect a number of large multi-access computer systems situated at universities and similar establishments. This network comprises a communication network employing small computers known as Interface Message Processors joined together by fairly high-speed data links capable of carrying information at 50,000 bits per second.The multi-access computers, known as "hosts", communicate with each other through the Interface Message Processors. In this system a user on-line to one of the host computers can be connected indirectly to some other host computer to obtain services or information that is not available on his own host machine.

4. Computer Utilities

The combination of large multi-access time-shared computers, operating singly or linked together, and telecommunication data links offer the prospect of computer utilities which will service the community with computer facilities on a scale comparable to the services currently provided by electric power, telephone, gas, water, radio and television utilities. Systems such as this have already been achieved on a relatively small scale, the MIT time-sharing system and the ARPA system to mention only two. Future computer utilities hold forth the potential of making the computer based services available to all sections and even all individuals of a community.

D. COMPUTERS IN ENGINEERING

1. General

Since their inception computers have been used for a wide variety of engineering applications. Initially the main use of the computer in this area was for the performance of complex and/or repetitive engineering calculations. In recent times however, the computer has been used increasingly as an engineering design tool. This trend is the result of several factors; firstly, increasingly widespread experience and knowledge of the use of the computer for solving engineering problems; secondly, the advent of the multi-access, time-shared computer system with its many advantages, of which user-computer interaction is probably the most important from an engineering design/problem-solving point of view; and thirdly, on-line graphical input/output terminals which enable the engineer to communicate with the computer by means of drawings and graphical symbols.

The interactive use of a computer for design problems is referred to as computer-aided-design. At the heart of a computer-aided-design system is a program designed for a particular engineering problem (such as electronic circuit analysis, or mechanical structure analysis) and designed to interact with the user through an on-line terminal. The user may begin by supplying the program with a basic description of the design problem including component configurations and parameters. The program will then perform a mathematical analysis of the specified circuit or structure and present the engineer with the results; after studying these the engineer may alter some component parameters, or add new components, etc., and then instruct the computer to repeat its analysis. The engineer will proceed to use the computer in this way until a satisfactory design is achieved. Then if the computer has been supplied with the appropriate programs the engineer may instruct it to produce an engineering specification of the design, including drawings, parts lists and performance data.

The importance of interactive graphical terminals stems from the fact that pictures and diagrams are basic to the way in which engineers and designers think about their problems and also because a great deal of information can be presented in a concise and easily assimilated form through the use of graphics. The use and application of interactive graphics is not restricted of course to engineering but exist throughout the whole range of scientific and engineering computing and is extending into many areas of commercial computer applications.

2. Graphical Input-Output Terminals

Computer graphic terminals display computer output in pictorial, and alphanumeric, form on a cathode-ray-tube screen similar in appearance to a TV screen. Information may be fed into these terminals either by means of a keyboard for alphanumeric data or by means of a *light-pen* device which is used to draw diagrams on the display screen or to identify, to the computer, specific graphic symbols or specific parts of a drawing on the display screen by pointing to them and touching them with the light-pen. The terminals are designed to detect the coordinate position of the light-pen as it moves across the screen and hence a drawing is stored in the computer as a set of coordinate point values. The computer displays a drawing by having a dot of light displayed at each coordinate point of the drawing. A picture is held on the screen because the computer redisplays it at a constant rate, usually about 20 to 25 times per second to avoid display flickering effects.

Provided the computer is fitted with the appropriate programs the user can proceed to delete, modify or add elements to the displayed graphic, and specify what these parts are, so the computer can take them into account when processing the problem represented by the graphic. The user can also command the computer to store the coordinate data representing the graphic in one of the user's data files so he can call the graphic up again when next he uses the computer.

Another interesting and valuable use of computer graphics is the generation of a sequence of pictures representing the solution to a specific problem for successive values of some variable, such as time, commencing with some specified reference value. In effect this produces a moving picture which displays the dynamic behaviour of the engineer's design or simulated model of some physical system under investigation. For example, the flow of water over a dam spillway following the opening of the dam sluice gates, or the pattern of air turbulence around an aerofoil of varying geometry.

The original developments in generalised computer graphics were carried out by Ivan E. Sutherland at the Massachusetts Institute of Technology, which culminated in Sutherland's Sketchpad system, completed in late 1962.

3. Computer Aided Design

When Sutherland began work on his Sketchpad system, the computer terminal that he had consisted of a cathode-ray-tube screen and a light-pen, these two facilities had existed for some time but little had been done to exploit their possibilities. It is interesting to repeat some observations about the Sketchpad system (Coon, 1966).

> Sutherland set out to develop a system that would make possible direct conversation between man and machine in geometric, graphical terms. In the course of the development of Sketchpad, he would invite people in to try out his system so that he could observe their reactions. Claude E. Shannon, Sutherland's advisor on his doctoral thesis, wanted to perform a geometric construction. Rather than work out the construction on the console screen, Shannon automatically turned to pencil and paper to make a preliminary sketch. This came as a disappointment to Sutherland, who intended the system to be so congenial to the user that it would not intrude on his thought processes. He thereupon disassembled his program and rewrote it. It went through several such revisions, and it stands today as a classic of well-considered human engineering.
>
> Using Sketchpad, I was able in one evening to set up and experiment with the following problems and constructions.

1. Evaluate a cubic polynomial equation. There is a simple geometric construction for polynomials of any degree and for real and complex values of the various terms. By manipulating the x variable with the light-pen I could cause y to vanish, thus "solving"the cubic equation.
2. Construct a general conic section, or second-degree curve, using the basic principles of projective geometry.
3. Draw and set in simulated motion a "four-bar" mechanical linkage. Although such linkages are simple in outward form, their analysis is troublesome and is still the subject of investigation.
4. Draw a pin-jointed structure (such as a bridge), displace one of the joints (as if loading the structure) and observe the "relaxation" that is thereupon carried out by the computer to minimise the energy of the system, thereby simulating the actual deflection.
5. Plot the potential field typical of the flow of an ideal fluid within a region of specified shape.

Now it is clear that these five problems are not much related to one another and that a conventional computer program written to deal with one of them would not be of the slightest use for any of the others. The computer did not contain a set of programs—one for each of the problems. Instead it had a flexible and quite general capability for performing a set of primitive geometric constructions and for applying a set of primitive geometric constraints. The computer played the role of an intelligent but innocent assistant and together the machine and I set up and solved the problems.

Computer-aided-design systems have been developed for a wide range of engineering applications. A computer-aided circuit design system, for example, might operate as follows. The computer would be supplied with a specially designed, interactive, circuit analysis program together with a data file containing specification and performance data on a range of electronic elements such as diodes, transistors, resistors, capacitors and so on. The circuit designer sits at a CRT graphical display with a light pen and a teletypewriter. The designer sets up a circuit schematic by pointing the light pen at a component selected from a list of components displayed on the screen and by then pointing the light pen to where he wants that component positioned in the schematic diagram displayed on the screen, where appropriate he enters the values for the component using the teletypewriter. The designer proceeds in this way, step by step, to complete a schematic diagram of the circuit that he wants to study. Having established his circuit, the designer will proceed to specify circuit input voltage or current signals either by defining them analytically, by drawing their waveform, or by specifying the name of some standard waveshape. Having done this the designer will specify the type of analysis and results that he wants from the computer program, such as a plot of voltage versus time or frequency, or the relative variation of voltages or current at specified points throughout the circuit. Upon obtaining the analysis results the designer may proceed to modify the circuit adding, deleting and altering component values and then get the computer to analyse the modified circuit. The designer will proceed in this way in a continuous cycle of analysis and modification until he arrives at a circuit having the performance characteristics that he requires.

A computer program for the interactive design of optical systems using a graphics terminal (Green, 1970) has been developed by the Beckman Corporation in California. This program enables a design engineer to specify, modify, test and adjust a computer simulated model of an optical diffractometer.

He can specify design parameters for mirrors and lenses, specify the groove spacing, the shape of the grating and the required light source. The computer will then calculate the path of light rays passing through the simulated optical system, and then display the output light beam on the CRT screen as if the display screen were a ground glass screen placed across the output of the diffractometer. The screen, in effect, can be stepped into or out from this simulated optical system displaying the resulting change in shape of the output beam, which should focus to a fine line for each wavelength when using an ideal light source. This interactive design system enables the engineer to test several different ideas and design methods in a short period of time. This could well have taken him many months to investigate under the old system of building and testing laboratory models.

The aerospace and the motor car industries are two major users of interactive, computer-graphics design systems, particularly for the design of motor car and aircraft body surfaces and shapes. In the motor car industry interactive graphics systems are used for jobs ranging from the checking of body surface measurements to the design of body-press tools and dies. In many cases the interactive design system is coupled to a large automatic drafting machine which can produce full scale detailed drawings of body surfaces.

E. COMPUTERS IN EDUCATION

1. General

The use of computers for education (Suppes, 1966) and training purposes is generally referred to as Computer Assisted Instruction (CAI). The most important aspect of computer assisted instruction is its potential capability to provide each and every student with a level of individualised instruction.

It is generally recognised that the most effective teaching method is personal tuition. The more that an educational situation can be adapted to the particular needs of individual students—each of whom has his own level of ability, and rate and style of learning—then the better the student's prospects of successful learning. The cost of providing sufficient teachers to establish anything like individualised instruction is economically prohibitive, however, the use of computers to assist teachers to achieve this objective appears to be a practical proposition. The computer could be used to supervise the simpler learning processes of each student, thereby releasing the teacher to deal with the special problems of particular students and to attend to more advanced types of training which cannot be provided by the computer. It appears that the range of subjects that may be "taught" by computer aided instruction is virtually unlimited. Programs that have been developed so far tend to concentrate on simple arithmetic or linguistic comprehension, however, programs have been prepared in botany, pathology, electrical engineering and many other subjects.

A major factor in the individualised tuition offered by computer based systems is that the learning program can be made up with many alternative strands of instruction in a particular subject and designed so that the computer monitors a student's performance and decides which strands are most suitable for each student. For example, if a student is weak in some particular aspect of the subject the computer will detect this and provide additional strands of instruction at appropriate levels. If the student is still unable to achieve a satisfactory grasp of that aspect of the subject the computer will notify the teacher who can then give the student special attention and tuition.

The major problem with computer aided instruction is its cost. Computer systems and terminal equipment are still too expensive, however, their cost is reducing and in the not too distant future it should come down to an acceptable level. In addition to equipment costs is the cost of preparing and programming courses of instruction, which is very high largely because the course designer has to plan for all eventualities—not only must he plan for correct responses but also for all the likely incorrect replies by the student and plan what course of action should be followed for each and every case. Apart from these factors research is needed into the effectiveness of various forms of computer aided instruction, and even into the basic learning and teaching processes, to establish a body of knowledge and experience upon which widescale computer assisted instruction can be successfully founded.

2. Computer Instruction Terminals

Personalised instruction involves a great deal of interaction between student and teacher, and so if a computer is to be used to assist in teaching it must be able to interact in a similar way. The on-line computer terminal is used to provide the basis for interactive communications between the teaching program and the student. These terminals are usually connected to a multi-access, time-shared computer system which can service a number of students simultaneously.

The terminals that are used take on a variety of forms ranging from the simple teletypewriter, which is the most common, to graphical CRT displays with keyboard and light-pens, to quite elaborate terminals which may also have computer controlled film or slide projectors and even a computer controlled audio device which the computer uses to speak to the student (these consist of recorded words on small sections of magnetic tape and the computer is programmed to select those words required to make up any particular sentence). The use of spoken messages is desirable at all educational levels, but it is particularly desirable for younger children who tend to learn at least as much by ear as they do by eye.

The desirable features of a terminal for computer aided instruction are that it should be programmed for interactive operations, it should consist of a cathode-ray-tube display for alphanumeric and graphic data, a keyboard, a light-pen which amongst other things enables students, particularly younger ones, to respond to questions by pointing the pen to touch the selected items on the screen, and a loud-speaker or earphones for computer generated audio messages. Other visual displays such as motion pictures or slide projectors operated under computer control are useful additions.

3. Computer Assisted Instruction

At the present time computer assisted instruction systems seem to fall into two broad categories; "drill and practice" type instruction systems which supplement the teacher's instruction, and "tutorial" systems the purpose of which is to take over from the teacher part of the task of the instruction of new concepts and principles. The following descriptions are based on work done at Stanford University and reported by Patrick Suppes (1966).

Research and experience have shown that students need a great deal of practice in order to master the algorithms and basic procedures of subjects like arithmetic and abstract algebra. A good deal of practice seems to be necessary for students to learn to master the basic algorithms with speed and accuracy. With the traditional "drill and practice" methods of instruction each and every student is presented with the same set of exercises from the same book for practice purposes. The major advantage with computer-assisted, drill and practice, systems is that once a number of alternative strands of drill and practice exercises have been set up the computer can be programmed to present students with exercises of varying degrees of difficulty and to select the appropriate level of difficulty for each student according to his past performance. In this way each student is presented with a set of exercises that are continuously tailored to his particular ability and rate of progress. The experience gained to date indicates that students achieve significantly higher levels of proficiency through the use of computer assisted, drill and practice instruction systems.

A system developed at Stanford University operates in the following way. The student sits down at a teletype terminal and types his name, the computer then searches a master file to obtain the student's record and the set of exercises to be presented to the student, which is determined on the basis of his performance the previous day.

> The computer types the first exercise. The [typewriter] carriage returns to a position at which the pupil should type his answer. At this point one of three things can happen. If the pupil types the correct answer, the computer immediately types the second exercise. If the pupil types a wrong answer, the computer types WRONG and repeats the exercise without telling the pupil the correct answer. If the pupil does not answer within a fixed time (in most cases 10 seconds), the computer types TIME IS UP and repeats the exercise. This second presentation of the exercise follows the same procedure regardless of whether the pupil was wrong or ran out of time on the first presentation. If his answer is not correct at the second presentation, however, the correct answer is given and the exercise is typed a third time. The pupil is now expected to type the correct answer. but whether he does or not the program goes on to the next exercise. As soon as the exercises are finished the computer prints a summary for the student showing the number of problems correct, the number wrong, the number in which time ran out and the corresponding percentages. The pupil is also shown his cumulative record up to that point, including the amount of time he has spent at the terminal.
>
> A much more extensive summary of student results is available to the teacher. By typing a simple code the teacher can receive a summary of the work of the class on a given day, of the class's work on a given concept, the work of any pupil and of a number of other descriptive statistics I shall not specify here. Indeed, there are so many questions about performance that can be asked and that the computer can answer that teachers, administrators and supervisors are in danger of being swamped by more summary information than they can possibly digest. We are only in the process of learning what summaries are most useful from the pedagogical standpoint.

Computer assisted tutorial type instruction systems are designed to teach students basic concepts and principles. These systems have been developed for subjects ranging from numerical analysis, basic arithmetic, and linguistic concepts to electrical engineering subjects. To illustrate, a system developed at Stanford University (Suppes, 1966) to teach young children the concepts associated with the words "top" and "bottom", and "first" and "last", and so forth, operates as follows.

1. The child uses his light pen to point to the picture of a familiar object displayed on the cathode-ray-tube screen.
2. The child puts the tip of his light pen in a small square box displayed next to the picture. (This is the first step in preparing the student to make the standard response to a multiple-choice exercise).
3. The words FIRST and LAST are introduced. (An instruction here is spoken, by the computer, rather than written. FIRST and LAST refer mainly to the order in which elements are introduced on the screen from left to right).
4. The words TOP and BOTTOM are introduced. (An instruction to familiarise the child with the use of these words might be PUT YOUR LIGHT PEN ON THE TOY TRUCK SHOWN AT THE TOP).
5. The two concepts are combined in order to select one of several things. (The first instruction might be PUT YOUR LIGHT PEN ON THE FIRST ANIMAL SHOWN AT THE TOP).

Again with such a computer assisted instruction system, tuition can be tailored to meet the learning characteristics of individual students.

While the example described is an elementary one the principles involved can be extended to provide tuition in a wide range of subjects such as solid-state physics, numerical methods, organic chemistry, transistor and solid state devices, circuit theory, thermo-dynamics, and so on. Hence, you will appreciate that computer based tuition systems offer rich potential for teaching purposes.

F. COMPUTERS IN INDUSTRY

1. General

One of the uses of computers in industry is for the control and automation of industrial processes. (Of course not all "automated" industrial plants are controlled by computers—it is possible to construct complex control systems based on continuous monitoring and feedback loops, without including computers, and this is the way in which most automated plants have been made up to recent times.)

In some industrial applications computers are used to carry out continuous calculations on data obtained from process monitoring instruments and on the basis of these calculations they inform a human operator about what needs to be done and adjusted to maintain correct process operations. In other cases the computer may be used to monitor and control one or more sub-processes, and in still other cases the computer may be used to control the routine operations of the whole production plant as is common in petroleum and chemical plants.

2. Monitoring and Control of Steel Production

Computers are playing an increasingly important role in the production of iron and steel. The behaviour of blast furnaces is somewhat unpredictable; not enough has been known quantitatively about what happens between the charging of the furnace—when iron ore, coke and limestone are fed in at the top—and the time, some hours later, when molten iron is cast from the bottom.

In recent years many steel companies have been involved in efforts to understand more precisely the complex physical and chemical processes that take place in the blast furnace. A great deal has been learnt in the past ten years or so by installing various types of sensors in the furnaces and by feeding the output of these sensors into a computer for detailed analysis. This process has lead to a more detailed understanding of blast furnace processes. For example, one steel company (*Computing Report,* 1970a), the Jones & Laughlin Steel Corporation in the U.S.A., has installed an array of instruments and sensors in one of its blast furnaces, including infra-red and thermal conductivity gas analysers, neutron gauges for determining coke moisture, dew cells for measuring gas moisture, and a wide variety of thermocouples, flow meters, and pressure gauges. Every 30 seconds these instruments monitor some 70 process variables including the pressure, temperature and composition of gases; the temperature, pressure and moisture content of the hot blast; the weight of the iron ore; the weight and moisture content of the coke; the stack temperature; the temperature and composition of the molten iron; and the composition of the slag.

According to Dr. Wakelin of the J & L Steel Corporation: "The prompt processing of this data by a computer and the immediate relay of information back to the furnace operator should enable the operator to make faster corrections for any irregularity in furnace operation. In the past he has had to base his decisions on limited information about what was happening inside the furnace. In fact he has had to depend entirely too much on analyses of the iron after it was cast, and sometimes he did not learn about an irregularity until hours after it had occurred. Now he will get much more complete data on all stages of the iron making process, enabling him to determine when the operation is not proceeding as desired and what should be done to bring it back into line."

3. Monitoring and Control of Nuclear Generating Plants

Nuclear generating plants are taking an increasingly important role in the generation of electric power (*Computing Report,* 1970b), it is estimated that by the year 2000 half of the electric power generated in the U.S.A. will be generated in plants powered by nuclear energy. A major advantage with nuclear fuel is that it is relatively cheap and a small amount of it can generate a vast amount of electricity—a cubic foot of uranium has the energy potential of 1.7 million tons of coal, 7.2 million barrels of oil, or 32 thousand million cubic feet of natural gas.

The nuclear plant at the Connecticut Yankee Atomic Power Company, Connecticut, U.S.A. consists of a reactor core which is housed in a steel vessel 41.6 feet high with walls 10.5 inches thick. Forty-five rods made of silver, indium and copper control the fission process by absorbing neutrons produced by the nuclear reaction. The rods are divided into four banks designated A, B, C and D. The rods in banks A and B, called the control rods, are lowered into the reactor core to decrease the rate of fission and are raised to increase it. The rods in banks C and D, called the shutdown rods, are used only to shut down the reactor completely.

The reinforced-concrete walls of the reactor and steam-generating complex are 4.5 feet thick, and the dome of the structure is 2.5 feet thick. Within the reactor, water is heated by fission to about 570 degrees Fahrenheit and is kept at a pressure of 2000 pounds per square inch to prevent boiling. The water circulates through a closed system of pipes called the primary loop, which passes through the steam generator. There, heat from the water in the primary loop vaporises water in a secondary loop. The steam produced in the secondary loop drives the turbine that is coupled to the plant's electrical generator.

Monitoring the operation of a nuclear generating plant is a complex task, necessitating the use of a digital computer to monitor and analyse critical operating parameters on an on-line basis, so that plant engineers are provided with plant status information 24 hours a day, 7 days a week. The computer used at the Connecticut plant is an IBM/1800 computer with 24,000 words of main core storage and two magnetic disk drive storage units, having a time-shared operating control system.

The 1800 regularly monitors 317 analogue signals and 237 digital signals. The analogue inputs come from 48 thermocouples that measure temperatures in the reactor core, from linear-voltage-differential transformers that indicate the position of each of the 45 reactor control rods, and from other instruments that measure core flux input, system temperatures, and so on.

Digital inputs come from contact sensors that indicate equipment status, such as whether a pump is on or off. The computer system scans 112 digital inputs every 12 milliseconds and compares the new input with that from the preceding scan. Should there be a change, a sequence-of-events message informs the plant operators about the time of the reading and the kind of change—high or low, on or off—and it identifies the piece of equipment involved.

Every ten seconds the computer system scans the analogue signals and stores the readings for subsequent calculation of one-minute averages. These values are later converted to five minute averages. This information is printed out in report form every hour on the hour, summarising plant operating conditions for the operators and engineers.

Of the 317 analogue monitoring signals, 182 have been assigned alarm limits—high, low or both. These signals are scanned every minute and any that exceed specified limits are immediately scanned again. If the signal has exceeded the assigned limit—for example, if a steam generator's water level is found to be outside of the preset high and low limits—a descriptive alarm message is printed in red on the message/trend printer, an alarm horn sounds in the control room and a message light flashes on the main control board. The alarm-scan program automatically issues a return-to-normal message when a point returns to a value within its alarm limits.

Thus the 1800 computer system provides a total plant monitoring system—in effect, it puts the entire plant at the operator's fingertips, and it provides continuous data for performance evaluation.

4. Control of Processing Plants

The control of chemical and petroleum processing plants are two areas in which computers are used to achieve very high levels of automation. Part of the reason for this is that the process ingredients are usually in liquid or powder form and hence their transfer is relatively easy to monitor and control and part of the reason is that the processing is performed by specially designed chemical plant to the exclusion of difficult to automate manual processing type operations.

The variables in a chemical process—temperature, pressure, flow, viscosity weight, volume, pH-factor, valve settings, and many others—are interrelated in a complicated way. Computers are used to monitor variable information from many measuring stations located at strategic places in the process plant, to perform the necessary calculation relating the variables to the particular process, and to send appropriate control signals to control devices—such as valve controls which regulate flow rates, heating controls, controlling temperature, and so on—in such a way as to maintain optimum operating conditions for optimum production rates and quality of product. Since there is essentially no limit to the complexity of the information with which a computer can deal, industrial engineers can now establish detailed analyses and control of plant processes so as to optimise these processes in a way that previously has not been possible.

5. Control of Machine Tools

Another industrial application of computers lies in the numerical control of machine tools, such as lathes, milling machines, drilling machines and so on (Coons, 1966). These machines employ various types of cutting tools to form metal parts of various shapes from stock material. The cutting tool is moved so as to cut some contoured shape out of sheet metal or heavier stock; conventional machine tools require the constant attention of a highly skilled operator particularly if the metal part is to have some complex two or three dimensional shape. Special machine tools have now been developed which can be controlled by a computer, the computer is supplied with a detailed numerical description of the metal part to be produced, which it uses to continuously position the cutting tool throughout the production of that part. The computer control of machine tools has resulted in high levels of precision and reproducibility of even the most complex shapes, and many parts can now be machined that would be prohibitively expensive to produce if a human operator had to control and monitor the operation of the machine tool.

Computers are also used in the production of computer circuit boards. At IBM for example the production process starts with an engineer and a graphical input/output terminal. The engineer uses a light-pen to draw a schematic circuit diagram of a required circuit and to draw a layout plan for the circuit board. The computer is programmed to take this detailed information and produce detailed production drawings for the circuit board. At a later stage a computer is used, for example, to control a drilling machine with a 24-spindle drill, which drills connection holes in the printed circuit board at rates up to 6000 holes per minute, the computer also controls an electron-beam "hole-tester" that spot tests hole diameters. At a still later stage in the production process a computer is used to test the completed circuit board, automatically performing hundreds of different tests for short circuits, impedance measurements, and logical functions on each board.

G. COMPUTERS IN MEDICINE

1. General

Computers are being put to work in a variety of ways in the medical field; they are being used by hospital administrations to automate the setting up and processing of administrative and medical records and accounts; they are being used to automate the operation of medical testing apparatus and the analysis of the results obtained from these instruments; they are being used for the continuous monitoring of the condition of patients, particularly those in intensive care wards; and they are being used to assist doctors in the diagnosis of illness.

2. Medical Administration

Hospitals are employing computer systems to improve their efficiency, to reduce clerical work loads and to minimise operating costs (*Data Processor*, 1969). Bed assignments, meal requirements, courses of medication, X-ray orders, surgery schedules, insurance account billings and outpatients' bills are just a few of the many procedures performed by computers.

Modern day hospitals are designed to accommodate communication lines, terminal installations and data processing facilities—the Loyola University Hospital in Maywood, Illinois, U.S.A. for example has installed a multi-access, time-shared IBM/360 computer system with remote terminals positioned at various locations throughout the hospital.

When a new patient is admitted the admissions clerk uses one of these terminals to set up a patient medical service record and an accounting record in master files held in the computer. Also using the terminal the clerk makes enquiries about room and bed availabilities; the computer prints out a list of available beds—based on the patient's preference (private or semi-private room) and the nature of his illness. Next the computer alerts a nursing station that the patient is coming.

During the patient's stay, the system is constantly handling information related to his care. It follows through for example on doctors' orders for medication by printing a patient's medication schedule at hourly intervals at the various nursing stations, nurses must then confirm both the receipt of the medicines and the fact that the patient received them. The same procedure is applied to lab tests and X-rays, the computer schedules these and prints out a daily schedule of tests and X-rays. The computer continues to print out reminder notices, until such time as a nurse notifies the system that the requirements have been completed.

The computer system is programmed to check against errors in treatment; each order entered into the system is compared with a comprehensive set of standards, any order that is not consistent with these standards is rejected, such as an overdose of drugs, and the originating terminal station is notified for corrective action. While the system performs these functions, and many others, it also keeps track of accounting information. Every morning the system produces an up-to-date account for each patient. It also produces a daily report for each of the hospital's resident and visiting doctors. The report lists every patient being looked after by the doctor, how long the patient has been in the hospital, every expiring medication order for the next 24 hours and other relevant information. The use of computers frees nurses and doctors from time consuming paper work and enables them to devote more time to the care and treatment of patients.

3. Medical Laboratories

Modern science has provided the medical scientist and technician with a wide range of equipment for performing medical tests, the present day trend is for this equipment to be connected to a computer and hence to automate its operation and the analysis of the results obtained from it.

For example, automatic blood analysing devices and computers have been combined to automate blood analyses. The computer performs mathematical and statistical processing on the output from the blood analyser, stores this data for future reference (perhaps for wide scale analyses) and prints the results. If a multi-access computer is being used then the results would be stored on the patient's medical record and printed out at the patient's nursing station. Because the computer provides rapid economic blood analyses the condition of a patient's blood can be assessed at regular intervals thus providing doctors with valuable information about trends in its condition, indicating for example whether or not a patient is responding to a particular course of treatment.

In the past it has required two or three days of tests including X-rays and chemical analysis to determine if a patient was suffering from a common kidney complaint —obstruction in the flow of blood to one of the kidneys. Current research (*Computing Report*, 1970c) indicates that this procedure may be reduced to a half-hour test, performed in the hospital outpatients department, and consequently at a fraction of the cost. The test involves an injection of radio-active iodine; the arrival of the iodine at the kidneys and its subsequent excretion are measured with a gamma-ray scintillating camera, which produces digital data about the iodine concentration by measuring the number of gamma-ray disintegrations per unit of time. This data is fed into a computer which processes it and produces a graph (called a renograph) of localised radio-activity (iodine concentration), as a function of time. For all practical purposes, the radio-active iodine is completely absorbed from the blood stream by the kidneys, hence its absorbtion rate is a valuable indicator of how well the kidneys are carrying out their basic function of removing metabolic wastes and other poisonous substances from the body circulation. The computer compares the data obtained with a standardised mathematical model of the kidneys and the circulating fluids that make up the vascular pool from which the kidneys extract the radio-active iodine. The computer proceeds to analyse the data from the camera comparing it with the model to detect any major differences. The computer (program) is able to estimate eight variables that have physiological significance, renal blood flow, for example, can be estimated to within two to four percent. Other parameters include the fraction of blood from the heart that passes through each kidney, and measures of obstruction and tubular reabsorption. These useful measurements are generally impossible to obtain through other clinical tests.

The computer is being used in a wide range of clinical laboratory tests providing rapid, economical and more thorough analyses to help in the detection of ailments and disease. With the help of multi-access computers these processes can be made widely available at relatively low cost in smaller hospitals, and even in the local doctor's surgery.

4. Patient Monitoring

Computers are also being used for the continuous monitoring of critically ill patients in intensive-care wards. For example, with coronary patients, bedside sensing devices are used to monitor such critical factors as arterial blood pressure, heart pumping rate and cardiac output. The output from these sensing devices is sampled by a computer at 10 to 20 second intervals. The computer compares these measurements with standards that are considered normal or acceptable under the circumstances, and nurses and doctors are automatically alerted by an alarm system when abnormal trends occur in the patient's condition. The computer may also accumulate historical data on any of these factors and present it in summary form (perhaps graphically) at regular intervals to assist doctors and nurses in their assessment of the patient's condition and his course of treatment.

5. Medical Diagnosis

Increasing use of the computer is being made in the decision making process that underlies diagnosis and treatment in clinical medicine (Taylor, 1970). When faced with the problem of diagnosing a patient's illness a doctor proceeds through a sequence of decisions, he examines the patient's symptoms selectively abandoning what he considers to be irrelevant symptoms (perhaps with the aid of laboratory tests) until he has gathered sufficient information and arrived at a set of symptoms which indicate the patient's probable illness, the appropriate course of treatment and the probability of its success. When diagnosing an illness a doctor uses items of a patient's history, physical examination, and laboratory tests. He must interpret these pieces of information and assess what group of diseases he thinks his patient is suffering from, he will then seek new information to prove or disprove this diagnosis. He revises his opinion about diagnosis and treatment on a continuous basis as new information comes to hand. There is a well-established theory of statistical influence which can be used to describe and model this process mathematically.

Diagnostic decisions are based, in part, on the previous knowledge of the diseases being studied, this knowledge is summarised in the form of "likelihoods" predicting the relative frequencies of test results in each disease. Another set of valuable predictive information is the prior probabilities of each disease, these are the incidence rates of the diseases in the population being studied. These probabilities vary in different populations, for example, tropical diseases are common around the equator but rare in England.

Decisions of this type form the basis of the whole of clinical medicine; the sequence of decisions which culminate in final diagnosis and choice of treatment can be analysed into a logical sequence of steps. Once this is done the decision making process can be restated in mathematical terms by making use of conditional probability theory and statistical decision theory. So by combining prior probabilities, likelihoods and test results and by using mathematical techniques the process of inference involved in the diagnostic process can be defined as a mathematical algorithm, and hence can be programmed and performed by a computer.

A computer diagnostic aid might operate as follows: the doctor after a preliminary diagnosis might select a general category of illness and a few basic symptoms and send this information to a computer through an on-line terminal. The computer will then decide upon the next "best test", print the name of the test on the terminal and wait for the doctor to enter the test results. After this the computer proceeds to select the next most informative test, it also prints out the probabilities of each of the most likely diseases at each step in this process. The delay between entering the results for each test and the computer's response is not more than a few seconds, which means that it is feasible to use this technique in real decision making situations in clinical practice. Experiments with this type of system as developed so far have indicated a diagnostic accuracy comparable to that of experienced medical specialists. Furthermore the system tends to minimise the number of tests needed for accurate diagnosis, it reduces the amount of information collected for each patient and hence lessens the burden of information with which the doctor has to cope.

Similar systems are also being used for preliminary screening of patients. For example in one such system people are asked to fill in a medical questionnaire of some 200 to 300 questions which they can do at home and then post the questionnaire to their doctor or hospital. Information from the questionnaire is entered into a computer which processes it and decides whether or not that person needs medical attention, whether it is urgently required, whether some clinical tests are required to complete the diagnosis, and so on. The computer will then schedule appointments with the doctor, or the hospital laboratory for tests, etc., it will send appointment details to the patient and schedules of appointments to doctors and hospital staff, and so on.

H. EXERCISES

1. Computers promise to be as pervading and ubiquitous as paper, printing, automobiles and electric power. What do you think should be the proper role of computers in society? What safeguards and protection procedures should be established for the average citizen?
2. How do you think computers could be used in libraries? Carry out a literature survey and write an essay on this topic. Design, in broad outline, a computer system for a library and define the major functions that it should perform. What would be the advantages, if any, of linking together the computers at various libraries throughout the country?
3. Carry out a literature survey investigating the use of computers in space flight. Do you think that it would have been possible for Major Yuri Gagarin to orbit the earth or for Commander Neil Armstrong to set foot on the moon if the computer had not yet been invented?
4. Take a street map of a section of a city or town and design a computer based traffic control system. Specify the objectives of such a system. Try and obtain information and statistics about typical traffic characteristics, densities and flow rates, if this is not possible state some estimates for these factors and work from these. Set up a mathematical model which can be used, firstly to simulate your system for testing and experimental purposes, and secondly which can be used for reference and/or control purposes in the final system.
5. Do you think that the use of computers will lead to a cashless society?—where, for example, when we purchase something from a store our specially prepared and coded credit card is put into a machine to identify us and our bank, the price of the purchase is then keyed into the same machine upon which our bank balance is automatically debited and the store's account credited by the price of the purchase.

REFERENCES

Coons, S. A. (1966). "The Uses of Computers in Technology," *Scientific American*, p. 177.

Computing Report, (1970a). "The Behaviour of Susan", p. 13, Vol. VI, No. 1.

Computing Report, (1970b). "Electricity for New England", p. 4, Vol. VI, No. 1.

Computing Report, (1970c). "A Test for the Kidneys", Vol. VI, No. 2.

Data Processor, (1969). "R for Hospitals", Vol. 1, XII, No. 4.

Fano, R. M. and Corbato, F. J. (1966) "Time-sharing on Computers", *Scientific American*, p. 129.

Green, R. E. (1970). "Interactive Computer Graphics", *Science Journal*, p. 67.

Pierce, J. R. (1966). "The Transmission of Computer Data", *Scientific American*, p. 145.

Suppes, P. (1966). "The Uses of Computers in Education", *Scientific American*, p. 207.

Taylor, T. (1970). "Computers in Medicine", *Science Journal*, Vol. 6, No. 10.

SUGGESTED READING MATERIAL

The best sources of readable material about computer systems and applications are periodicals and journals, of which the following are recommended.

Scientific American (U.S.A.) In particular the special issue on Computers, Sept. 1966.

Science Journal (U.K.) In particular the special issue on Computers, Oct. 1970.

Datamation (U.S.A.); *Computers and Automation* (U.S.A.); *Data Processing* (U.K.); *New Scientist* (U.K.); *Data Processor* (U.S.A.); *Computing Report* (U.S.A.); *I.E.E.E. Spectrum* (U.S.A.).

CHAPTER 25

Electric Power and Computers

R. E. Vowels

"We can hope that the wealth possible through automation will enable us to achieve Bagrit's 'really high civilisation'. This will not be measured in terms of the merchant's gold, but only in our intellectual and artistic achievements. The threat of automation is only as real as we, by our own lack of preparedness, make it ourselves."

J. B. Davenport, in the epilogue to Automation: Threat or Promise?. *Published by Sun Books, 1969 as an account of a symposium held by the Australian and New Zealand Association for the Advancement of Science.*

A. INTRODUCTION

The linking of "power" and "computers" might seem a little unusual since it is only within the last 15 years that the theory of discrete control systems has been actively developed. Only comparatively recently have digital computers been used for control purposes and it would be correct to say that their application to power systems has barely commenced when viewed in the light of their true potential. "Systems Engineering" and the "Control of Large Systems" are topical subjects destined to play important roles in the electric power industry.

History has shown that large systems evolve or grow from small beginnings. Their control by man has in the past developed through experience. At various stages of the growth of the system crises arise and problems which emerge must be solved before the system can be safely enlarged. It is therefore relevant to make some mention of the development of power systems and the problems which have arisen.

B. EARLY HISTORY

The first alternating current system in America using transformers was installed at Great Barrington in Massachusetts in 1886 by William Stanley. The Siemens alternator rated at 500 volts, 12 amperes was imported from England by George Westinghouse. He secured all rights in the U.S.A. for the Gaulard and Gibbs transformer. Stanley designed and constructed a number of transformers wound for a 500 volt primary and a 100 volt secondary, six of which were installed in the town of Great Barrington. It is interesting to note that the distance from the generator to the centre of the town was 4000 feet. Since the electric power industry of today is predominantly alternating current we could regard the Great Barrington installation as the beginning of the modern electric power system or "integrated power pool."

C. SOURCES OF ENERGY

The sources of energy for the generation of electric power may be classified as follows:

1. Solar Energy:
   (a) Directly by photocells or steam raising for the generation of electricity.
   (b) Photo-synthesis; to produce combustible vegetable* and fossil fuels such as coal, petroleum and natural gas.
   (c) Wind power.
   (d) Precipitation hydro-electric power.
2. Tidal hydraulic power.
3. Geothermal power.
4. Chemical energy (not involving combustion).
5. Nuclear energy—fission or fusion.

1. Solar Energy

Schemes have been developed to focus the rays of the sun by large mirrors on a boiler for steam raising. However, it has been estimated that solar radiation falling on the whole of the land surface could satisfy all energy requirements if conversion could be effected at an average efficiency of 2.5 per cent, but unfortunately the technology for its achievement does not exist today. Even if it did exist, the cost would probably be prohibitive.

Attempts were made at Abidjan on the West Coast of Africa to utilise the difference in the temperature of the sea water at the surface and at a depth of a few hundred feet. Wind generators have been installed in various parts of the world but it has been concluded that our present technology does not enable us to convert the large amounts of kinetic energy in the movement of the earth's atmosphere.

2. Geothermal Power

Geothermal power remains largely an unknown quantity. However geothermal stations exist in various parts of the world; for example, at Taupo in the North Island of New Zealand, approximately 200 megawatts (MW) are generated from a geothermal source.

3. Fuel Cells

Chemical energy in the form of fuel cells may yet assume great importance for small sources of power especially in the field of transportation but for economic and other reasons it will not play a major role in large-scale power generation.

4. Hydro-electric Power

Hydro-electric power is recognised as an important source in many countries, including Sweden, the U.S.A., Canada, Britain, Australia, etc. In Tasmania, for example, the entire generating plant has been hydro-electric, but provision is being made for an alternative source of power to meet emergent emergencies. However, in the long-range future, it has been recognised that precipitation hydraulic power can provide only a small proportion of the total energy requirement.

5. Pumped Storage

Pumped storage hydro-electric plant can be designed in such a manner that it can be brought on the line rapidly so that it can be considered a relatively quick-release source of reserve power to meet the peak load demands of power authorities. The construction of pumped storage hydro-electric plants, favoured by the topography of the area, may offset the relatively high fixed costs of other plants such as nuclear stations. During off-peak load periods electric power may be used to pump water to a suitable storage at a higher level and the resultant potential energy converted by the

* The Townsville Regional Electricity Board, Queensland, Australia, accepts generation of electric power from sugar mills using begasse (the residue of crushed sugar cane) as fuel.

hydro-electric plant as required to meet the peak energy demands of the system. The combination of say nuclear and pumped storage hydro-electric plant provides an efficient base load, economical peaking capacity, and an effective spinning reserve in addition to many recreational benefits.

Following extensive engineering economic studies a 1000 MW pumped storage hydro-electric plant at Northfield Mountain in Franklin Country, Massachusetts, U.S.A. will be the largest of its kind in the world when completed in 1971.

Pumped storage is also used in connection with tidal power schemes to meet peak demands or to extend the hours of energy supply.

6. Tidal Power

Tidal power has been recognised as an important source of energy for many years. Coastal dwellers have been trying for centuries to harness the energy resulting from the ebb and flow of the tide. Paddle wheels set in motion by the sea water have served to work grain crushers. Such low capacity tidal mills were to be seen on the Breton coasts of France as early as the 12th century. One or two were in operation until recently in the Rance River estuary near St. Malo and elsewhere in Brittany. The tide rises and falls twice in every 24 hours and 50 minutes which is the apparent period of rotation of the moon. It reaches a maximum every two weeks when the moon and the sun act in conjunction; that is, when these two bodies are pulling together we have what is termed a "spring" tide.

Unfortunately, there is a relatively limited number of locations throughout the world where the tide amplitude is large enough to justify a tidal power station. For example, in the Bay of Fundy on the Atlantic Coast of Canada the tide amplitude exceeds 50 feet and a series of tidal plants has been proposed at the Passamaquoddy site.

Alaska has sites with tidal amplitudes exceeding 33 feet. In the Gulf of California near the mouth of the Colorado River the tidal variations reach 30 feet. The Severn estuary has a tidal amplitude in excess of 45 feet and Argentina, India, Korea and parts of the northern coast line of Australia have good sites as does the Russian north coast. In France there are several good locations with a tidal range of 40 feet or more; for example, the coast of Brittany.

The planetary system sets the rhythm of the seas, but only the sea, with the constraints of the topography of land masses, decides the nuances and values of this cadence. The simplest tidal power plant involves damming an estuary or other suitable location, admitting tidal water to the basin thus formed, closing the sluice gates until the tide has eased and then allowing the basin water to flow back to the sea through turbines to generate power. More sophisticated schemes involve turbining during filling and pumping in off-tide periods to raise or lower the level of the main or adjacent basins. Various schemes have been devised to increase the period over which power is available.

Figures 25.1 and 25.2 illustrate some of the methods of operation. Both methods may be improved by pumping (a) at high tide as in Fig. 25.3 and (b) at low tide as in Fig. 25.4. In the first case of pumping at high tide, the level of the basin can be raised when the level of the sea and the basin are approximately equal. Relatively little energy is needed to raise the water level but when the tide has fallen this "pump" water falls through a greater head. Similarly at low tide relatively little energy is required to lower the basin level but the energy benefits at full tide are readily apparent.

Many other schemes and refinements have been devised. Combinations of these schemes permit energy to be extracted at times more suited to the industrial demand for power. The turbining and pumping functions have been combined into a single

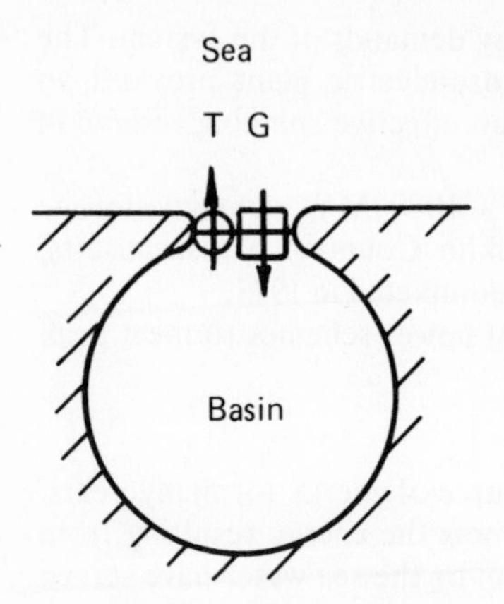

FIG. 25.1 Single Way Operation on Emptying.

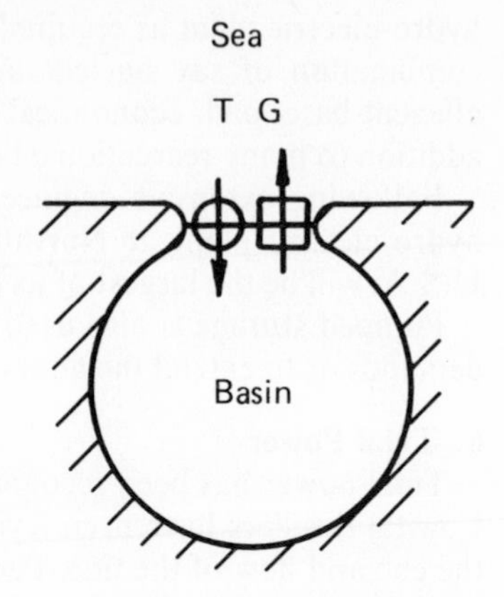

FIG. 25.2 Single Way Operation on Filling.

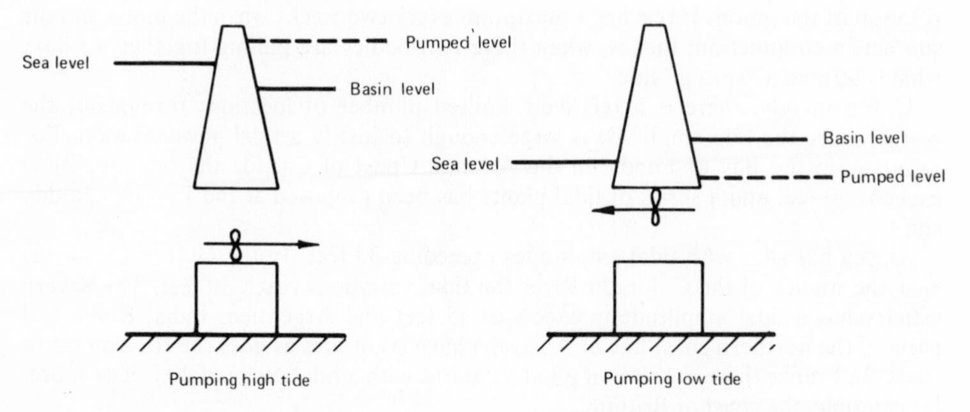

FIG. 25.3 Pumping, High Tide.

FIG. 23.4 Pumping, Low Tide.

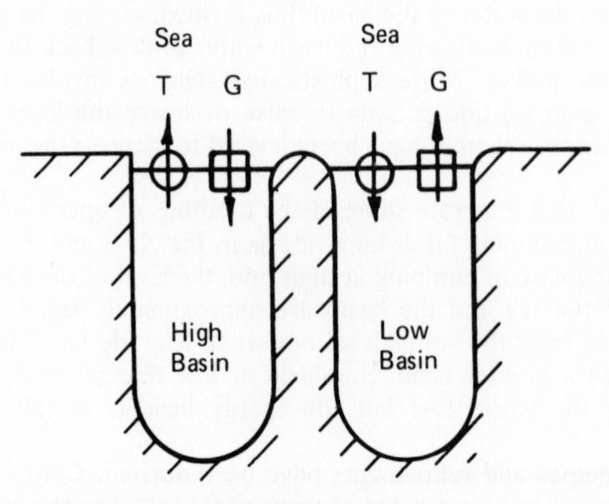

FIG. 25.5 Double Way Operation. This scheme extends the period over which energy can be supplied to the electrical network. T = Turbine, G = Gate.

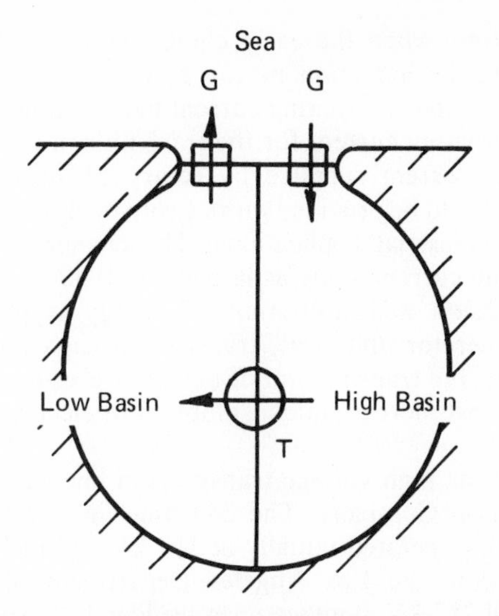

FIG. 25.6 Sesqui Operation. This cycle makes use of two successive spring tides to pass the water three times through the turbines, with two intermediate pumpings. T = Turbine, G = Gate.

piece of machinery which can function as a turbine with water flowing in either direction. In other words its unique design enables it to perform the same tasks as four separate machines. The first tidal power plant was the Rance River project which was constructed by Electricité de France. It is located about two miles upstream of the town of St. Malo. The tidal variations are of the order of 40 feet and twenty-four 10 MW units are being installed, the first of which went into operation in 1966. The average head of water is about 25 feet.

At the Passamaquoddy site in the mouth of the Bay of Fundy it is planned to build two 500 MW plants.

7. Pattern of Energy Sources

The changing pattern of energy sources is highlighted by the fact that in 1947, coal supplied roughly half of all energy consumed; by 1965 it had dropped to about one-fifth of the total and this proportion continues to fall.

Although the consumption of petroleum, natural gas and coal will continue to increase absolutely, these fossil fuels will in the future account for a relatively smaller proportion of the total energy produced. The hopes for the future rest largely on nuclear energy.

D. GENERATOR UNIT SIZE

The size of generator units has continued to increase rather rapidly over the past 20 years and naturally the question arises as to whether there is an optimum size of machine for a given power system.

Investigation has shown that the most economical method of power system expansion is to add generation units of 7 to 10 per cent of the size of the system. This assumes that the investment cost per kW of generating units continues to decrease with size and the forced outage rate for large generators remains at the present level. There is no indication that the maximum economic size of units has been reached and machines of 1000 to 1200 MW are now in service.

E. TRANSMISSION OF POWER

The main factor determining the adoption of alternating current has been the ease with which it can be converted from one voltage to another by means of a "transformer". At the time when the early choice between alternating and direct current was being made, the induction motor as invented by Tesla provided the prospect of an extremely simple alternating current industrial machine. In spite of the decision in favour of alternating current for the transmission of electric power, many cities operated on direct current supplied by rotary converters and mercury arc rectifiers. Variable speed could be provided more easily by direct current machines for lift drives and certain industrial applications. The changeover of the Australian capital cities to alternating current came as late as the 1930-45 period. A key role in the generation, transmission and utilisation of energy is played by the power transformer. As with generator unit sizes, transformer capacities have increased to over 1000 MVA. Without the transmission of power at ever increasing voltage there could not have been the vast increase in consumption of electric power for industrial expansion.

Historically, the first long high voltage transmission line was built in 1914 by the Southern California Edison Company. The 243 mile line from the Sierra Nevada Mountains to Los Angeles operated initially at 150 kV and in 1924 the voltage was raised to 220 kV. In 1936 the Los Angeles Department of Water and Power constructed the 265 mile 287 kV Boulder lines feeding Los Angeles. Sweden's 593 mile 380/420 kV line was completed in 1952. More recently, the Hydro-Quebec of Canada has built a 700/735 kV 400 mile system to develop the vast hydro-electric potential of Northern Canada and it is believed that alternating current (a.c.) voltages will reach 1000 kV in the not too distant future.

Naturally, the energy sources near the load centres were the first to be developed, and very short transmission lines were required. However, it soon became necessary to build power stations at considerable distances from the industrial towns and longer transmission distances were involved. Problems arose over the stability of the electrical system and higher voltages were adopted to increase the capacity of the transmission lines.

Economical extra-long distance point-to-point transmission lines are becoming difficult to design because of the high cost of the compensation or resonance equipment required to maintain stability. Two systems, inter-connected through an a.c. line must run in synchronism, that is the rotors of their generators must remain fixed with certain angular limits with respect to each other. Sufficient stability is provided only if this angular displacement or the phase shift does not exceed 30 electrical degrees. On a 60 cycle per second (c/s) or Hertz (Hz) system, a 250 mile line with surge impedance loading has a phase angle of approximately 30°. Because of stability problems, to go beyond this length of transmission line requires complete compensation for the miles in excess of 250 with consequent cost penalties.

However, communications engineers encounter long lines and are aware of the interesting phenomenon which occurs for electrical lengths between one half and threequarter wave length (or 180° to 270°). The system is as stable as one operating in the first quadrant, namely 0° to 90°. The most attractive feature of the half-wave length system is that the cost per unit length decreases as the line length increases. The transition point between the conventional and the half-wave length system occurs at a distance of 900 miles.

Thought is being given to the transmission of several gigawatts* (GW) of power over distances of thousands of kilometers. Using classical extra high voltage (e.h.v.) a.c. transmission the line voltage has to be higher than 750 kV. It has already been predicted that a.c. transmission line voltages will reach 1000 kV by 1980, 1500 kV by 1990 and 2000 kV by the year 2000.

* 1 gigawatt (GW) is equivalent to 1000 MW

Alternatives to the e.h.v. alternating current transmission line are:

(i) microwave lines using oversized circular wave guides in the T E_{01} mode at frequencies from 3 to 10 GHz;

(ii) superconducting conductors carrying direct current of 100 kA with current densities of 20 kA/cm^2 and 40 kV between conductors; and

(iii) e.h.v. d.c. transmission at $\pm$ 400 kV to earth.

All these alternatives need careful consideration and each has its own special problems.

The microwave transmission of power through a waveguide is clearly limited by the maximum electric field which the medium filling waveguide can withstand before breakdown. However, by considering large pipes, gasfilled at higher than atmospheric pressure, the power handling capacity can be quite high; for example a five foot diameter pipe, air filled can carry approximately 70 GW of power. The advantages of microwave power transmission systems are considerable and merit further consideration. However, the basic problem lies in the provision of suitable generators. At this stage it is difficult to predict whether such a system would be more economically feasible than other methods mentioned above.

Recent progress in cryogenics makes the earlier visions of super-conducting power transmission increasingly attractive. One proposal for such a cable is a twin line carrying a direct current of 100 kA with a voltage of $\pm$ 20 kV between the conductors. The two conductors are enclosed in a common tube 12 cm. in diameter with the temperature held constant at 4.2° K by means of liquid helium.

Much attention has been paid to the use of extra high voltage direct current (d.c.). In principle such lines have no stability limitations. The cost of the basic d.c. transmission line is always less than that of the a.c. line. For submarine installations the difference is particularly pronounced as the sea may be used as an earth return. For a given capacity, an overhead d.c. transmission line costs about two-thirds of the a.c. line. Usually it is not a simple case of a d.c. source with d.c. transmission to a d.c. load; in general the d.c. line connects two a.c. systems.

A summary of high voltage d.c. transmission systems in operation is given in Table 25.1.

It is worth noting that the ground is an excellent return conductor for a d.c. line. With a.c., the inductive effect forces the ground current to follow the line route near the surface. By contrast, the d.c. ground current follows the path of least resistance, penetrating the good conduction interior of the earth and the only notable ground effects are in the vicinity of the earthing electrodes.

High voltage d.c transmission would have been adopted much earlier had it been possible to construct suitable high voltage d.c. generators and/or motors. Important research with this end in view was carried out by Noel Felici at Grenoble, France. He concluded that the Wimshurst machine which had been known for over a century was not efficiently designed and proceeded to develop a new form of electrostatic machine. He predicted that the Felici generator might ultimately produce enough power at a voltage suitable for d.c. power transmission. At the receiving end, the line could be coupled to the a.c. network through an electrostatic motor driving a conventional synchronous alternator. However, it appears unlikely that the conversion equipment as developed by A.S.E.A. of Sweden (see Figs 25.7 and 25.8) will be replaced by rotating machines in the near future.

During the last few years efforts have been directed towards the development of solid state high voltage d.c. terminal equipment in the form of thyristors. The feasibility of connecting thyristors in series to reach extra high voltage levels has been demonstrated. The main tasks confronting the groups studying thyristor applications were the problems involved in gating series strings of thyristors as well as the

Table 25.1
Summary of High Voltage Direct Current Transmission Systems

Location	Installation Date	Rating kV	Rating MW	Type of Transmission and Distance	Remarks
Gotland (Sweden)	1954	100	20	60 miles—cable	Single conductor Swedish mainland to Gotland Island.
English Channel (France-England)	1961	±100	160	40 miles—cable	Single conductor interconnecting English and French systems.
Donbass-Volgograd (U.S.S.R.)	1964	±400	750	300 miles—overhead line	
New Zealand (Cook Strait)	1965	±250	600	25 miles—cable 360 miles—overhead line	Single conductor transmitting power from the South to the North Island.
Japan (50.60 c/s)	1965	2×125	300		
Konti-Skan	1965	250	250	54 miles—cable 59 miles—overhead line	Connecting Denmark and Sweden.
Sardinia—Italy via Corsica	1966	200	200	61 miles—cable 217 miles—overhead line	

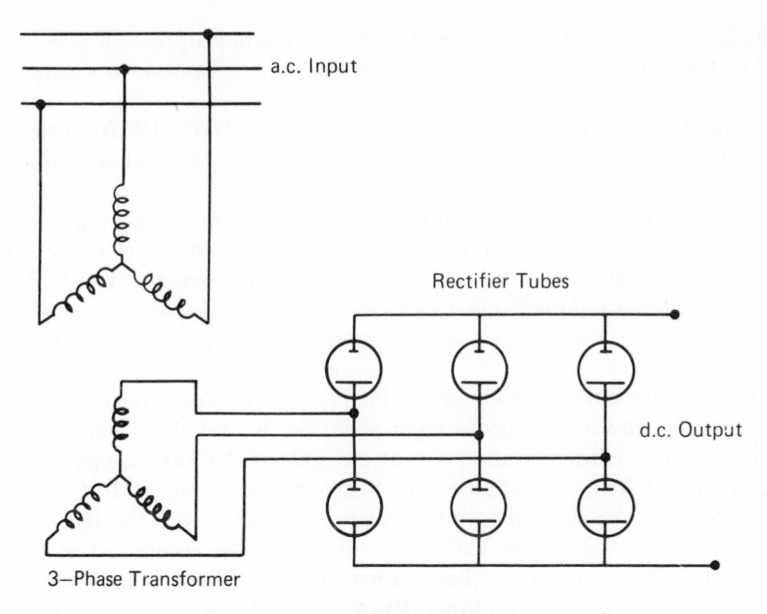

FIG. 25.7 A Six Pulse Two-way Converter.

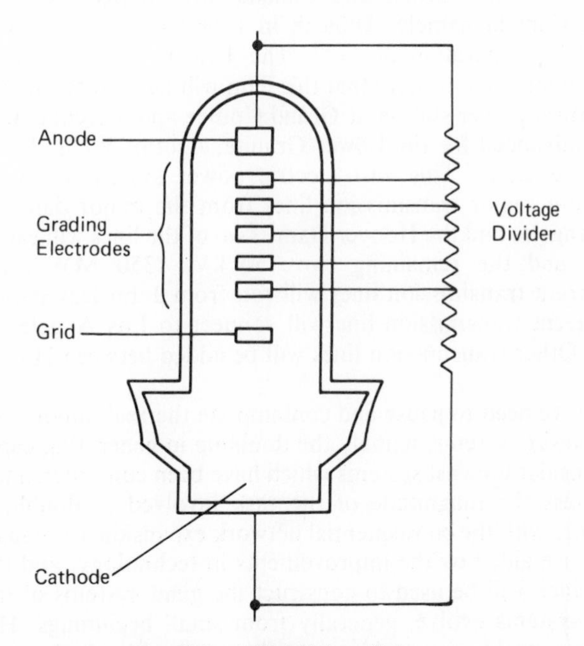

FIG. 25.8 A High Voltage Valve.

problems of equal voltage distribution during switching. It is now possible to utilise the advances in solid state control and logic in order to take full advantage of the fast response times of thyristors as compared with the mercury arc devices which convert from a.c. to d.c. or invert from d.c. to a.c. With these developments the future of extra high voltage (e.h.v.) d.c. interties appears to be assured.

As a result of research and satisfactory experience in operation, high voltage d.c. lines are becoming recognised as the method of interconnecting large power systems. The Germans have begun work on a 400 kV d.c. system to transmit 600 MW 75 miles from Berlin to a terminal on the Elbe. Design work is in progress for a ±750 kV 5250 MW d.c. transmission inter-tie from Kazakhstan to Tambov a distance of 2500 kilometres and it is expected to be in operation in 1971.

The giant d.c. inter-tie of two 800 kV lines rated at 2700 MW from the Columbia River to the Hoover Dam and Los Angeles will have the highest voltage, largest capacity and be the longest in the world.

Electricity consumption has been doubling every eight to ten years and in most countries a doubling of the generation capacity is being planned for the next ten years. Although an annual growth rate of 7 per cent is normal, the U.S.S.R. electrical energy production was 11 per cent greater in 1965 than 1964, and the vast land masses of European and Asiatic Russia are being inter-connected by long distance extra high voltage inter-ties.

The tremendous developments of the Pacific North West Coast are typical of the pressures on the power industry brought about by the doubling every ten years. The Bonneville Power Administration serves an area of 290,000 square miles with 21 existing dam sites. The main source of its hydro-electric power is the Columbia River. The harnessing of the river did not commence until 1933 with the Rock Island Dam near Wenatchee, Washington. The Bonneville Dam was completed in 1939 and the Grand Coulee in 1941. Since then numerous dams have been constructed. The potential capacity of the Columbia River is far from exhausted and under an agreement between the U.S.A. and Canada three dams have been or are being constructed in Canada namely, Duncan in 1968, Arrow Lakes in 1969 and Mica scheduled to be in operation in 1973. The U.S.A. will construct Libby Dam in Montana. It is interesting to note that this dam will back water 42 miles into Canada. Apart from a third power station at Grand Coulee and a second at Bonneville Dam, plans are well advanced for the Lower Granite, Asotin, Wells, Boundary, and High Mountain Sheep dams. This vast electric power system is to have an inter-tie consisting of four major transmission lines from the major dams on the Columbia River to Los Angeles and the Hoover Dam. Two of the lines will each be 500 kV, 1000 MW capacity and the remaining two 750 kV, 1350 MW capacity each. The alternating current transmission lines will run from John Day dam to Los Angeles. One direct current transmission line will connect to Los Angeles and the other to Hoover Dam. Other transmission lines will be added between Hoover Dam and Los Angeles.

At this point we need to pause and contemplate the real import of the rapid rate of expansion of power systems, namely the doubling in generating capacity in less than ten years.* Consider the vast systems which have been constructed to the present date and try to assess the magnitude of the task involved in doubling the generating capacity together with the consequential network expansion accompanying it.

At least we are aided by the improvements in technology, and the increase in the size of units which will be used to construct the giant systems of the future. As was stated earlier systems evolve, generally from small beginnings. However, with the expansion come problems; problems such as the North East power failure of November 9-10, 1965 which involved the massive outage of the states of New York, Vermont, Massachusetts, Connecticut and New Hampshire in the U.S.A. and Ontario in Canada. The period of the outage ranged from up to 2 hours in New Hampshire to a maximum of 18½ hours in parts of New York. So seriously did President Johnson regard the occurrence that he immediately appointed a Federal Power Commission (F.P.C.). The urgency of the situation was underlined by the fact that in less than one month the Commission submitted its report entitled "North East Power Failure", dated 6th December, 1965, to the President.

* Doubling in ten years would correspond to a steady growth rate of 7.2 per cent per annum.

The origin of the incident was traced to a "back-up protective relay". The particular relay setting was reassessed and the critical level at which it was set to function was increased in January, 1963. This relay operated normally and isolated an unfaulted line. The power flow on the disconnected line shifted to the remaining four lines each of which became loaded beyond the critical level and tripped.

Quoting from the F.P.C. report:

> With the dropping of the lines to Toronto, the power being generated at the Beck plant and at PASNY'S Niagara plant which had been serving the Canadian loads around Toronto amounting to approximately 1500 MW, reversed, and was superimposed on the lines to the south and east of Niagara. It was this tremendous thrust upon the transmission system in Western New York state which exceeded its capacity and caused it to break up.

The consequent chain reaction shut down the system.

It must be realised that apart from the losses incurred by industry due to failure of supply and to the business community, such a catastrophic failure may cause damage to equipment. In the North East power outage, failure of the electric pumps supplying lubricating oil caused considerable damage to generator shaft bearings. Consequently, the reconnection of the system presented a major problem. Even if equipment had not been damaged, reconnection would still have presented a difficult problem.

Hydro-electric stations have a distinct advantage over thermal stations in that they can usually be started with power other than emergency batteries. The time lags in placing thermal stations on load are considerable. Above all it must be appreciated that the coordination of the re-energising of such a vast inter-connected system is an immense problem in itself. It must now be apparent that the protection and control of large integrated power pools must soon be performed by digital computers. The task is too complex for individuals and needs urgent attention.

The Federal Power Commission's report contained nineteen recommendations including the need for auxiliary power sources, stability studies, fully coordinated power pools, mixed generation (hydro-electric, pumped storage, coal, etc.), internal load shedding and improved communication facilities.

Following the North East power failure a new philosophy for the analysis and design of large electric power inter-connected systems needs to be evolved. Disaster control procedures providing successive lines of defence need to be evaluated and implemented. The principles of load shedding, distribution of spinning reserves and system separation must be applied in a more positive manner, with a view to their eventual control by a digital computer which can assess a complex situation and make vital decisions far more quickly than a human operator.

F. NUCLEAR POWER

At the time of writing, it is some eleven years since the first nuclear power plant at Shippingport, Pennsylvania, U.S.A. first supplied electric power to the public on a regular basis. Progress from that date was slow, but in 1965 there was a rapid increase in the number of orders for nuclear stations.

It would appear that during the 1970s about one half of all the new power stations in the United States will be nuclear. Such power projects are already under contract as a construction period of 5-6 years is required from the date of authorisation. The significant number of recent commitments to construct nuclear plants is clear evidence of their acceptance by power authorities. The acceptance of light water reactors for current contracts has served to speed the progress of advanced reactor projects. The sodium cooled liquid-metal fast breeder reactor (LMFBR) is receiving the highest priority in the development programme of a number of countries.

It is predicted that rapid developments will take place within the next ten years. The accelerated rate of construction of nuclear power stations will give great impetus to the applications of nuclear power to the desalting of sea water, the production of chemicals and metals, the synthesis of ammonia for fertilisers, the electrolytic production of aluminium and the production of magnesium from sea water by electrolysis.

G. LINEAR MOTION ELECTRICAL MACHINES

Recently Professor E. R. Laithwaite has drawn attention to some interesting phenomena arising from the study of linear motion electric machines. He points out that in almost a hundred years of rotating machine development the shape of the magnetic circuit has remained virtually unchanged. However, linear motor development has been characterised by the extent of the diversity of shapes that become possible. The field over an open-sided linear motor contains a rotating component which can be used to spin rotors in a manner similar to a rack and pinion action. Laithwaite observed that a steel washer rolled along a linear stator in the opposite direction to the motion of the magnetic field. These types of machine are virtually unexplored and it is possible that electrical machine development could still be in its infancy. Such machines could well play an important role in power development, either as generators or motors.

H. MAGNETO-HYDRODYNAMIC GENERATORS

The conventional alternator has been used for power generation for over 80 years. However, it must be recognised that the effective further development of the steam cycle as used in turbines is nearing an end. Additional gains from increases in temperature and pressure are limited in their economic achievement. Furthermore, the trend during recent years to atomic sources calls for new and improved forms of power generation. The adoption of a working fluid which becomes an integral part of the electric generation process would be a distinct advancement.

The principle adopted in magneto-hydrodynamic (MHD) or magneto-plasmadynamic generation is relatively simple. A stream of electrically conducting fluid flows through a transverse magnetic field. As a result an electric field is induced in a direction mutually perpendicular to both the flow and the magnetic field and from this induced electromotive force (e.m.f.) power may be drawn. Thus the MHD generator is analogous in action to the conventional generator. Actually the simple MHD generator produced low voltage direct current and various schemes have been proposed for the production of alternating current. Research experiments producing MHD power have been carried out at the Massachusetts Institute of Technology and a number of universities. To carry this experimentation beyond a certain stage required rather expensive large scale equipment but power prototypes have now been successfully constructed, for example, the AVCO-Everett Research Laboratory in the U.S.A. has produced a maximum power of 1.5 MW burning oxygen and kerosene/alcohol. The same laboratory has also constructed a Mark V version with an output of 40 MW.

Other research is directed toward determining the potential of the liquid-metal concept for commercial power systems. Binary cycles are being considered such as the coupling of a liquid metal primary cycle with conventional steam secondary cycle. The primary cycle would consist of a liquid metal generator tied directly to a liquid metal cooled reactor. It has been suggested that efficiencies of 60 per cent may be feasible.

I. AUTOMATIC CONTROL

The application of the principles of Automatic Control to items of equipment in power systems is being extended as the systems grow in complexity. Quantities such as voltage, frequency, time, active and reactive power flow, etc., have been controlled for many years. With the trend towards large boiler and turbo-generator units and the parallel development of instrumentation telemetry, control by on-line analogue-digital or hybrid computers is essential to achieve optimum operating conditions. The operations of each power station need to be optimised automatically, in order to achieve the best performance. In fact, large economies can be effected in integrated power systems, by optimising the many variable functions with computer control.

Some of the power system functions which become an essential part of the automatic processes are:

(1) Load forecasting:
Load predictions 24 hours in advance.

(2) Generator incremental cost:
The determination of current incremental power costs for generators using fixed data or power plant computation of heat rate for economic despatch.

(3) Economic despatch:
The computation of the most economic loadings for all units in operation based on generator incremental cost curves and transmission losses. It is achieved when the generators within the area are loaded to equal incremental costs of delivered power.

(4) Generator selection:
The selection of the most economic generator units for the increase or decrease in load. Dynamic programming achieves the best solution.

(5) Interchange accounting:
The determination of the optimum amounts of power interchange and the statement of account for the inter-connected operation.

(6) System security and protection:
A check of overload transmission circuits and relay schedules. In due course protective relays will be replaced by computers.

(7) Load-frequency control:
The loading of generators to maintain frequency and interchange flows at desired levels. The settings should be calculated every five seconds or less to follow load dynamics.

(8) Spinning Reserve:
The computation of the amount of immediately available reserve necessary to minimise the risk of power shortage and at the same time optimising costs.

Digital computers are being used extensively for the design, analysis and operation of power systems. All the relevant system design parameters may be stored and software programs arranged to determine the performance and behaviour of the system. The updating of the system parameters can be carried out by the computer for the addition of new transmission lines, inter-ties, additional power stations, substations, industrial loads, etc.

Furthermore, it must be stressed that with the dynamic growth of already complex integrated power pools there is an urgent need for all operations including protection to be controlled by computers.

Initially we were concerned with the stability of systems having two or three power stations and load centres inter-connected by transmission lines. With the growth of the power network the concepts have changed to such an extent that each power station had to be replaced by an "area" representing in itself a power system. These areas are inter-connected by e.h.v. tie lines; this is the new concept of the "stability problem". Each area needs to be controlled within itself and should be responsible for the absorption of its own load changes within its perimeter. This is termed "area control". The way in which these areas behave when e.h.v. inter-ties connect them to form an integrated power pool is most important.

We need to study the dynamics of large systems to discover the significant parameters and the properties of the system for control purposes. It all becomes part of the new form of the stability problem, and the system must remain stable under sudden changes such as the inadvertent loss of component parts under fault conditions.

It should not be imagined that a single computer will control the whole of a power system but we may expect inter-connected computer systems communicating with and interacting with each other. Computers located in power plants will report to an "area computer" or controller. The area computers would in turn report to the "integrated power pool" computer. In the future, the decisions for control and operation will become far too complex for human controllers. Furthermore, these decisions and actions need to be taken quickly to ensure optimum system performance.

Already computer controlled power plants have achieved considerable economies in both power generation and equipment maintenance. It is expected that the economic savings to be achieved by computer control of a large system far outweigh the cost of the computer system.

The total investment in the power industry in the next ten years justifies an intensive effort on the part of the new breed of power engineer to solve the problems ahead as the systems grow to the next stage.

J. EXERCISES

1. Comparing a.c. and d.c. transmission, it has been stated that direct current transmission becomes economical when the savings on transmission lines exceed the higher cost of the terminals. What is the measure of the saving in transmission costs?

Let us assume that the same power W is to be transmitted at the same insulation level; that is, the same peak voltage E to earth is to be adopted in each case. The a.c. line to neutral voltage is $\frac{E}{\sqrt{2}}$ (root mean square value).

Assuming that the line currents are balanced, then the three phase a.c. power is expressed by

$$W = \frac{3E}{\sqrt{2}} I_{ac} \cos\phi \tag{25.1}$$

and for a d.c. line with a voltage $\pm E$ above or below earth (that is, one line is $+E$ above earth and the other $-E$ below earth) the power is

$$W = 2EI_{dc}. \tag{25.2}$$

If the power factor is $\cos\phi = 1$,

$$\text{then} \quad \frac{3EI_{ac}}{\sqrt{2}} = 2EI_{dc} \tag{25.3}$$

$$\text{and} \quad I_{dc} = \frac{3}{2\sqrt{2}} I_{ac}. \tag{25.4}$$

Assuming that the same current density will apply for the a.c. line, then the ratio

$$\frac{\text{d.c. copper per line}}{\text{a.c. copper per line}} = \frac{3}{2\sqrt{2}}$$

and, since there are two d.c. conductors and three a.c. conductors

$$\frac{\text{total d.c. line copper per unit length}}{\text{total a.c. line copper per unit length}} = \frac{3 \times 2}{2\sqrt{2} \times 3} = \frac{1}{\sqrt{2}} = .707$$

Apart from the saving in conductors for d.c. transmission, there will be a saving in insulation and transmission line power costs.

K. PROBLEMS FOR DISCUSSION

1. What are the advantages of pumped-storage in relation to:
(a) tidal power schemes and
(b) hydro-electric plants?
2. What are the limitations of extra-high voltage long distance alternating current power transmission lines? Are they likely to be superseded by direct current transmission or wave guides?

SUGGESTED READING MATERIAL

LAITHWAITE, E. R. "Rack and Pinion Motors", *Electronics and Power,* p. 251. Institution of Electrical Engineers, July, 1970.

Northeast Power Failure, Federal Power Commission, December 6, 1965, United States Government Printing Office.

LAMM, U. "Long Distance Power Transmission" *Science and Technology*, No. 52, pp. 50-53. April, 1966.

RINCLIFFE, R. G. "Planning and Operation of a Large Power Pool", *I.E.E.E. Spectrum*, Vol. 4, No. 1, pp. 91-96, January, 1967.

CHAPTER 26

Modelling and Simulation

R. M. Huey

***"In the very beginning of science
the parsons, who managed things then,
Being handy with hammer and chisel,
made gods in the likeness of men;
Till Commerce arose, and at length
some men of exceptional power
Supplanted both demons and gods by
the atoms, which last to this hour."***

James Clerk Maxwell, ca. 1875.

A. USE OF MODELLING IN SCIENCE AND ENGINEERING

In this book we have endeavoured to acquaint the reader with the concepts of information theory, signals and systems theory. We have described something of the way in which the development of materials science has enabled modern electronics engineering to achieve the hardware developments of the preceding two decades, which have resulted in the construction of powerful computer facilities, new telecommunications systems and the application of electronics to a host of purposes—many of which had not been dreamt of a few years before.

We have tried to interweave with our examples and references to actual systems, some of the ideas and concepts which have been the starting point of these new systems. This is of course an essential point, that new uses of available scientific knowledge must start from an idea or concept. It is a generally accepted idea that the scientific method can be described in simple and brief terms as "*We make an hypothesis and then test it*". The testing of a new hypothesis consists in devising an experiment whose results will clearly show agreement or disagreement with the hypothesis or theory. For this reason it is quite usual to speak of scientists as being either theoreticians or experimentalists. Of course, the one person at different times may act in the rôle of either theoretician or experimentalist. Others may elect to act out only one rôle.

Science is sometimes spoken of as the discovery of what exists. Engineering, on the other hand, may be spoken of as the creation of what has never been. The two activities are closely related. The engineer depends on the scientist for the knowledge and data which he uses in the creation of new designs. The scientist depends on the engineer to construct some of the equipments and systems with which to probe the secrets of nature. A man or woman trained in either science or engineering may step over into the other field or may sometimes work in both areas. It is not always easy to draw a dividing line between the two fields in a way that is applicable to all circumstances and at all times.

In this book we have spoken of models and modelling, mainly in the sense that a given situation, process or system may be represented by a mathematical model. In order to show something of the difference between the activities of the scientist and the engineer we should indicate the different way in which each profession makes use of its models.

The scientist will make a model which represents his new theory or hypothesis, usually in a mathematical form. He then uses this model in the pursuit of knowledge and will devise experiments from which measurements can be used to verify or to discredit the predictions of the mathematical model. He is investigating what actually exists.

The engineer on the other hand is required to create *new* systems or structures. He too will use a model (very often in a mathematical form) by which he may compute or predict the behaviour of the system or structure *before* it has ever been constructed. On the basis of his faith in the applicability of his chosen model and the accuracy of his computations, he produces the instructions to which the system or structure is made. Engineering judgment is required in choosing the configuration of the system which it is desired to create and the choice of sensible (i.e., applicable) models to represent its behaviour.

An equally high degree of creative skill is needed in each profession, and as we have said above, it is not infrequent for individuals to be engaged in a mixture of science and engineering.

With this preamble, which has served to point out that making models of real world situations or systems is a very important part of both science *and* engineering, let us go on to summarise some of the things which have been said about models and modelling.

B. KINDS OF MODELS

Following the division given by Harold Chestnut (1965) we will say that models may be classified in three ways (of course, there are other schemes in which the whole field of modelling could be divided into a number of sub-fields, however, the three set out below serve excellently for the immediate purpose).

1. Iconic Models

The adjective iconic is derived from the same root as the word icon meaning image or something which *looks* like the thing it is representing, although it may be larger or smaller, faster or slower than the original. Another word used with a rather similar meaning is homologue, meaning a model which has the same proportions, and is also of the same *kind.* The word homomorphic may also be seen, from the Greek root *morphe* = form.

2. Analogue Models

The noun analogue is here used as an adjective where the original meaning, of a similar or parallel word or thing, has been somewhat altered to emphasise the notion of a transformation (according to some specified set of rules) in going from the original to the model. Usually the transformation is governed by the simple rule that the *same form of mathematical equation* is used, but that the variables will represent different quantities in the original and in the model, and that the coefficients in the equations may differ in some predetermined way. The graph of one quantity *versus* another is a very simple analogue model. Many other graphical representations are analogue models, and so also are many hardware models (e.g., electric circuits to represent heat flow problems; streamline fluid flow to represent a magnetic field).

3. Symbolic Models

In this case both the elements comprising the real situation and the inter-relationships or interactions between the elements are represented by symbols in the model. The symbols may be those of mathematics or logic (i.e., alpha-numeric characters, with relationships such as equals, does not equal, is greater than, is less than, *if* A *then* B, *if* A *and* B *then* C, etc.) or they may be of other less conventional kinds (e.g., the detective in a crime film wears a certain kind of hat or coat; the baddies wear dark clothes etc.; the sexpot in a play may be identified by her bust measurement of at least 38—or in metric units 96; consider too, how much information is conveyed by the wearing of a political or club badge). In science and engineering the sort of symbols involved may well be those of sequential logical procedure or of the mathematics of probability or group theory, in addition to the readily recognised symbolic models which make use of everyday algebra and calculus.

C. USE OF MODELS

The above definitions of models are quite useful but they give very little indication, taken by themselves, of what such models are used for. Mathematical and logical models may be "solved" in the sense that if the essential specified data is inserted (i.e., as if the inputs were connected to the system which is being modelled) then going through the necessary mathematical or logical procedure can yield a "solution" which will indicate the behaviour of the system when those particular inputs exist. This "solution" is no doubt interesting but by itself it may not tell us a great deal. Usually many such solutions will be needed to yield the sort of information needed for either a scientific inquiry or an engineering task.

Several cases exist and we can make a simplified list of these to illustrate the last statement.

1. Analysis of an Existing System

By an *existing* system is usually meant a situation where we are either unable or not desirous to alter the parameters (i.e., the properties) of the system. In this case the problem of analysis may be stated in three steps.

(a) Choice of a suitable model, i.e., one which will display the variations in behaviour of the system in a manner which is related to our mental queries, and which is fine enough in detail to tell us what we wish to know but coarse enough in detail to be economical to use.
(b) Choice of a technique and provision of the means for "solving" the model. The technique must be intelligible to those who are required to use it, the people concerned must agree to devote their efforts to this task and the means must be provided (pencil, paper, slide rules, text books, access to digital computers or to analogue computers, office space, etc., etc.).
(c) The range of those inputs, initial (starting) conditions and boundary conditions which are of interest must be decided and these, with any other relevant data, must be inserted and the actual solution carried through to the end.

Repetition of the solution for as many inputs and conditions as desired must then be carried through and the solutions set out in some intelligible tabular or graphical format.

The repetition of the process of solution is sometimes known as "running" the problem. It is also sometimes known as "simulation". Simulation in this fashion may be less expensive (particularly is this the case for complicated systems) than conducting a series of trials on an actual system. More importantly, simulation can also be carried out even if an actual system does *not yet exist,* and hence it provides a valuable and economical means of testing the performance of an intended system.

2. Design of a Future System

Simulation of a model for the system which has been conceived by a designer is a very helpful process. The simulation may be run either on an iconic model (e.g., a small scale hydraulic model of a new harbour to predict its behaviour under the action of tides, wind and waves), *or* an analogue model together with the relevant mathematical equations (e.g. an electric circuit analogue using resistance and capacitance to simulate the thermal behaviour of a fireproof masonry wall, with the aid of the relevant circuit equations), or a symbolic model (e.g., a task scheduling network including some conditional or logical branch-points, together with the very simple arithmetic of adding up total times for each possible path through the network—a very simple but tedious calculation and therefore well suited to being a computer task).

The simulation may be "run" by using an actual experiment yielding observed values as the "solution", by hand computation, by digital computation, by setting up an analogue computer using electronic circuits and observing results typically displayed on a chart recorder (a form of galvanometer which makes a permanent record on a moving chart) or an oscilloscope (which may be viewed or photographed).

If we choose a given set of inputs (which will remain fixed during the simulation or run), but choose a number of different values for the parameters of the system we may view the behaviour of the system over a range of parameter values and decide which one gives *optimum performance*.

The function of designing (sometimes called synthesis) therefore involves rather more sophistication than the analysis of a known system. The steps involved in designing may be set out as:

1. Statement of the objective or (usually) objectives, in a way which will allow the subsequent steps to be carried out. This will also require a definition at this or a later stage of what we mean by *optimum performance*.
2. Choice of a structure or description for a system which appeals to the designer as being a likely candidate for *satisfactory performance*. (In real life a designer may choose several different *structures*, run a simulation for each and allow them to compete with each other before *choosing* the one which seems best; at that early stage quite coarse models might be used).
3. Given a structure (or structures) which the designer wishes to investigate, he must next make a model (or perhaps more than one model if a number of radically differing properties are to be investigated, i.e., multiple objectives for the one system). Each model must then be simulated for a range of its own parameters and the one with optimum performance *chosen*.
4. Having chosen a design, it is usual to carry out some sort of trial. This may involve going back to the original specification and checking whether it was sufficiently complete as well as accurate. Alternatively, a further simulation with wider terms of reference or the actual construction of a prototype might be envisaged, leading up to a simulated or an actual *field trial*. This last stage is an important one and will usually generate modifications which must be fed back into the design and construction procedure. The user (customer) of the equipment would be involved in this final stage.

3. Improvement of an Existing System

Modelling may also be the technique whereby the behaviour of an existing system may be improved. Just as in the original design stage a suitable model (i.e., one which is sensitive in relation to the desired investigation) must be chosen, the means for simulation must be provided together with means for gathering essential data from the operating system, and a number of simulated runs made for different values of those inputs and parameters *which can be manipulated*. Simulations to determine the optimum value of some component which we are not allowed to alter, would be a waste of time. Improvement of an existing system may well present its greatest difficulties in the collection of data and in gaining authorisation to manipulate the system. The works manager of a steel mill producing a large volume of steel products each day will be hard to persuade that he should allow one of his engineers to alter the inputs or environment of the system in a way which might possibly reduce production over the test period involved. In these circumstances simulation is a vital tool, provided that access can be had both to the means for simulation and the necessary data (i.e., inputs and parameters of the chosen model). If such information is not available it must be created either by experimental measurement or by computation.

D. CONCLUSION

In conclusion we would like to quote from Harold Chestnut (1965).

> Simulation and modelling complement each other. The model provides a basic understanding of the process to be studied. The simulation shows how the model of the process will operate under the conditions for which the simulation is run. Both simulation and models may be analytical and/or experimental in nature.
>
> Although models and simulation can prove very valuable if properly used, they seldom yield more worthwhile results than the data or information given to them. They are an aid to good system design; they are not a replacement for it.

It seems quite obvious that the system designer (and also the system user) will be called upon to exercise a great deal of creativity and skill.

SUGGESTED READING MATERIAL

CHESTNUT, H. *Systems Engineering Tools,* pp. 126 and 141, John Wiley and Sons, New York, 1965.

KOESTLER, A. *The Act of Creation*, Danube edition, Hutchinson, 1969.

This is quite an interesting book for browsing; see Appendix II in which he refers to the quotation, at the beginning of this chapter, where Maxwell indulges in his hobby of verse making with a poem satirising both sides in a controversy of his time which raged between the Victorian theologians and the exponents of a materialist philosophy based on a solely deterministic view of the universe: rendered with some entertaining comment. It is said to have been inspired by the narrowness of the views presented by an exponent for the materialists, John Tyndall in his presidential address to the British Association, at a meeting in Belfast.

CHAPTER 27

Conclusion: And What of the Future?

A. E. Karbowiak and
R. M. Huey

IT'S UP TO US
"In the past, men could shrug their shoulders in the face of most of the evils of life because they were powerless to prevent them . . . Now there is no one to blame but ourselves. Nothing is any longer inevitable. Since everything can be accomplished, everything must be deliberately chosen. It is in human power for the first time to achieve a level of human welfare exceeding our wildest imaginings or to commit race suicide, slowly or rapidly. The choice rests only with us."

Jerome D. Frank
quoted in Fabun Dynamics of Change *(Prentice-Hall, 1967).*

In his study of science, D. J. de Solla Price (1963) observes that science is *immediate*. By this he means that most discoveries in science are of recent origin. To make this point clearer an examination of numerical estimates might be helpful. The population of the world today counts about 3,000 million, a number of the order of 3×10^9. From various studies which have been carried out one obtains a figure of 60,000 million for the total number of people who lived on this earth. This, therefore, indicates that only about 5 per cent of all those who ever lived are alive today, and it will continue to be so that most of those who ever lived are dead today. This indicates that human population is *not immediate*.

In contrast, irrespective of which idea we have for measuring the *amount* of science we find that most of scientific discovery is of recent origin. Indeed, *science is immediate* to the 80-90 per cent level. If we were able to analyse the progress of technology numerically (no one has yet done so, in a universally applicable way), we would find it to be even more immediate than science.

Scientific knowledge doubles every 15 or 20 years, and many disciplines—electronics included—grow even faster than that. One must accept therefore, the basic fact of life that no single human being can possibly keep up with all the details at this rate of development. Hence, the need for a greater degree of specialisation in narrower fields, and a greater need for team efforts. Systems engineering will now be recognised as a way of achieving objectives even when confronted with many unknowns. The branches of science and technology will be recognised as specialised disciplines: complex, maybe, but not something to be afraid of, not something to run away from. On the contrary, scientific and technological knowledge must be organised and used by those who *care* for the future as well as for the present.

In order to assign value judgments and hence priorities, to establish objectives for the world of tomorrow we must be able to identify models for social systems. Something of what we mean by this has been well said by John Rose (1967).

> The long-winded and fatuous descriptions, in the 1950s, of the progress of automation look rather shabby in the cold light of reality. Thus a pamphlet issued by the American National Union of Manufacturers in the 1950s contained the following high-sounding phrases, 'Guided by electronics, powered by atomic energy, geared to the smooth effortless workings of automation, the magic carpet of our free economy heads for distant and undreamed of horizons. Just going along for the ride will be the biggest thrill on earth.' This 'magic-carpet' view of advanced technology has to be abandoned and human factors considered. The ride was rather bumpy, the road to Shangri-La being littered with vast sums of money wasted on gadgets, and economic recessions, topped by new social problems of unprecedented magnitude and complexity.
>
> The need is, therefore, for an appreciation of the proper use of automation to be considered in the context of the economic and social changes, which may be brought about by the Cybernetic Revolution.
>
> In fact, the ball is right back in man's court. As Professor Gabor puts it: 'Until now Man has been up against Nature, from now on he will be up against his own nature' (Gabor, 1963). Perhaps the answer to this problem lies in education, as elaborated by Professor Gabor in the 1963 Thomson Lecture and by Sir Leon Bagrit in his Reith Lectures in 1964.

The one conclusion has been reached by so many different people of vastly differing backgrounds who have pursued vastly differing routes from almost equally differing starting points, that we are driven inescapably to the conviction that it must be true. Let us quote from a biologist (who is an optimist in a gentle way) and an engineer (who is a realist and needs consciously to avoid falling into pessimism).

In the epilogue of a book by Sir Macfarlane Burnet (1968) based on the Boyer lectures in 1966, the famous Australian biologist Macfarlane Burnet reminds us "Dr. Albert Schweizer summed up his philosophy in a single phrase 'reverence for life' ". This philosophy informs much of Burnet's thinking and it is from this belief that he reaches the conclusion (ibid, 1968).

> In one way or another we *will* reach the stage of a single community of human beings making a sane use of planetary resources. This is physically and biologically possible in a hundred years. It could take 10,000 and be repeatedly thwarted by recurrent wars and collapses of civilisations everywhere but it must come eventually.

Denis Gabor (1963) has written:

> "At the present juncture of history we can no longer afford to have uncommon men with demoniac power and the morals of adolescent gangsters . . . In plain words 'rounding this corner' means that history as we know it will come to an end; history in terms of wars and conquests. It means also the coming of *'post-historic man'*; of the common man living for his own happiness, not for the glorification of his rulers, and of the uncommon man who finds his fulfilment in enriching the life of the common man."

The joint authors of this chapter (who are the editors of this book) share these beliefs and commend them to their readers.

REFERENCES

de Solla Price, D. J. *Little Science, Big Science*. Chapter 1, and conclusion of Chapter 4. Columbia University Press, 1963 (or paperback 1965).

Rose, J. *Automation: it's Uses and Consequences*. Contemporary Science Paperback series, Oliver and Boyd, 1967.

BURNET, M. *Biology and the Appreciation of Life*. Sun Books, Melbourne, 1968. The Boyer lectures are presented by the Australian Broadcasting Commission. See also the chapter on "Adaption and Change".

GABOR, D. *Inventing the Future*. Secker & Warburg, 1963. p. 130; Pelican, 1964.

Problems for Discussion

1. COSTS OF RUNNING A MOTOR CAR

Prepare your own estimates for running a motor car, and in discussion with other students compare your ideas with theirs.

You might assume that the motor car you buy costs $4,000. The costs will then be made up of the following.

(i) Fixed costs (e.g., capital costs and depreciation).

(ii) Variable costs which might be proportional to the mileage covered (e.g., fuel, servicing, wear and tear, storage of vehicle).

On the assumption that the variable costs are proportional to the mileage covered, draw a graph of costs against time for several values of assumed number of miles covered per year (say, 5,000, 10,000 and 20,000). Take into consideration as many items as you can think of, including that the capital ($4,000) will have depreciated to say, $400 in ten years at an assumed realistic rate and that the interest on the capital might be 7%, that maintenance, storage, registration and insurance costs must be covered.

Finally, try to answer questions such as:

(a) what is the true cost of owning the vehicle in dollars per month or per mile,

(b) do the costs of running the vehicle vary with age of the vehicle,

(c) how much higher are the costs of running the vehicle as compared with costs of using public transport or a taxi?

2. PATTERNS OF GAMES

In a game of draughts there are at the beginning of the game 12 black and 12 white pieces. Omitting the existence of queens, if it is known at a certain time in the game that there are 4 white pieces and 3 black left on the board, calculate how much information needs to be given (at most) to reconstruct the position of all pieces?

Solution:

The are $4 \times 8 = 32$ black squares accessible to all pieces. Assuming that the 4 white and 3 black pieces can be placed without constraints the number of ways in which this can be done is the number of placings (combinations) of 7 objects in 32 spaces. Therefore, the total number of ways, N, is given by

$$N = C_7^{32} = \frac{32 \times 31 \times 30 \times 29 \times 28 \times 27 \times 26}{1 \times 2 \times 3 \times 4 \times 5 \times 6 \times 7}$$

$$\approx \frac{29^7}{5040}$$

$$\approx 29^7 \times 2^{-12}$$

Taking logarithms to the base of two we have approximately

$$\text{lgg}\, N = 34 - 12 = 22 \text{ bits}.$$

The information needed is therefore 22 binary decisions. In actual fact, the information is less that that because we left out of consideration constraints of the game. Discuss their effect.

3. A PROBLEM IN HUMAN PERCEPTION

With some training a human being can commit to memory (over a limited period of time) English words at a rate of about five words per minute. Assuming that an average word is five letters long and that each letter carries one bit of information, the rate of committing things to memory works out to be less than one bit per second.

However, it is known from the study of reading habits that most literate people can be trained to read, and to retain the information read (such as a story or a novel), at the rate of five to ten words per second, which is equivalent to about 50 bits per second.

Discuss the apparent conflict that exists between the two estimates and suggest possible ways of accounting for the discrepancy.

Try to think of other examples, such as the rate with which television data can be absorbed by a viewer and compare this rate with that of the television data transmitted. What conclusions can you draw?

4. FEEDBACK CONTROL

The system of Fig. 12.10 can be designed to ensure that the response R follows very closely (at least in its steady-state value) the reference input I.

What value of the feedback path gain H must be chosen to ensure this?

How large must the loop gain GH be made to ensure that R is held

(i) to within 10% of I,
(ii) to within 1% of I,
(iii) to within 0.1% of I?

5. SEMICONDUCTION

A crystal of silicon contains 10^{16} cm^{-3} donors and 10^{15} cm^{-3} acceptors, calculate the electron density at room temperature. (Necessary data is given in Chapter 16 Section E, problem 1).

6. VISUAL PERCEPTION

(a) With reference to the design of a visual perception machine which is freely moving discuss the merits of a system which requires continual eye tremor to operate (like eyes) versus one which cannot operate adequately in the presence of movement (like cameras).

(b) With reference to the design of a visual perception machine, discuss in the light of your knowledge of computers and animals the relative merits of a scanning system, like a TV raster versus numerous parallel paths as in living systems.

7. COMMUNICATION SYSTEMS

A human voice occupies effectively a bandwidth of about 4 kHz but for a satisfactory reproduction of orchestral music about 10 kHz bandwidth is needed. This rate, in terms of data, is equivalent to 10^4 sites/second. If we assume that the dynamic range needed is equivalent to 16,000 levels (approximately 2^{14} and therefore equivalent to 14 bits per site) the channel data rate becomes 1.4×10^5 bits/sec.

Now consider the following proposition. At the transmitter end there is a large store of recorded music and a similar store is available at the receiving end. To transmit a given symphony an instruction of the type "select Beethoven's Choral Symphony production number DK3782 and reproduce at the recommended level" is sent over a channel. In response to such an instruction the computer at the receiving end performs the requested actions and as far as the listener is concerned, he might be convinced that his service is provided through the use of a high quality communication channel. In fact the objectives are achieved without such facilities.

Discuss the problem in detail and resolve any discrepancies. What conclusions can be drawn from this?

Appendix I

SYSTEMS OF UNITS

Most electrical engineers and scientists today make use of the rationalised meter-kilogram-second-coulomb RMKSC system of units, a system which is completely compatible with, and forms a part of the S.I. (scientifique internationale) system of units. The English speaking countries which still employ the English system of units have mostly committed themselves to a rather gradual programme of metrication, in which the units of measurement for everyday commerce in weights and sizes will be made very largely compatible with the S.I. system, i.e., metrication will imply use of a meter-kilogram system in everyday affairs.*

Recently it has been standard practice to refer to the MKS (meter-kilogram-second) system and indeed a very large number of units (covering mechanical systems of all sorts) may be based on these three basic quantities. If we wish to consider electricity and magnetism it is necessary to permit a further fundamental quantity, which has been chosen as the coulomb of electric charge. It should be noted that the factor 4π (the number of steradians in a complete spherical solid angle in three dimensions) will inevitably appear in some equations of electrical and magnetic actions. It may be cancelled out in some cases by suitable choice of the definition (and hence the size) of various derived units for electric and magnetic quantities. The rationalised system† which has been adopted, succeeds in eliminating the factor 4π from the majority (but not all) of frequently used electromagnetic equations.

Table A.1 below lists symbols approved by the Institution of Electrical and Electronics Engineers, New York‡ for publication in their journals.

In addition Table A.2 below lists accepted prefixes which can be applied to these units in order to scale them down to size by steps of 10^3 or 10^{-3} respectively. In addition scaling prefixes (less often used) for 100, 10, 0.1 and 0.01 are given. Any prefix may be used with any unit, e.g., kilowatt 1 kW = 1000W; gigahertz 1 GHz = 10^9 Hz; picofarad 1 pF = 10^{-12} F; nanohenry 1 nH = 10^{-9} H; millitesla 1 mT = 10^{-3}T.

The prefixes hecto—deka—deci—and centi—are not widely used except in decibel and centimetre.

* One fundamental difference is that metric units are based on the kilogram as a unit of force (i.e., weight) while the RMKSC and S.I. systems are based on the kilogram as a unit of mass.

† In years gone by there were quite fierce arguments about whether it would be preferable to use a rationalised or a so-called natural (but unrationalised) system. The name rationalised was coined by Heaviside in the 1890s, but his habit of making insulting remarks in public about many fellow scientists and engineers of his time may have helped to delay until the next generation, the recognition of the undoubted virtues of the RMKS system of units.

‡ see IEEE Spectrum, August 1965, pp. 111-115. Note that certain quantities such as temperature (measured in degrees Kelvin, °K) have been omitted from the list.

TABLE A.1
The RMKSC Units

Unit	Quantity	Symbol
ampere	current	A
coulomb	charge	C
farad	capacitance	F
henry	inductance	H
hertz	frequency	Hz
joule	energy	J
kilogram	mass	kg
meter	length	m
newton	force	N
ohm	resistance	Ω
radian	angle	rad
second	time	s
siemens*	conductance	S
tesla	magnetic flux density	T
volt	potential	V
watt	power	W
weber	magnetic flux	Wb

* An acceptable alternative name is mho (ohm spelled backward).

The initials RMKSA may also be encountered. The A here refers to ampere, and implies an alternative way of defining the units necessary in electricity and magnetism. The question of whether the coulomb or the ampere is the more desirable *fundamental* unit for this purpose has not yet been fully resolved.

Appendix 2

VALUES OF CERTAIN MATHEMATICAL FUNCTIONS AND PHYSICAL CONSTANTS

TABLE A.2
Recommended Magnitude Prefixes

Prefix	Exponent of 10	Symbol
tera-	+12	T
giga-	+9	G
mega-	+6	M
kilo-	+3	k
hecto-	+2	h
deka-	+1	da
	0	
deci-	−1	d
centi-	−2	c
milli-	−3	m
micro-	−6	μ
nano-	−9	n
pico-	−12	p
femto-	−15	f
atto-	−18	a

TABLE A.3
Values of Certain Physical Constants

Constant	Symbol	Value	Units
Speed of light (in vacuo)	c	2.998×10^8	ms^{-1}
Electron charge	e	1.6021×10^{-19}	C
Electron rest mass	m	9.109×10^{-31}	kg
Planck's constant	h	6.626×10^{-34}	Js
Boltzmann's constant	k	1.3805×10^{-23}	$J(°K)^{-1}$
Energy per electron-volt	eV	1.6021×10^{-19}	$J(eV)^{-1}$
Energy per ° K	ek	1.1605×10^{4}	$°K(eV)^{-1}$

C, coulomb; J, joule; ° K, degrees Kelvin; kg, kilogram; m, metre; s, second.

TABLE A.4
Powers and Inverse Powers of Two

2^n	n	2^{-n}
1	0	1.0
2	1	0.5
4	2	0.25
8	3	0.125
16	4	0.062 5
32	5	0.031 25
64	6	0.015 625
128	7	0.007 812 5
256	8	0.003 906 25
512	9	0.001 953 125
1 024	10	0.000 976 562 5
2 048	11	0.000 488 281 25
4 096	12	0.000 244 140 625
8 192	13	0.000 122 070 312 5
16 384	14	0.000 061 035 156 25
32 768	15	0.000 030 517 578 125
65 536	16	0.000 015 258 789 062 5
131 072	17	0.000 007 629 394 531 25
262 144	18	0.000 003 814 697 265 625
524 288	19	0.000 001 907 348 632 812 5
1 048 576	20	0.000 000 953 674 316 406 25
2 097 152	21	0.000 000 476 837 158 203 125
4 194 304	22	0.000 000 238 418 579 101 562 5
8 388 608	23	0.000 000 119 209 289 550 781 25
16 777 216	24	0.000 000 059 604 644 775 390 625
33 554 432	25	0.000 000 029 802 322 387 695 312 5
67 108 864	26	0.000 000 014 901 161 193 847 656 25
134 217 728	27	0.000 000 007 450 580 596 923 828 125
268 435 456	28	0.000 000 003 725 290 298 461 914 062 5
536 870 912	29	0.000 000 001 862 645 149 230 957 031 25
1 073 741 824	30	0.000 000 000 931 322 574 615 478 515 625

$\mathrm{lgg}\,x = \log_2 x = 3.322 \log_{10} x$

TABLE A.5
Powers and Inverse Powers of e (e = 2.71828 . . .)

e^n	n	e^{-n}
1.000	0	1.000
1.105	0.1	.905
1.221	0.2	.819
1.350	0.3	.741
1.492	0.4	.670
1.649	0.5	.607
1.822	0.6	.549
2.014	0.7	.497
2.226	0.8	.449
2.460	0.9	.407
2.718	1.0	.368
4.482	1.5	.223
7.389	2	.135
20.09	3	.0498
54.60	4	.0183
148.4	5	.00674
22026	10	.000045

$\text{lin}\, x = \log_e x = 2.3026 \log_{10} x$

Appendix 3

MATRIX ALGEBRA

1. Definitions

Matric algebra is a method of manipulating groups of symbols. In this algebra the symbols are grouped in a special array called a matrix.

An n × m matrix has n rows and m columns, with the elements laid out as follows for a matrix A:

$$A = \begin{bmatrix} a_{11} & a_{12} & \cdots & a_{1m} \\ a_{21} & a_{22} & \cdots & a_{2m} \\ \bullet & \bullet & & \bullet \\ a_{n1} & a_{n2} & \cdots & a_{nm} \end{bmatrix}$$

where the elements a_{ij}, $i = 1,2, \ldots n$; $j = 1,2, \ldots m$, are algebraic symbols denoting numbers or functions and where, in the case

$n = m$, the matrix is called a square matrix,
$m = 1$, the matrix is called a column vector,
$n = 1$, the matrix is called a row vector.

The rules of matrix algebra provide for addition (and subtraction) and multiplication of matrices, provided certain conditions relating to the numbers of rows and columns of the two matrices are met.

2. Addition

Provided two matrices have the same numbers of rows and columns they may be added as in the following example.

if $A = \begin{bmatrix} a & b \\ c & d \end{bmatrix}$ and $B = \begin{bmatrix} e & f \\ g & h \end{bmatrix}$

then $A + B = \begin{bmatrix} a+e & b+f \\ c+g & d+h \end{bmatrix}$, also is a 2×2 matrix.

3. Multiplication

The product, A, B, may be calculated provided the number of columns of A is equal to the number of rows of B. For instance if

$$A = \begin{bmatrix} a & b \\ c & d \end{bmatrix}$$

and

$$B = \begin{bmatrix} e \\ f \end{bmatrix}$$

we note that A has 2 columns and B has 2 rows. The product is

$$C = AB = \begin{bmatrix} ae + bf \\ ce + df \end{bmatrix}$$

which we note is a column vector in this case. In general, the c_{ij} th element of the product is obtained by taking the sum of the products of the elements of the *i*th row of A with those of the *j*th column of B.

Thus, if

$$A = \begin{bmatrix} a_{11} & \ldots a_{1m} \\ \vdots & \vdots \\ a_{n1} & a_{nm} \end{bmatrix}, \quad \text{an } n \times m \text{ matrix,}$$

and

$$B = \begin{bmatrix} b_{11} & \ldots b_{1r} \\ \vdots & \vdots \\ b_{m1} & b_{mr} \end{bmatrix}, \quad \text{an } m \times r \text{ matrix,}$$

then an element of the product matrix is

$$c_{ij} = [a_{i1} \ldots a_{im}] \begin{bmatrix} b_{1j} \\ \vdots \\ b_{mj} \end{bmatrix} = a_{i1} b_{1j} + \ldots + a_{im} b_{mj}$$

and the product matrix C has dimension $n \times r$.

For solving sets of algebraic and differential equations it is convenient to retain these arrays and carry out the algebraic manipulation by matrix methods and thus greatly reduce the number of symbols to be carried.

For a further discussion the reader may care to refer to the text by Hohn given in the reference below.

4. Reference

HOHN, F. E., *Elementary Matrix Algebra,* 2nd Ed. Macmillan Publishing Co., New York, 1964.

Subject Index

Abacus, 53
Acceptor impurities, 154, 163
Across-variable, 122
Active components or devices, 33, 123
Acuity, visual, 248
Adder, 116, 118
Address (in computer), 195, 196, 198 et seq., 205, 225
Algorithms, for programs, 212
Aluminium, 166
Amplifier, linear, 45, 47
, switched, 46-7
Analogue, 63, 279, 329
computer, 51
data, 19-20, 93
Analysis, 113, 330
AND, definition of, 169, 172
Argument (of a function), 180
Arithmetic operations, 168, 172, 173 et seq.
Arsenic, 152
Association law, 172
Automatic control, *see System, Control*
Avalanche in *p-n* junction, 165

Bandwidth, 104, 107, 120, 286
Bar pattern, 102
Binary addition, 169, 172, 177
arithmetic, 173-4, 177, 179, 182
coded character, 189
decimal, 188
devices, 170-1, 188
multiplication, 178, 182
variable, 169
Bit, 21
, definition of, 24, 27
Bipolar transistors, 166
"Black box", 36
Block diagrams, 41 et seq., 115 et seq.
Boltzmann's constant, k, 15
Boolean algebra, 172
Bonds, inter-atomic, 32, 143 et seq.
Boron, 154 et seq., 163
Buffer register, 197
Byte, 202, 206

Cable, balanced-pair, 286, 288
, coaxial, 289
, open, 286
Capacity (for information), 98, 105
Channel, capacity of, 98, 106
Characteristic curves, 47-8
Chequer-board pattern, 98-9
Circuit diagram, 42, 121
Circuit theory, 120 et seq.
Commutation law, 172
Complex frequency, s, 115
Compilers, 231, 235
Complementary function, 44, 114, 115
Computer, 51 et seq., 66, 293
-aided design, 300, 301-2
-assisted instruction, 303
central processing unit, 189
, control of power systems by, 325 et seq.
control unit, 54, 192
function unit, 191
, industrial process control, 306 et seq.
input/output units, 194
, interactive, 295, 300
, medical use of, 309 et seq.
-s, network of, 289, 297 et seq.
, organisation of, 179, et seq. 183, 186
, parallel, in living systems, 66 et seq.
register unit, 192
, sequential or iterative, 66 et seq.
storage unit (see memory), 189
, time-shared, 294 et seq.
, MIT system, 296
Conductivity, electric, 144-6, 147 et seq.
, of meals and semi conductors, 114, 147, 148, 150
, super-, 150, 319
Conservation laws, 38, 122
Constraints, 40, 90, 98, 100, 101, 102
Contacts, metallic, 155, 163
Control forces, 267, 271
Control interval, 274
Coupling of systems, 94, 133
Convolution, 114
Core storage memory, 180 et seq.
Correlation, in vision, to detect movement, 255
Cost-benefit analysis, example of, 131-2
Cost-flow diagram, 131
Covalence bonds, 146
CPM (Critical Path Method), 129
Crystalline structure, 142
Crystals, ionic, 143
, metallic, 144
, molecular, 145
Curie temperature, 150

Decimal numbers, 174 et seq.
, binary coded, 188
Defects in crystal structure, 147
Depletion layer, 155, 158
Design (or synthesis), 113, 330
Determinants, 43
Deterministic (signals), 64

Dielectric constant of silicon, 153
Differential equations, 43 et seq., 64
, partial, 134 et seq.
Differential operators, 43-4
Diffusion, of atoms in a crystal, 150
, dependence on temperature
of solid-state in silicon, 150
Digital computer, 51 et seq., 168 et seq.
Digital data (see also binary), 20
transmission, 298
Directed line, the, 118
Distributed systems, 134
Distribution law, 172
Disturbance, 86-7, 266
Dollars, as conservable quantity, 38, 131
Donor impurities, 153, 162
Dynamics, 37, 41 et seq., 49, 64, 111 et seq., 266

e, powers of, 340
Electrons as charge carriers, 147
in metals, 144
, valence, 145
Encoding, 20, 107
Energy (see also Power), 13 et seq., 266
flow diagram, 130-1
Engineers, compared with scientists, 326 et seq.
Entropy, 79, 89, 109
Epitaxial, layer of n-Si, 167
process, 167
Equivocation, 285
Error correcting, 104
detecting, 103
Etching, selective, 163, 166
Excitation, 43
Execution (of program), 196-8
Extrinsic semiconductors, 152
Eye (of the crab), 248 et seq.

Feedback, 79
control system, 86, 119
, visual, 256
Flip-flop, 46, 171, 180
Flowchart, computing, 202, 213

Gain, 46
Germanium transistor, 34, 161
Guided waves, 290, 319

Hartley, definition of, 27
Hexadecimal numbers, 178
Holes, concept of, 148
as charge carriers, 148 et seq.
Homologue, 63, 329
Hydrofluoric acid, 166

Iconic model, 329
Identification, 113
Image (of a transformation), 126
Induction motor, 318
Inequalities, 40
Information, 13, 19 et seq., 98
, extraction of, 282 et seq.
theory, 19 et seq., 98 et seq.
, transportation of, 282 et seq.
Inputs (outputs and signals), 38, 39, 42, 116, 188, 238, 266, 295, 300-1
Instructions, 195
, branch, 200, 207
, machine, 195, 205, 230
Integrated circuits, 28, 32, et seq., 180
, fabrication of, 161 et seq.
, oxide-masking in, 163
, planar technology for, 165
Interconnections, 167
Intrinsic semiconductors, 152
Ionic crystals, 143
Isolation, using reverse bias
on epitaxial layer, 167

Karnaugh map, 173

Language, Algol, 222
as a set of symbols, 108
, assembly, 223
, binary, 108•
, Cobol, 222, 228
, computer, 57, 221 et seq.
, Fortran, 222, 227
, high level, 221
, machine, 202
, Snobol, 222, 230
, syntax of programming, 231
, correction of, 296
Laplace transform, 114 et seq., 136
Learning machines, 88, 92
Linearisation, 36, 47
Linear motion electrical machines, 324
Linear systems, etc., 43 et seq.
Logic, 169 et seq.
Logical circuits, 170, 180, 187
Logical functions, 168 et seq., 187
, definition (AND OR
NAND NOR), 169
variable, 169

Magneto-hydrodynamic generator, 324
Management, 71 et seq.
, bureaucratic, 76
, systems approach to, 78
theory, administrative, 75
, modern, 77
, scientific, 75
Managers, 70, 72

Material, 266
Matrix algebra, 42, 272, 340
Maya Indians, 173
Measurements, 61, 266, 276, 329
Mechanical, circuit theory, 123
-electric circuit analogues, 123
Memory, 55, 68, 74, 179, 180 et seq., 187, 194
Memory systems, characteristics and cost of, 180
, reading from and writing into, 181
Metallic crystals, 144
Microelectronics (see integrated circuit), 28
Microwave link, 289
"Mill" (in computer), 54
Models, 7, 35, 37, 42, 62, 64, 67, 72, 86, 111, 127-8, 131, 135, 273, 276, 280, 328 et seq., 332
Modulation, 20, 29
Morse code, 21-2
Movement perception, 253
Multiplexing, 286

NAND, definition of, 169
Nerve impulses, 248 et seq.
response as a graded potential, 252
Nervous system, 46, 66, 248 et seq.
Network, 37
chart, 42, 128
Nit, definition of, 27
Noise, 15, 16, 106
NOR, definition of, 169
Number, representation of, 173 et seq.
negative, 175

Octal numbers, 175, 178
Operand, 187 et seq., 197
Operations research, 35, 72
Optimisation, 45, 72, 87, 131, 273, 275, 285, 325
OR, definition of, 169
Organisation, social (*see management*)
Original (of a transformation), 126
Outputs (inputs and signals), 38, 39, 42, 43, 86, 115 et seq., 188, 238, 266, 300-1

Partial fractions, 125
Particular integral, 44, 114, 115
Passive components or devices, 123
Pattern generating capacity, 18 et seq.
recognition, 92
Patterns and form (see also waveform, semantic, sequence), 19 et seq., 24
Performance index, 273, 277
Periodic table, 146, 153
PERT, 128
Phasor algebra, 112, 126, 136
Phosphorus, 153
Photo-resist, 166
Physical constants, 339
Planck's constant, *h*, 15
p-n junction, 31, 46, 155
, breakdown of, 165
capacitance, 155, 158
, charge density near, 155
, current-voltage characteristic, 157
passivation, 165
, potential barrier at, 155, 165
saturation current, 157
Power (see also Energy), 13 et seq., 313 et seq.
source, geothermal, 314
, nuclear, 323
, solar, 314
, tidal, 315
Predictors, 88
Prefixes, 338
Probabilistic, (signals), 64
Probability, 108
Program, ICES and ECAP, 222
, job control, 238
, supervisor, 240
Propagation time, 134
Punched cards, 193-4
Push-pull, 47

Quine-McCluskey algorithm, 173

Radix, 174 et seq.
conversion, 175
Recombination of holes and electrons, 149
Recovery (inverse analysis), 113
Redundancy, 100
Register, 179
, shift, 180
Reliability, 31, 32
Remote sensing (other than sight or sound), 290
Repeaters, 286
Representative point (in state space), 269
Resilient environment, 94
Resource allocation, 70
Response (see outputs)

Scientists, 70, 72, 283, 328
328
Semantic content, 89, 90, 291
Semiconductors, 29 et seq., 149
, extrinsic, 152
, intrinsic, 152
Sensitivity, 120

Sequences of symbols, 108, 229
Set, capacity of, 22
Shift register, 180
Signals, 43, 111, 119
  , damped sinusoidal, 115, 126
  exponential, 125, 126
  sinusoidal, 44, 91, 111 et seq., 126
Silicon, 28 et seq., 146, 148
  , dielectric constant of, 153
  dioxide, 143, 163, 166
Simulation, 62, 280, 330, et seq.
Simultaneous equations, solution by
  block diagrams, 234
"Sketchpad", 301
Social groups, 71
Software, system, 221 et seq.
Solar batteries, 32, 34
Source of information, discrete and
  continuous, 105
  , controlled, 123
  , driving, 123
Spectrum, optical, 290
  , radio frequency, 290
Spinning reserves, 315, 323
Stability, 45
State, 67
  estimation, 276
  space, 268 et seq.
Statics of systems, 41, 64, 127 et seq.
Steady state, 44, 64, 79, 111, 115, 127
Stochastic process, 85
Store (storage) in computers, 54, 56, 168,
  237
Stored programs, 196, 237
Superconductivity, 150, 319
Switching circuits, 169 et seq.
Systems, 6 et seq., 28, 35 36 et seq., 78, 329
  et seq.
  , adaptive, 86
  , biological, 66 et seq.
  , communication, 60, 282, 285 et seq.
  , control, 59
  , distributed, 134
  , economic, 280
  , electric power, 276, 313 et seq.
  , global, 94
  , human, 71 et seq.
    , closed or open, 78 et seq.
    , subsystems of, 80 et seq.
  , hydraulic, 279
  , large, 266 et seq.
  , radar, 282
  , satellite attitude, 273
  , satellite launch, 276
  , social, 333-4
  , variable, 89
  , vehicular traffic, 278

Task network, 128
Technology, 5 et seq., 69, 73, 333
Telegrapher's equation, 135
Through-variable, 122
Thyristors, 319
Time lag, 266
Time model, 128
Trajectory (in state space) 269, 271, 275-6
Transducer, 29, 45, 49, 123-4, 267, 291
Transfer function, 67, 114
Transformer, 124, 313, 317
Transient response, 44, 86
Transistors, 29 et seq., 123
  , bipolar, 166
  , current distribution in, 159
  , field effect, (FET), 167
  , junction, 34, 158
  , point contact, 34
  , potential distribution in, 158-9
Transmission line, 134, 318 et seq.
Truth table, 169
Two, powers of, 339

Units, systems of, 337

Vacuum tubes, 29 et seq., 123
Valence bonds and electrons, 143 et seq.
Variables, logical or binary, 169
Vectors (see also matrix algebra), 43, 272
Venn diagram, 24

Waveform, continuous, 105
  , equivalence to bar-pattern, 25
  , information capacity
    per unit time, 26
Word-cell, 204

Name Index

Babbage, 53
Bagrit, 334
Bauchon, 53
Bell, 137
Boole, 169
Braiwerd, 293
Burks, 55
Burnet, 334-5

Charles XII, 174

de Morgan, 169
de Solla Price, 333

Eckert, 54, 293

Falcon, 53
Fayol, 74
Felici, 319
Ferranti, 137

Gabor, 334-5
Gauss, 137
Goldstine, 55, 293

Harriott, 174
Hartley, 21
Hertz, 137
Hollerith, 54

Jaques, 53

Kahn, 80
Katz, 80
al-Khowarizmi, 174, 212
Kelvin, 135, 137

Leibnitz, 53, 174

Marconi, 137
Mauchley, 54, 293
Maxwell, 137
Morse, 286
Muller, 53

Nyquist, 21

Pascal, 51, 53
Piave, 137
Pupin, 137

Rose, 333

Shannon, 21, 169, 301
Shepherd, 137
Siemens, 137
Stanley, 313
Stibitz, 54
Sutherland, 301

Taylor, 74
Tesla, 137, 318
Thomas, 53

von Neumann, 55, 293

Wakelin, 306
Weber, 137
, Ernst, 137
, Max, 74
Westinghouse, 313
Wheatstone, 286
Wiener, 21
Wilkes, 155, 294